UNTIL THE
END OF TIME

UNTIL THE END OF TIME

Mind, Matter, and Our Search
for Meaning in an Evolving Universe

Brian Greene

RANDOM HOUSE
LARGE PRINT

Published in the United States of America by Random House Large Print, in association with Alfred A. Knopf, a division of Penguin Random House LLC, New York, and distributed in Canada by Penguin Random House Canada Limited, Toronto.

Cover photograph by shaunl/E+/Getty Images
Cover design by Chip Kidd

The Library of Congress has established a Cataloging-in-Publication record for this title.

ISBN: 978-0-593-17172-1

www.penguinrandomhouse.com/large-print-format-books

FIRST LARGE PRINT EDITION

Printed in the United States of America
10 9 8 7 6 5 4 3

This Large Print edition published in accord with the standards of the N.A.V.H.

For Tracy

Contents

Contents

Preface

"I do mathematics because once you prove a theorem, it stands. Forever."[1] The statement, simple and direct, was startling. I was a sophomore in college and had mentioned to an older friend, who for years had taught me vast areas of mathematics, that I was writing a paper on human motivation for a psychology course I was taking. His response was transformative. Until then, I hadn't thought about mathematics in terms even remotely similar. To me, math was a wondrous game of abstract precision played by a peculiar community who would delight at punch lines turning on square roots or dividing by zero. But with his remark, the cogs suddenly clicked. **Yes,** I thought. **That is the romance of mathematics.** Creativity constrained by logic and a set of axioms dictates how ideas can be manipulated and combined to reveal unshakable truths. Every right-angled triangle drawn from before Pythagoras and on to eternity satisfies the famous theorem that bears his name. There are no exceptions. Sure, you can change the assumptions and find yourself exploring new realms, such as triangles drawn on a curved surface like the skin of a basketball, which can upend Pythagoras's conclusion. But fix your assumptions, double-check your work, and your result is ready to be chiseled in stone. No climbing to the mountain-top, no wandering the desert, no triumphing over the underworld. You can sit comfortably at a desk and use paper, pencil, and a penetrating mind to create something timeless.

The perspective opened my world. I had never really asked myself **why** I was so deeply attracted to mathematics and physics. Solving problems, learning how the universe is put together—that's what had always captivated me. I now became convinced that I was drawn to these disciplines because they hovered above the impermanent nature of the everyday. However overblown my youthful sensibilities rendered my commitment, I was suddenly sure I wanted to be part of a journey toward insights so fundamental that they would never change. Let governments rise and fall, let World Series be won and lost, let legends of film, television, and stage come and go. I wanted to spend my life catching a glimpse of something transcendent.

In the meantime, I still had that psychology paper to write. The assignment was to develop a theory of why we humans do what we do, but each time I started writing, the project seemed decidedly nebulous. If you clothed reasonable-sounding ideas in the right language it seemed that you could pretty much make it up as you went along. I mentioned this over dinner at my dorm and one of the resident advisors suggested I take a look at Oswald Spengler's **Decline of the West.** A German historian and philosopher, Spengler had an abiding interest in both mathematics and science, no doubt the very reason his book had been recommended.

The aspects responsible for the book's fame and scorn—predictions of political implosion, a veiled espousal of fascism—are deeply troubling and have since been used to support insidious ideologies, but I was too narrowly focused for any of this to register. Instead, I was intrigued by Spengler's vision of an all-encompassing set of principles that would reveal hidden patterns playing out across disparate cultures, on par with the patterns articulated by calculus and Euclidean geometry that had transformed understanding in physics and mathematics.[2] Spengler was talking my language. It was inspiring for a text on history to revere math and physics as a template for progress. But then came an observation that caught me thoroughly by surprise: "Man is the only being that knows death; all others become old, but with a consciousness wholly limited to the moment which must seem to them eternal," knowledge that instills the "essentially human fear in the presence of death."

Spengler concluded that "every religion, every scientific investigation, every philosophy proceeds from it."[3]

I remember dwelling on the last line. Here was a perspective on human motivation that made sense to me. The enchantment of a mathematical proof might be that it stands forever. The appeal of a law of nature might be its timeless quality. But what drives us to seek the timeless, to search for qualities that may last forever? Perhaps it all comes from our singular awareness that we are anything but timeless, that our lives are anything but forever. Resonating with my newfound thinking on math, physics, and the allure of eternity, this felt on target. It was an approach to human motivation grounded in a plausible reaction to a pervasive recognition. It was an approach that didn't make it up on the fly.

As I continued to think about this conclusion, it seemed to promise something grander still. Science, as Spengler noted, is one response to the knowledge of our inescapable end. And so is religion. And so is philosophy. But, really, why stop there? According to Otto Rank, an early disciple of Freud who was fascinated by the human creative process, we surely shouldn't. The artist, in Rank's assessment, is someone whose "creative impulse . . . attempts to turn ephemeral life into personal immortality."[4] Jean-Paul Sartre went farther, noting that life itself is drained of meaning "when you have lost the illusion of being eternal."[5] The suggestion, then, threading its way through these and other thinkers who followed, is that much of human culture—from artistic exploration to scientific discovery—is driven by life reflecting on the finite nature of life.

Deep waters. Who knew that a preoccupation with all things mathematics and physics would tap into visions of a unified theory of human civilization driven by the rich duality of life and death?

Well, OK. I'll take a breath as I remind my long-ago sophomore self not to get too carried away. Nonetheless, the excitement I felt proved more than a passing wide-eyed intellectual wonderment. In the nearly four decades since, these themes, often simmering on a mental back burner, have stayed with me. While my day-to-day work has pursued unified theories and cosmic origins, in ruminating on the larger significance of scientific advances I have found myself returning repeatedly to questions of time and the limited allotment we are each given. Now, by training and temperament, I'm skeptical of one-size-fits-all explanations—physics is littered with unsuccessful unified theories of nature's forces—only more so if we venture into the complex realm of human behavior. Indeed, I have come to see my awareness of my own inevitable end as having considerable influence but not providing a blanket explanation for everything I do. It's an assessment, I imagine, that to varying degrees is common. Still, there is one domain in which mortality's tentacles are particularly evident.

Across cultures and through the ages, we have placed significant value on permanence. The ways we have done so are abundant: some seek absolute truth, others strive for enduring legacies, some build formidable monuments, others pursue immutable laws, and others still turn with fervor toward one or another version of the everlasting. Eternity, as these preoccupations demonstrate, has a powerful pull on the mind aware that its material duration is limited.

In our era, scientists equipped with the tools of experiment, observation, and mathematical analysis have blazed a new trail toward the future, one that for the first time has revealed prominent features of the eventual if still far-off landscape-to-be. Although obscured by mist here and fog there, the panorama is becoming sufficiently clear that we cogitating creatures can glean more fully than ever before how we fit into the grand expanse of time.

It is in this spirit, in the pages that follow, that we will walk the timeline of the universe, exploring the physical principles that yield orderly structures from stars and galaxies to life and consciousness, within a universe destined for decay. We will consider arguments establishing that much as human beings have limited life spans, so too do the very phenomena of life and mind in the universe.

Indeed, at some point it is likely that organized matter of any kind will not be possible. We will examine how self-reflective beings contend with the tension entailed in these realizations. We emerge from laws that, as far as we can tell, are timeless, and yet we exist for the briefest moment of time. We are guided by laws that operate without concern for destination, and yet we constantly ask ourselves where we are headed. We are shaped by laws that seem not to require an underlying rationale, and yet we persistently seek meaning and purpose.

In short, we will survey the universe from the beginning of time to something akin to the end, and through the journey explore the breathtaking ways in which restless and inventive minds have illuminated and responded to the fundamental transience of everything.

We will be guided in the exploration by insights from a variety of scientific disciplines. Through analogies and metaphors, I explain all necessary ideas in nontechnical terms, presuming only the most modest background. For particularly challenging concepts, I provide brief summaries that allow you to move on without losing the trail. In the endnotes I explain finer points, spell out particular mathematical details, and provide references and suggestions for further reading.

Because the subject is vast and our pages limited,

I have chosen to walk a tight path, pausing at various junctures I consider essential for recognizing our place within the larger cosmological story. It is a journey powered by science, given significance by humanity, and the source of a vigorous and enriching adventure.

UNTIL THE
END OF TIME

1

THE LURE OF
ETERNITY

Beginnings, Endings, and Beyond

In the fullness of time all that lives will die. For more than three billion years, as species simple and complex found their place in earth's hierarchy, the scythe of death has cast a persistent shadow over the flowering of life. Diversity spread as life crawled from the oceans, strode on land, and took flight in the skies. But wait long enough and the ledger of birth and death, with entries more numerous than stars in the galaxy, will balance with dispassionate precision. The unfolding of any given life is beyond prediction. The final fate of any given life is a foregone conclusion.

And yet this looming end, as inevitable as the setting sun, is something only we humans seem to notice. Long before our arrival, the thunderous

clap of storm clouds, the raging might of volcanoes, the tremulous shudders of a quaking earth
surely sent scurrying everything with the power to
scurry. But such flights are an instinctual reaction
to a present danger. Most life lives in the moment,
with fear born of immediate perception. It is only
you and I and the rest of our lot that can reflect on
the distant past, imagine the future, and grasp the
darkness that awaits.

It's terrifying. Not the kind of terror that makes
us flinch or run for cover. Rather, it's a foreboding
that quietly lives within us, one we learn to tamp
down, to accept, to make light of. But underneath
the obscuring layers is the ever-present, unsettling
fact of what lies in store, knowledge that William
James described as the "worm at the core of all our
usual springs of delight."[1] To work and play, to
yearn and strive, to long and love, all of it stitching
us ever more tightly into the tapestry of the lives
we share, and for it all then to be gone—well, to
paraphrase Steven Wright, it's enough to scare you
half to death. Twice.

Of course, most of us, in the service of sanity,
don't fixate on the end. We go about the world
focused on worldly concerns. We accept the inevitable and direct our energies to other things. Yet
the recognition that our time is finite is always with
us, helping to shape the choices we make, the challenges we accept, the paths we follow. As cultural
anthropologist Ernest Becker maintained, we are

under a constant existential tension, pulled toward the sky by a consciousness that can soar to the heights of Shakespeare, Beethoven, and Einstein but tethered to earth by a physical form that will decay to dust. "Man is literally split in two: he has an awareness of his own splendid uniqueness in that he sticks out of nature with a towering majesty, and yet he goes back into the ground a few feet in order blindly and dumbly to rot and disappear forever."[2] According to Becker, we are impelled by such awareness to deny death the capacity to erase us. Some soothe the existential yearning through commitment to family, a team, a movement, a religion, a nation—constructs that will outlast the individual's allotted time on earth. Others leave behind creative expressions, artifacts that extend the duration of their presence symbolically. "We fly to Beauty," said Emerson, "as an asylum from the terrors of finite nature."[3] Others still seek to vanquish death by winning or conquering, as if stature, power, and wealth command an immunity unavailable to the common mortal.

Across the millennia, one consequence has been a widespread fascination with all things, real or imagined, that touch on the timeless. From prophesies of an afterlife, to teachings of reincarnation, to entreaties of the windswept mandala, we have developed strategies to contend with knowledge of our impermanence and, often with hope, sometimes with resignation, to gesture toward eternity. What's new in our age is the remarkable power of science to tell a lucid story not only of the past, back to the big bang, but also of the future. Eternity itself may forever lie beyond the reach of our equations, but our analyses have already revealed that the universe we have come to know is transitory. From planets to stars, solar systems to galaxies, black holes to swirling nebulae, nothing is everlasting. Indeed, as far as we can tell, not only is each individual life finite, but so too is life itself. Planet earth, which Carl Sagan described as a "mote of dust suspended on a sunbeam," is an evanescent bloom in an exquisite cosmos that will ultimately be barren. Motes of dust, nearby or distant, dance on sunbeams for merely a moment.

Still, here on earth we have punctuated our moment with astonishing feats of insight, creativity, and ingenuity as each generation has built on the achievements of those who have gone before, seeking clarity on how it all came to be, pursuing coherence in where it is all going, and longing for an answer to why it all matters.

Such is the story of this book.

Stories of Nearly Everything

We are a species that delights in story. We look out on reality, we grasp patterns, and we join them into narratives that can captivate, inform, startle, amuse, and thrill. The plural—narratives—is utterly essential. In the library of human reflection, there is no single, unified volume that conveys ultimate understanding. Instead, we have written many nested stories that probe different domains of human inquiry and experience: stories, that is, that parse the patterns of reality using different grammars and vocabularies. Protons, neutrons, electrons, and nature's other particles are essential for telling the reductionist story, analyzing the stuff of reality, from planets to Picasso, in terms of their microphysical constituents. Metabolism, replication, mutation, and adaptation are essential for telling the story of life's emergence and development, analyzing the biochemical workings of remarkable molecules and the cells they govern. Neurons, information, thought, and awareness are essential for the story of mind—and with that the narratives proliferate: myth to religion, literature to philosophy, art to music, telling of humankind's struggle for survival, will to understand, urge for expression, and search for meaning.

These are all ongoing stories, developed by thinkers hailing from a great range of distinct disciplines. Understandably so. A saga that ranges from quarks to consciousness is a hefty chronicle. Still, the different stories are interlaced. **Don Quixote** speaks to humankind's yearning for the heroic, told through the fragile Alonso Quijano, a character created in the imagination of Miguel de Cervantes, a living, breathing, thinking, sensing, feeling collection of bone, tissue, and cells that, during his lifetime, supported organic processes of energy transformation and waste excretion, which themselves relied on atomic and molecular movements honed by billions of years of evolution on a planet forged from the detritus of supernova explosions scattered throughout a realm of space emerging from the big bang. Yet to read Don Quixote's travails is to gain an understanding of human nature that would remain opaque if embedded in a description of the movements of the knight-errant's molecules and atoms or conveyed through an elaboration of the neuronal processes crackling in Cervantes's mind while writing the novel. Connected though they surely are, different stories, told with different languages and focused on different levels of reality, provide vastly different insights.

Perhaps one day we will be able to transit seamlessly between these stories, connecting all products of the human mind, real and fictive, scientific and imaginative. Perhaps we will one day invoke a unified theory of particulate ingredients to explain the overwhelming vision of a Rodin and the myriad responses **The Burghers of Calais** elicits from those who experience it. Maybe we will fully grasp how the seemingly mundane, a glint of light reflecting from a spinning dinner plate, can churn through the powerful mind of a Richard Feynman and compel him to rewrite the fundamental laws of physics. More ambitious still, perhaps one day we will understand the workings of mind and matter so completely that all will be laid bare, from black holes to Beethoven, from quantum weirdness to Walt Whitman. But even without having anything remotely near that capacity, there is much to be gained by immersion in these stories—scientific, creative, imaginative—appreciating when and how they emerged from earlier ones playing out on the cosmic timeline and tracing the developments, both controversial and conclusive, that elevated each to their place of explanatory prominence.[4]

Clear across the collection of stories, we will find two forces sharing the role of leading character. In chapter 2 we will meet the first: **entropy.** Although familiar to many through its association with disorder and the often-quoted declaration that disorder is always on the rise, entropy has subtle qualities that allow physical systems to develop in a rich variety of ways, sometimes even appearing to swim against the entropic stream. We will see important examples of this in chapter 3, as particles in the aftermath of the big bang seemingly flout the drive to disorder as they evolve into organized structures like stars, galaxies, and planets—and ultimately, into configurations of matter that surge with the current of life. Asking how that current switched on takes us to the second of our pervasive influences: **evolution.**

Although it is the prime mover behind the

gradual transformations experienced by living systems, evolution by natural selection kicks in well before the first forms of life start competing. In chapter 4, we will encounter molecules battling molecules, struggles for survival waged in an arena of inanimate matter. Round upon round of molecular Darwinism, as such chemical combat is called, is what likely produced a series of ever more robust configurations ultimately yielding the first molecular collections we would recognize as life. The details are the stuff of cutting-edge research, but with the last couple of decades of stupendous progress, the consensus is that we are heading down the right track. Indeed, it may be that the dual forces of entropy and evolution are well-matched partners in the trek toward the emergence of life. While that might sound like an odd coupling— entropy's public rap veers close to chaos, seemingly the antithesis of evolution or of life—recent mathematical analyses of entropy suggest that life, or at least lifelike qualities, might well be the **expected** product of a long-lived source of energy, like the sun, relentlessly raining down heat and light on molecular ingredients that are competing for the limited resources available on a planet like earth.

Tentative though some of these ideas currently are, what's certain is that a billion or so years after the earth formed it was teeming with life developing under evolutionary pressure, and so the next phase of developments is standard Darwinian fare.

Chance events, like being hit by a cosmic ray or suffering a molecular mishap during the replication of DNA, result in random mutations, some with minimal impact on the organism's health or welfare but others making it more or less fit in the competition for survival. Those mutations that enhance fitness are more likely to be passed on to descendants because the very meaning of "more fit" is that the trait's carrier is more likely to survive to reproductive maturity and produce fit offspring. From generation to generation, qualities that enhanced fitness thus spread widely.

Billions of years later, as this long process continued to unfold, a particular suite of mutations provided some forms of life with an enhanced capacity for cognition. Some life not only became aware, but became aware of being aware. That is, some life acquired conscious self-awareness. Such self-reflective beings have naturally wondered what consciousness is and how it arose: How can a swirl of mindless matter think and feel? Various researchers, as we will discuss in chapter 5, anticipate a mechanistic explanation. They argue that we need to understand the brain—its components, its functions, its connections—with far greater fidelity than we now do, but once we have that knowledge, an explanation of consciousness will follow. Others anticipate that we are up against a far greater challenge, arguing that consciousness is the most difficult conundrum we have ever encountered, one that will require radically new perspectives regarding not just mind but also the very nature of reality.

Opinions converge when assessing the impact our cognitive sophistication has had on our behavioral repertoire. Across tens of thousands of generations during the Pleistocene, our forebears joined together in groups that subsisted through hunting and gathering. In time, an emerging mental dexterity provided them with refined capacities to plan and organize and communicate and teach and evaluate and judge and problem-solve. Leveraging these enhanced abilities of the individual, groups exerted increasingly influential communal forces. Which takes us to the next collection of explanatory episodes, those focused on developments that made us us. In chapter 6 we examine our acquisition of language and subsequent obsession with the telling of stories; chapter 7 probes a particular genre of stories, those that foreshadow and transition into religious traditions; and in chapter 8 we explore the long-standing and widespread pursuit of creative expression.

In seeking the origin of these developments, both common and sacred, researchers have invoked a wide range of explanations. For us, an essential guiding light will continue to be Darwinian evolution, applied now to human behavior. The brain, after all, is but another biological structure evolving via selection pressures, and it is the brain that informs what we do and how we respond.

Over the past few decades, cognitive scientists and evolutionary psychologists have developed this perspective, establishing that much as our biology has been shaped by the forces of Darwinian selection, so too has our behavior. And thus in our trek across human culture we will often ask whether this or that behavior may have enhanced the prospects for survival and reproduction among those who long ago practiced it, promoting its wide propagation throughout generations of descendants. However, unlike the opposable thumb or upright gait—inherited physiological features tightly linked to specific adaptive behaviors—many of the brain's inherited characteristics mold predilections rather than definitive actions. We are influenced by these predispositions but human activity emerges from a comingling of behavioral tendencies with our complex, deliberative, self-reflective minds.

And so a second guiding light, distinct but no less important, will be trained on the inner life that comes hand in hand with our refined cognitive capacities. Following a trail marked by many thinkers, we will come to a revealing vista: with human cognition we surely harnessed a powerful force, one that in time elevated us to the dominant species worldwide. But the mental faculties that allow us to shape and mold and innovate are the very ones that dispel the myopia that would otherwise keep us narrowly focused on the present. The ability to manipulate the environment thoughtfully provides the capacity to shift our vantage point, to hover above the timeline and contemplate what was and imagine what will be. However much we'd prefer it otherwise, to achieve "I think, therefore I am" is to run headlong into the rejoinder "I am, therefore I will die."

Mildly put, the realization is disconcerting. Yet most of us can take it. And our survival as a species attests to our brethren having been able to take it too. But how do we do it?[5] According to one line of thought, we tell and retell stories in which our place in a vast universe migrates to center stage, and the possibility of our being permanently erased is challenged or is ignored—or, simply put, is just not in the cards. We craft works in painting, sculpture, movement, and music in which we wrest control of creation and invest ourselves with the power to triumph over all things finite. We envision heroes, from Hercules to Sir Gawain to Hermione, who stare down death with a steely resolve and demonstrate, albeit fancifully, that we can conquer. We develop science, providing insights into the workings of reality that we transform into powers earlier generations would have reserved for gods. In short, we can have our cognitive cake—the nimbleness of thought that, among much else, reveals our existential predicament—and enjoy eating it too. Through our creative capacities we have developed formidable defenses against what would otherwise have been debilitating disquiet.

All the same, because motives don't fossilize, tracing the inspiration for human behavior can be a knotty undertaking. Perhaps our creative forays, from the stags at Lascaux to the equations of general relativity, emerge from the brain's naturally selected but overly active ability to detect and coherently organize patterns. Perhaps these and related pursuits are exquisite but adaptively superfluous by-products of a sufficiently large brain released from full-time focus on securing shelter and sustenance. As we will discuss, theories abound but unassailable conclusions are elusive. What lies beyond question is that we imagine and we create and we experience works, from the Pyramids to the Ninth Symphony to quantum mechanics, that are monuments to human ingenuity whose durability, if not whose content, point toward permanence.

And with that, having considered cosmic origins, explored the formation of atoms, stars, and planets, and swept across the emergence of life, consciousness, and culture, we will cast our sights toward the very realm that for millennia, literally and symbolically, has both stimulated and quelled our cosmic anxiety. We will look, that is, from here to eternity.

Information, Consciousness, and Eternity

Eternity will be a long time coming. A lot will happen along the way. Breathless futurists and

Hollywood sci-fi spectaculars envision what life and civilization will be like over spans that while significant by human standards pale in comparison to cosmic timescales. It is an entertaining pastime to extrapolate from a short stretch of exponential technological innovation to future developments, but such predictions are likely to differ profoundly from how things will actually unfold. And that's over relatively familiar durations of decades, centuries, and millennia. Over cosmic timescales, predicting these sorts of details is a fool's errand. Thankfully, for most of what we will explore here, we will find ourselves on more solid ground. My intent is for us to paint the future of the universe with rich colors but only with the broadest of strokes. And with that level of detail, we can portray the possibilities with a reasonable degree of confidence.

An essential recognition is that there is little emotional equanimity to be gained from leaving a trace on a future bereft of anyone there to notice. The future we tend to envision, even if only implicitly, is one that's populated by the kinds of things we care about. Evolution will surely drive life and mind to take on a wealth of forms supported by a range of platforms—biological, computational, hybrid, and who knows what else. But regardless of the unpredictable details of physical composition or environmental backdrop, most of us imagine that in the vastly distant future, life of some stripe, and intelligent life more particularly, will exist and it will think.

And this raises a question that will ride along with us throughout the journey: Can conscious thought persist indefinitely? Or might the thinking mind, like the Tasmanian tiger or the ivory-billed woodpecker, be something sublime that rises up for a period but then goes extinct? I'm not focused on any individual consciousness, so the question has nothing to do with wished-for technologies—cryogenic, digital, whatever—capable of preserving a given mind. Instead, I am asking whether the phenomenon of thought, supported by a human brain or an intelligent computer or entangled particles floating in the void or any other physical process that proves relevant, can persist arbitrarily far into the future.

Why wouldn't it? Well, think about the human incarnation of thought. It arose in conjunction with a fortuitous set of environmental conditions explaining why, for example, our thinking takes place here and not on Mercury or on Halley's comet. We think here because the conditions here are hospitable to life and thought, which is why deleterious changes to earth's climate are so distressing. What's not at all obvious is that there is a cosmic version of such consequential but parochial concerns. By thinking of thought as a physical process (an assumption we will examine), it is not surprising that thought can take place only when certain stringent environmental conditions are met, whether on earth in the here and now or somewhere else in the there and then. And so as we consider the broad-brush evolution of the universe, we will determine whether the evolving environmental conditions across space and time can support intelligent life indefinitely.

The assessment will be guided by insights from research in particle physics, astrophysics, and cosmology that allow us to predict how the universe will unfold over epochs that dwarf the timeline back to the bang. There are significant uncertainties, of course, and like most scientists I live for the possibility that nature will slap down our hubris and reveal surprises we can't yet fathom. But focusing on what we've measured, on what we've observed, and on what we've calculated, what we'll find, as laid out in chapters 9 and 10, is not heartening. Planets and stars and solar systems and galaxies and even black holes are transitory. The end of each is driven by its own distinctive combination of physical processes, spanning quantum mechanics through general relativity, ultimately yielding a mist of particles drifting through a cold and quiet cosmos.

How will conscious thought fare in a universe experiencing such transformation? The language for asking and answering this question is provided once again by entropy. And by following the entropic trail we will encounter the all-too-real possibility that the very act of thinking, undertaken by any entity of any kind anywhere, may be thwarted by an unavoidable buildup of environmental waste: in the distant future, anything that thinks may burn up in the heat generated by its own thoughts. Thought itself may become physically impossible.

While the case against endless thought will be based on a conservative set of assumptions, we will also consider alternatives, possible futures more conducive to life and thinking. But the most straightforward reading suggests that life, and intelligent life in particular, is ephemeral. The interval on the cosmic timeline in which conditions allow for the existence of self-reflective beings may well be extremely narrow. Take a cursory glance at the whole shebang, and you might miss life entirely. Nabokov's description of a human life as a "brief crack of light between two eternities of darkness"[6] may apply to the phenomenon of life itself.

We mourn our transience and take comfort in a symbolic transcendence, the legacy of having participated in the journey at all. You and I won't be here, but others will, and what you and I do, what you and I create, what you and I leave behind contributes to what will be and how future life will live. But in a universe that will ultimately be devoid of life and consciousness, even a symbolic legacy—a whisper intended for our distant descendants—will disappear into the void.

Where, then, does that leave us?

Reflections on the Future

We tend to absorb findings about the universe intellectually. We learn some new fact about time or unified theories or black holes. It momentarily tickles the mind, and if sufficiently impressive, it sticks. The abstract nature of science often leads us to dwell on its content cognitively, and only then, and then only rarely, does that understanding have a chance of touching us viscerally. But on the occasions when science does conjure both reason and emotion, the result can be powerful.

Case in point: Some years ago when I began to think about scientific predictions regarding the far future of the universe, my experience was mostly cerebral. I absorbed relevant material as a fascinating but abstract collection of insights entailed by the mathematics of nature's laws. Still, I found that if I pressed myself to **really** imagine all life, all thought, all struggle, and all accomplishment being a fleeting aberration on an otherwise lifeless cosmic timeline, I absorbed it differently. I could sense it. I could feel it. And I don't mind sharing that the first few times I went there, the journey was dark. Through decades of study and scientific research, I've often had moments of elation and wonder, but never previously had results in mathematics and physics overwhelmed me with a hollow dread.

Over time, my emotional engagement with

these ideas has refined. Now, more often than not, contemplating the far future leaves me with a feeling of calm and connection, as if my own identity hardly matters because it has been subsumed by what I can only describe as a feeling of gratitude for the gift of experience. Since, more than likely, you don't know me personally, let me put this in context. I'm open-minded with a sensibility that demands rigor. I come from a world in which you make your case with equations and replicable data, a world in which validity is determined by unambiguous calculations that yield predictions matching experiments digit by digit, sometimes as far as a dozen places beyond the decimal point. So the first time I had one of these moments of calm connection—I happened to be at a Starbucks in New York City—I was deeply suspicious. Perhaps my Earl Grey was tainted with some bad soy milk. Or perhaps I was losing my mind.

On reflection, neither was the case. We are the product of a long lineage that has soothed its existential discomfort by envisioning that we leave a mark. And the more lasting the mark, the more indelible its imprint, the more a life seems to be a life that mattered. In the words of philosopher Robert Nozick—but they could just as easily have come from George Bailey—"Death wipes you out . . . To be wiped out completely, traces and all, goes a long way toward destroying the meaning of one's life."[7] Especially for those, like me, without a traditional religious orientation, an emphasis on not being "wiped out," a relentless focus on endurance, can pervade everything. My upbringing, my education, my career, my experiences have all been informed by it. During every stage, I've gone forward with an eye trained on the long view, on seeking to accomplish something that would last. There is no mystery why my professional preoccupation has been dominated by mathematical analyses of space, time, and nature's laws; it is hard to imagine another discipline that more readily keeps one's day-to-day thoughts focused on questions that transcend the moment. But scientific discovery itself casts this perspective in a different light. Life and thought likely populate a minute oasis on the cosmic timeline. Though governed by elegant mathematical laws that allow for all manner of wondrous physical processes, the universe will play host to life and mind only temporarily. If you take that in fully, envisioning a future bereft of stars and planets and things that think, your regard for our era can appreciate toward reverence.

And **that** is the feeling I had experienced at Starbucks. The calm and connection marked a shift from grasping for a receding future to the feeling of inhabiting a breathtaking if transient present. It was a shift, for me, compelled by a cosmological counterpart to the guidance offered through the ages by poets and philosophers, writers and artists, spiritual sages and mindfulness teachers, among countless others who tell us the simple but surprisingly subtle truth that life is in the here and now. It's a mind-set that is hard to maintain but one that has infused the thinking of many. We see it in Emily Dickinson's "Forever—is composed of Nows"[8] and Thoreau's "eternity in each moment."[9] It is a perspective, I've found, that becomes all the more palpable when we immerse ourselves in the full expanse of time—beginning to end—a cosmological backdrop that provides unmatched clarity on how singular and fleeting the here and now actually is.

The purpose of this book is to provide that clarity. We will journey across time, from our most refined understanding of the beginning to the closest science can take us to the very end. We will explore how life and mind emerge from the initial chaos, and we will dwell on what a collection of curious, passionate, anxious, self-reflective, inventive, and skeptical minds do, especially when they notice their own mortality. We will examine the rise of religion, the urge for creative expression, the ascent of science, the quest for truth, and the longing for the timeless. The deep-seated affinity for something permanent, for what Franz Kafka identified as our need for "something indestructible,"[10] will then propel our continued march toward the distant future, allowing us to assess the prospects for everything we hold dear, everything constituting reality as we know it, from planets and stars, galaxies and black holes, to life and mind.

Across it all, the human spirit of discovery will shine through. We are ambitious explorers seeking to grasp a vast reality. Centuries of effort have illuminated dark terrains of matter, mind, and the cosmos. During millennia to come, the spheres of illumination will grow larger and brighter. The journey so far has already made evident that reality is governed by mathematical laws that are indifferent to codes of conduct, standards of beauty, needs for companionship, longings for understanding, and quests for purpose. Yet, through language and story, art and myth, religion and science, we have harnessed our small part of the dispassionate, relentless, mechanical unfolding of the cosmos to give voice to our pervasive need for coherence and value and meaning. It is an exquisite but temporary contribution. As our trek across time will make clear, life is likely transient, and all understanding that arose with its emergence will almost certainly dissolve with its conclusion. Nothing is permanent. Nothing is absolute. And so, in the search for value and purpose, the only insights of relevance, the only answers of significance, are those of our own making. In the end, during our brief moment in the sun, we are tasked with the noble charge of finding our own meaning.

Let us embark.

2

THE LANGUAGE OF TIME

Past, Future, and Change

On the evening of January 28, 1948, nestled between a performance of the Schubert Quartet in A minor and a presentation of English folk songs, BBC Radio broadcast a debate between one of the most potent intellectual forces of the twentieth century, Bertrand Russell, and Jesuit priest Frederick Copleston.[1] The topic? The existence of God. Russell, whose innovative writings in philosophy and humanitarian principles would earn him the 1950 Nobel Prize in Literature, and whose iconoclastic political and social views would earn him a pink slip from both Cambridge University and the City College of New York, provided numerous arguments for questioning, if not rejecting, the existence of a creator.

One line of thought that informed Russell's position is relevant to our exploration here. "So far as scientific evidence goes," Russell noted, "the universe has crawled by slow stages to a somewhat pitiful result on this earth and is going to crawl by still more pitiful stages to a condition of universal death." With such a bleak outlook, Russell concluded, "if this is to be taken as evidence of purpose, I can only say that the purpose is one that does not appeal to me. I see no reason, therefore, to believe in any sort of God."[2] The theological thread will be stitched into later chapters. Here, I want to focus on Russell's reference to scientific evidence for a "universal death." It comes from a nineteenth-century discovery with roots as humble as its conclusions are profound.

By the mid-1800s, the Industrial Revolution was in full swing and across a landscape of mills and factories the steam engine had become the workhorse driving production. Nevertheless, even with the critical leap from manual to mechanical labor, the efficiency of the steam engine—the useful work performed compared to the quantity of fuel consumed—was meager. Roughly 95 percent of the heat generated by burning wood or coal was lost to the environment as waste. This inspired a handful of scientists to think deeply about the physical principles governing steam engines, seeking ways to burn less and get more. Over the course of many decades their research gradually led to an iconic result that has become justly famous: **the second law of thermodynamics.**

In (highly) colloquial terms, the law declares that the production of waste is unavoidable. And what makes the second law vitally important is that while steam engines were the catalyst, the law is universally applicable. The second law describes a fundamental characteristic inherent in all matter and energy, regardless of structure or form, whether animate or inanimate. The law reveals (loosely, again) that everything in the universe has an overwhelming tendency to run down, to degrade, to wither.

Stated in these everyday terms you can see where Russell was coming from. The future seemingly holds a continued deterioration, a relentless conversion of productive energy into useless heat, a steady draining, so to speak, of the batteries powering reality. But a more precise understanding of the science reveals that this summary of where reality is headed obscures a rich and nuanced progression, one that has been under way since the big bang and will carry onward to the far future. It is a progression that helps explain our place in the cosmic timeline, clarifies how beauty and order can be produced against a backdrop of degradation and decay, and also offers potential ways, exotic though they may be, to sidestep the bleak end Russell envisioned. As it is this very science, involving concepts such as entropy, information, and energy, that will guide much of our journey, it is worth our while to spend a little time understanding it more fully.

Steam Engines

Far be it from me to suggest that the meaning of life will be found lurking in the sweaty depths of a clamorous steam engine. But understanding the steam engine's capacity to absorb heat from burning fuel and use it to drive recurrent motion in a locomotive's wheels or a coal mine's pump proves indispensable to grasping how energy—of any sort and in any context—evolves over time. And the way energy evolves has a deep impact on the future of matter, mind, and all structure in the universe. So let's descend from the lofty realms of life and death and purpose and meaning to the incessant chugging and clanking of an eighteenth-century steam engine.

The scientific basis of the steam engine is simple but ingenious: Water vapor—steam—expands when heated and so pushes outward. A steam engine harnesses this action by heating a canister filled with steam that is capped by a snuggly fitting piston free to slide up and down along the canister's inner surface. As the heated steam expands, it pushes forcefully against the piston, and that outward thrust can drive a wheel to turn, or a mill to grind, or a loom to weave. Then, having expended energy through this outward exertion, the steam cools and the piston slides back to its initial position, where it stands ready to be pushed when the steam is heated again—a cycle that will repeat so long as there is burning fuel to heat the steam anew.[3]

While history records the steam engine's central role in the Industrial Revolution, the questions it raised for fundamental science were just as significant. Can we understand the steam engine with mathematical precision? Is there a limit to how efficient its conversion of heat into useful activity can be? Are there aspects of the steam engine's basic processes that are independent of the details of mechanical design or materials used and thus speak to universal physical principles?

In puzzling over these issues, the French physicist and military engineer Sadi Carnot launched the field of thermodynamics—the science of heat, energy, and work. You wouldn't have known it from sales of his 1824 treatise, **Reflections on the Motive Power of Fire**.[4] But while slow to catch on, his ideas would inspire scientists over the course of the following century to develop a radically new perspective on physics.

A Statistical Perspective

The traditional scientific perspective, handed down in mathematical form by Isaac Newton, is that physical laws provide ironclad predictions for how things move. Tell me the location and velocity of an object at a particular moment, tell me the forces that are acting upon it, and Newton's equations do the rest, predicting the object's subsequent trajectory. Be it the moon pulled by earth's gravity or a baseball you just whacked toward center field, observations have confirmed that these predictions are spot-on accurate.

But here's the thing. If you took high school physics, perhaps you will recall that when we analyze the trajectories of macroscopic objects we generally, if quietly, invoke a great many simplifications. For the moon and the baseball we ignore their internal structure and imagine that each is just a single massive particle. It's a coarse approximation. Even a grain of salt contains about a billion billion molecules, and that's, well, a grain of salt. Yet as the moon orbits we generally don't care about the jostling motion of one or another molecule inhabiting the dusty Sea of Tranquility. As the baseball soars, we don't care about the vibration of one or another molecule residing in its cork core. The overall movement of the moon or the baseball as a whole is all we're after. And for that, applying Newton's laws to these simplified models does the trick.[5]

These successes highlight the challenge faced by nineteenth-century physicists concerned with steam engines. The hot steam pushing against the engine's piston comprises an enormous number of water molecules, perhaps a trillion trillion particles. We can't ignore this internal structure as we do in our analysis of the moon or the baseball. It is the motion of these very particles—slamming into the piston, bouncing off its surface, hitting the walls of the container, streaming back toward the piston again—that lies at the heart of the engine's workings. The problem is that there is no way that anyone, anywhere, however smart they may be and however formidable the computers they may use, can calculate all of the individual trajectories followed by such an enormous collection of water molecules.

Are we stuck?

You might think so. But as it turns out, we are saved by a change in perspective. Large collections can sometimes yield their own powerful simplifications. It is surely difficult, impossible really, to predict exactly when you will next sneeze. However, if we broaden our view to the larger collection of all humans on earth, we **can** predict that in the next second there'll be roughly eighty thousand sneezes worldwide.[6] The point is that by shifting to a statistical perspective, earth's large population becomes the key—not the obstacle—to predictive power. Large groups often display statistical regularities absent at the level of the individual.

An analogous approach for large groups of atoms and molecules was pioneered by James Clerk Maxwell, Rudolf Clausius, Ludwig Boltzmann, and many of their colleagues. They advocated jettisoning detailed consideration of individual trajectories in favor of statistical statements describing the average behavior of large collections of particles. They showed that this approach not only makes calculations mathematically tractable, but the physical properties it can quantify are the very ones that matter most. The pressure pushing on a steam engine's piston, for instance, is hardly affected by the precise path followed by this or that individual water molecule. Instead, the pressure arises from the average motion of the trillions upon trillions of molecules that slam into its surface each second. **That's** what matters. And **that's** what the statistical approach allowed the scientists to calculate.

In our current era of political polls, population genetics, and big data more generally, the shift to a statistical framework might not sound radical. We've grown accustomed to the power of statistical insights extracted from studying large groups. But in the nineteenth and early twentieth centuries, statistical reasoning was a departure from the rigid precision that had come to define physics. Bear in mind, too, that up through the early years of the twentieth century there were still well-respected scientists who challenged the existence of atoms and molecules—the very basis of a statistical approach.

Notwithstanding the naysayers, it didn't take long for statistical reasoning to prove its worth. In 1905, Einstein himself quantitatively explained the jittery motion of pollen grains suspended in a glass of water by invoking the continual bombardment by H_2O molecules. With that success, you had to be one heck of a contrarian to doubt the existence of molecules. What's more, a growing archive of theoretical and experimental papers revealed that conclusions based on statistical analyses of large collections of particles—describing how they bounce around containers and thereby exert pressure on this surface, or acquire that density, or relax to that temperature—matched data so exquisitely that there was simply no room to question the explanatory power of the approach. The statistical basis for thermal processes was thus born.

This was all a great triumph and has allowed physicists to understand not only steam engines but also a broad range of thermal systems—from earth's atmosphere, to the solar corona, to the vast collection of particles swarming within a neutron star. But how does this relate to Russell's vision of the future, his prognostication of a universe crawling toward death? Good question. Hang tight. We're getting there. But we still have a couple of steps to go. The next is to use these advances to shed light on the quintessential quality of the future: it differs profoundly from the past.

From This to That

The distinction between past and future is at once basic and pivotal to human experience. We were born in the past. We will die in the future. In between, we witness innumerable happenings that unfold through a sequence of events that, if considered in reverse order, would appear absurd. Van Gogh painted **Starry Night** but could not then lift the swirling colors through reverse brushstrokes, restoring a blank canvas. The **Titanic** scraped along an iceberg and ripped open its hull but could not then reverse engines, retrace its path, and undo the damage. Each one of us grows and ages but we cannot then turn back the hands of our internal clocks and reclaim our youth.

With irreversibility being so central to how things evolve, you would think we could easily identify its mathematical origin within the laws of physics. Surely, we should be able to point to something specific in the equations that ensures that although things can transform from **this** to **that,** the math forbids them from transforming from **that** to **this.** But for hundreds of years the equations we've developed have failed to offer us anything of the sort. Instead, as the laws of physics have been continually refined, passing through the hands of Newton (classical mechanics), Maxwell (electromagnetism), Einstein (relativistic physics), and the dozens of scientists responsible for quantum physics, one feature has remained stable: the laws have steadfastly adhered to a complete insensitivity to what we humans call future and what we call past. Given the state of the world right now, the mathematical equations treat unfolding toward the future or the past in exactly the same way. While that distinction matters to us, profoundly so, the laws shrug at the difference, assessing it as of no greater consequence than a stadium's game clock ticking off time elapsed or time remaining. Which means that if the laws allow for a particular sequence of events to occur, then the laws necessarily permit the reverse sequence too.[7]

As a student, when I first learned about this, it struck me as just shy of ludicrous. In the real world we don't see Olympic divers popping out of pools feetfirst and landing calmly on springboards. We don't see shards of stained glass jumping up from the floor and reassembling into a Tiffany lamp. Clips from films run in reverse are amusing for the very reason that what we see projected differs so thoroughly from anything we experience. And yet, according to the math, the events depicted in reverse-run clips are fully in keeping with the laws of physics.

Why then is our experience so lopsided? Why

do we only ever see events unfold in one temporal orientation and never the reverse? A key part of the answer is revealed by the notion of **entropy,** a concept that will be essential to our understanding of the cosmic unfolding.

Entropy: A First Pass

Entropy is among the more confusing concepts in fundamental physics, a fact that has not diminished the cultural appetite for freely invoking it to describe everyday situations that have evolved from order to chaos or, more simply, from good to bad. As colloquial usage goes, this is fine; at times, I've invoked entropy that way too. But as the scientific conception of entropy will guide our journey—and also lies at the heart of Russell's dark vision of the future—let's tease out its more precise meaning.

Start with an analogy. Imagine you vigorously shake a bag containing a hundred pennies and then dump them out on your dining room table. If you found that all hundred pennies were heads, you'd surely be surprised. But why? Seems obvious, but it's worth thinking through. The absence of even a single tail means each of the hundred coins, randomly flipping, bumping, and jostling, must hit the table and land heads up. All of them. That's tough. Getting that unique outcome is a tall order. By comparison, if we consider even a slightly different outcome, say in which we have a single tail (and the other 99 pennies are still all heads), there are a hundred different ways this can happen: the lone tail could be the first coin, or it could be the second coin, or the third, and so on up to the hundredth coin. Getting 99 heads is thus a hundred times easier—a hundred times more likely—than getting all heads.

Let's keep going. A little figuring reveals that there are 4,950 different ways we can get two tails (first and second coins tails; first and third tails; second and third tails; first and fourth tails; and so forth). A little more figuring and we find that there are 161,700 different ways to have three of the coins come up tails, almost 4 million ways to have four tails; and about 75 million ways to have five tails. The details of the numbers hardly matter; it's the overall trend I'm driving at. Each additional tail allows for a far larger collection of outcomes that fit the bill. Phenomenally larger. The numbers peak at 50 tails (and 50 heads), for which there are about a hundred billion billion billion possible combinations (well, 100,891,344,545,564,193,334,812,497,256 combinations).[8] Getting 50 heads and 50 tails is therefore about a hundred billion billion billion times more likely than getting all heads.

That's why getting all heads would be shocking.

My explanation relies on the fact that most of us intuitively analyze the collection of pennies much as Maxwell and Boltzmann advocated analyzing a container of steam. Just as the scientists turned a cold shoulder to analyzing the steam molecule by molecule, we typically don't evaluate a random collection of pennies coin by coin. We hardly care or notice if the 29th penny is heads or the 71st is tails. Instead, we look at the collection as a whole. And the feature that catches our attention is the number of heads compared to the number of tails: Are there more heads than tails or more tails than heads? Twice as many? Three times as many? Roughly equal amounts? We can detect significant changes in the ratio of heads to tails, but random rearrangements that preserve the ratio—like flipping the 23rd, 46th, and 92nd coins from tails to heads while also flipping the 17th, 52nd, and 81st coins from heads to tails—are virtually indistinguishable. Consequently, I divvied up the possible outcomes into groups, each containing those configurations of coins that pretty much look the same, and I enumerated the membership of each group: I counted the number of outcomes with no tails, the number of outcomes with 1 tail, the number of outcomes with 2 tails, and so on, up to the number of outcomes with 50 tails.

The key realization is that these groups do not have equal membership. Not even close. That made it obvious why you'd be shocked for a random shake of the pennies to yield no tails (a group with precisely 1 member), slightly less shocked for a random shake to yield one tail (a group with

100 members), a touch less shocked still to find two tails (a group with 4,950 members), but you'd yawn if the shake yields a configuration that's half heads and half tails (a group with roughly one hundred billion billion billion members). The greater the membership in a given group, the more likely it is that a random outcome will belong to that group. Group size matters.

If this material is new to you, you may not realize that we have now illustrated the essential concept of entropy. The entropy of a given configuration of the pennies is the size of its group—the number of fellow configurations that pretty much look like the given configuration.[9] If there are many such look-alikes, the given configuration has high entropy. If there are few such look-alikes, the given configuration has low entropy. All else being equal, a random shake is more likely to belong to a group with higher entropy since such groups have more members.

This formulation also connects with the colloquial uses of entropy I referenced at the outset of this section. Intuitively, messy configurations (think of a chaotic desktop piled high with scattered documents, pens, and paper clips) have high entropy because a great many rearrangements of the constituents all pretty much look the same; randomly rearrange a messy configuration and it still looks messy. Orderly configurations (think of a pristine desktop with all documents, pens, and paper clips neatly placed in their designated positions) have low entropy because very few rearrangements of the constituents look the same. As with the pennies, high entropy beckons because messy arrangements far outnumber orderly ones.

Entropy: The Real Deal

The pennies are particularly useful because they illustrate the approach scientists developed for dealing with the voluminous collection of particles constituting physical systems, whether water molecules flitting to and fro in a hot steam engine or air molecules drifting across the room in which you are now breathing. As with the pennies, we ignore the details of individual particles—whether any one particular molecule of water or air happens to be here or there is of little consequence—and instead group together those configurations of the particles that pretty much look the same. For the pennies, the criterion for look-alikes invoked the ratio of heads to tails because typically we are indifferent to the disposition of any particular coin, and generally take note only of the configuration's overall appearance. But what does "pretty much look the same" mean for a large collection of gas molecules?

Think about the air now filling your room. If you're like me and the rest of us, you couldn't care less whether this molecule of oxygen is flitting by the window or that molecule of nitrogen is bouncing off the floor. You care only that each time you inhale there is an adequate volume of air to meet your needs. Well, there are a couple of other features you likely care about too. If the air's temperature was so hot that you scorched your lungs, you'd be unhappy. Or if the air's pressure was so high (and you hadn't equalized it with the air already in your eustachian tubes) that you burst your eardrums, you'd be unhappy too. Your concern, then, is with the air's volume, the air's temperature, and the air's pressure. Indeed, these are the very macroscopic qualities that physicists from Maxwell and Boltzmann on through today care about too.

Accordingly, for a large collection of molecules in a container, we say that different configurations "pretty much look the same" if they fill out the same volume, have the same temperature, and exert the same pressure. Much as with the pennies, we group together all look-alike configurations of the molecules and say that each member of the group gives rise to the same **macrostate.** The entropy of the macrostate is the number of such look-alikes. Assuming you are not just now turning on a space heater (affecting temperature) or putting up an impermeable room divider (affecting volume), or pumping in additional oxygen (affecting pressure), the ever-evolving configuration of air molecules flitting to and fro in the room you are now inhabiting all belong to the same group—they all pretty much look the same—as they all yield the very same macroscopic features you are currently experiencing.

The organization of particles into groups of look-alikes provides an extraordinarily powerful schema. Just as randomly tossed pennies are more likely to belong to a group with greater membership (with higher entropy), so too for randomly bouncing particles. The realization is as straightforward as its implications are far-reaching: Whether the bouncing particles are in a steam engine, in your room, or anywhere else, by understanding the typical features of the most commonplace configurations (those that belong to the groupings with the greatest membership), we can make predictions about the system's macroscopic qualities—the very qualities we care about. These are statistical predictions, to be sure, but ones with a fantastically high probability of being accurate. And the kicker is that we achieve all this while avoiding the insurmountable complexity of analyzing the trajectories of an absurdly large number of particles.

To carry out the program we therefore need to sharpen our ability to identify commonplace (high entropy) versus rare (low entropy) particle configurations. That is, given the state of a physical system, we need to determine whether there are many or few rearrangements of the constituents that would leave the system looking pretty much the same. As a case study, let's visit your steam-filled bathroom just after you've taken a long hot shower. To determine the steam's entropy, we need to count the number of configurations of the molecules—their possible positions and their possible speeds—that all have the same macroscopic properties, i.e., have the same volume, same temperature, and same pressure.[10] Carrying out the count mathematically for a collection of H_2O molecules is more challenging than the analogous count for a collection of pennies, but is something most physics majors learn to do by their sophomore year. More straightforward, and more enlightening too, is working out how volume, temperature, and pressure qualitatively affect entropy.

Volume first. Imagine that the flitting H_2O molecules are tightly clustered in one tiny corner of your bathroom, creating a dense knot of steam. In this configuration, the possible rearrangements of the positions of the molecules will be sharply curtailed; as you move the H_2O molecules around, you have to keep them within that knot or else the modified configuration **will** look different. By comparison, when the steam is evenly spread throughout your bathroom, the game of molecular musical chairs is far less constrained. You can exchange the positions of molecules near the vanity with those floating by the light fixture, those near the shower curtain with those hovering by the window, and yet, overall, the steam will look the same. Note too that the bigger your bathroom, the greater the number of locations you have for sprinkling around the molecules, which also increases the number of rearrangements available. The conclusion, then, is that smaller and tightly clustered configurations of molecules have lower entropy, while larger and evenly spread configurations have higher entropy.

Temperature next. At the level of molecules, what do we mean by temperature? The answer is well-known. Temperature is the average speed of a collection of molecules.[11] Something is cold when the average speed of its molecules is low and it is hot when the average speed is high. So determining how temperature affects entropy is tantamount to determining how the average molecular speed affects entropy. And much as we found with molecular positions, a qualitative assessment is close at hand. If the temperature of the steam is low, the allowed rearrangements of the molecular speeds will be comparatively few in number: to keep the temperature fixed—and thus ensure that the configurations all pretty much look the same—you have to offset any increase in the speeds of some molecules by a suitable decrease in the speeds of others. But the burden of having low temperature (low average molecular speed) is that you don't have a lot of room to decrease the speeds before hitting rock bottom, zero. The available range of possible molecular speeds is thus narrow, and so your freedom to rearrange the speeds is limited. By comparison, if the temperature is high, your game of musical chairs once again revs up: with a higher average, the range of molecular speeds—some larger than the average and some smaller—is much wider, providing greater latitude for mixing up the speeds while preserving the average. More rearrangements of the molecular speeds that all pretty much look the same means that higher temperature generally entails higher entropy.

Finally, pressure. The pressure of the steam on your skin or on the bathroom walls is due to the impact of streaming H_2O molecules that slam into these surfaces: each molecular impact exerts a tiny push, and so the greater the number of molecules the higher the pressure. For a given temperature and volume, pressure is thus determined by the total number of steam molecules in your bathroom, a quantity whose consequences for entropy can be worked out with the greatest of ease. Fewer H_2O molecules in your bathroom (you took a shorter shower) means fewer rearrangements are possible, and so entropy is lower; more H_2O molecules (you took a longer shower) means more rearrangements are possible, and so entropy is higher.

To summarize: Having fewer molecules, or having lower temperature, or filling a smaller volume results in lower entropy. Having more molecules, or having higher temperature, or filling a larger volume results in higher entropy.

From this brief survey, let me underscore one way of thinking about entropy, lacking in precision but providing a useful rule of thumb. You should expect to encounter high-entropy states. Because such states can be realized by a great many different arrangements of the constituent particles, they're typical, pedestrian, easily configured, a dime a dozen. By contrast, if you encounter a low-entropy state it should command your attention. Low entropy means there are far fewer ways the given macrostate can be realized by its microscopic ingredients, and so such configurations are hard to come by, they're unusual, they're carefully arranged, they're rare. Step out of a long hot shower and find the steam uniformly spread throughout your bathroom: high entropy and totally unsurprising. Step out of a long hot shower and find the steam all clustered in a perfect little cube hovering in front of the mirror: low entropy and extraordinarily unusual. So unusual, in fact, that were you to encounter such a configuration you should be extremely skeptical of the explanation that you've simply come upon one of those unlikely things that on occasion happen. That **could** be the explanation. But I'd bet my life it isn't. Just as you'd suspect there's a reason beyond mere chance that a hundred pennies on your dining room table are all heads (such as someone judiciously flipped over each coin that landed tails), you should seek an explanation beyond mere chance for any low-entropy configurations you encounter.

This reasoning applies even to the seemingly mundane, like coming across an egg or an anthill or a mug. The orderly, crafted, low-entropy nature of these configurations calls out for an explanation. That the random motion of precisely the right particles could just happen to coalesce into an egg or an anthill or a mug is conceivable, but far-fetched. Instead, we're motivated to find more convincing explanations, and of course we don't have far to search: the egg and the anthill and the mug each arise from particular forms of life arranging the otherwise random configuration of particles in the environment to yield orderly structures. How life is able to produce such exquisite order is a theme we will address in later chapters. For now, the lesson is simply that low-entropy configurations should be viewed as a diagnostic, a clue that powerful organizing influences may be responsible for the order we've encountered.

In the late 1800s, armed with these ideas, many of his own devising, Austrian physicist Ludwig Boltzmann believed he could address the question that launched this section of our discussion: What distinguishes the future from the past? His answer relied on a quality of entropy articulated by the second law of thermodynamics.

Laws of Thermodynamics

While entropy and the second law enjoy a great many cultural references, public nods to the first law of thermodynamics are less common. Yet to fully grasp the second law it's good to grasp the first law first. As it turns out, the first law is widely known too, but under an alias. It's the law of energy conservation. Whatever energy you have at the beginning of a process is the same energy you'll have at the end of the process. You must be fastidious in your energy accounting, including all forms into which an initial cache of energy may have transformed, such as kinetic energy (energy of motion), or potential energy (stored energy, as in a stretched spring), or radiation (energy carried by fields, like the electromagnetic or gravitational fields), or heat (the random jittery motion of molecules and atoms). But if you keep track carefully, the first law of thermodynamics ensures that the energy balance sheet will balance.[12]

The second law of thermodynamics focuses on entropy. Unlike the first law, the second is not a law of conservation. It is a law of growth. The second law declares that over time there is an overwhelming tendency of entropy to increase. In colloquial terms, special configurations tend to evolve toward ordinary ones (your carefully pressed shirt becomes creased and wrinkled) or order tends to descend into disorder (your organized garage degenerates into a haphazard mess of tools, storage boxes, and sporting equipment). While this depiction provides fine intuitive imagery, Boltzmann's statistical formulation of entropy allows us to describe the second law with precision and, just as important, gain a clear understanding of why it's true.

It comes down to a numbers game. Consider again the pennies. If you carefully arrange the pennies so they are all heads, a low-entropy configuration, and then subject them to a little shaking and jostling, you expect to get at least a few tails, a higher-entropy configuration. If you shake them further, it's conceivable that you'll get back to all heads, but that would require the jostling to be just right, to be so perfectly attuned that it flips back only those few coins that happened to be tails. That's extraordinarily improbable. It is fantastically more likely that the jostling will instead flip a random collection of the pennies. Some of the few coins that were tails might revert to heads, but of the coins that were heads, many more will become tails. So straightforward logic—no fancy math, no unduly abstract ideas—reveals that if you begin with all heads, random shaking will drive an increase in the number of tails. An increase, that is, of entropy.

The progression toward a greater number of tails will continue until we reach a roughly 50-50 heads-tails split. At that point, the jostling will tend to flip about as many heads to tails as tails to heads, and so the pennies will spend most of their time migrating among the members of the most populous, highest-entropy groups.

What's true for the pennies is true more generally. Bake bread and you can be sure that the aroma will shortly fill rooms far from the kitchen. At first, the molecules released as the bread bakes are clustered near the oven. But those molecules will gradually disperse. The reason, similar to our explanation with the pennies, is that there are many more ways for the aroma molecules to spread compared with ways for them to cluster. It is thus overwhelmingly more probable that through random bumping and jostling the molecules will waft outward as opposed to clump inward. The low-entropy configuration of molecules clustered near the oven thus naturally evolves toward the higher-entropy state in which they are spread throughout your house.[13]

Saying it yet more generally, if a physical system is not already in the highest-entropy state available, it is overwhelmingly likely that it will evolve toward it. The explanation, illustrated well by the bread's aroma, rests on the most basic reasoning: because the number of configurations with more entropy is enormously greater than those with less entropy (by the very definition of entropy), the odds are enormously larger that random jostling— the relentless bumping and vibrating of atoms and molecules—will drive the system toward higher entropy, not lower. The progression will continue until we reach a configuration with the highest entropy available. From that point onward the jostling will tend to drive the constituents to migrate among the (typically) gargantuan number of configurations of the highest-entropy states.[14]

That's the second law of thermodynamics. And that's why it is true.

Energy and Entropy

The discussion might lead you to think that the first and second laws are thoroughly distinct. After all, one focuses on energy and its conservation, the other on entropy and its growth. But there's a deep connection between them, highlighting a fact implicit in the second law that we will return to repeatedly: all energy is not created equal.

Consider, as an example, a stick of dynamite. Because all the energy stored in the dynamite is contained in a tight, compact, orderly chemical package, the energy is easy to harness. Place the dynamite where you want its energy deposited and light the fuse. That's it. Post-explosion, all of the dynamite's energy still exists. That's the first law in action. But because the dynamite's energy has been transformed into the rapid and chaotic motion of widely dispersed particles, harnessing the energy is now extremely difficult. So, although the total amount of energy doesn't change, the character of the energy does.

Before the explosion, we say that the dynamite's energy is high quality: it's concentrated and easy to access. After the explosion, we say that the energy is low quality: it's spread out and difficult to utilize. And since the exploding dynamite fully abides by the second law, going from order to disorder—from low entropy to high entropy—we associate low entropy with high-quality energy and high entropy with low-quality energy. Yes, I know. It's a lot of highs and lows to keep track of. But the conclusion is pithy: whereas the first law of thermodynamics declares that the quantity of energy is conserved over time, the second law declares that the quality of that energy deteriorates over time.

Why then is the future different from the past? The answer, apparent from what we've now developed, is that the energy powering the future is of lower quality than that powering the past. The future has higher entropy than the past.

Or at least that is what Boltzmann proposed.

Boltzmann and the Big Bang

Boltzmann was surely onto something. But there is a subtle clarification to the second law whose implications, truth be told, took some time even for Boltzmann to appreciate fully.

The second law is not a law in the traditional sense. The second law does **not** absolutely preclude entropy from decreasing. It declares only that such a decrease is unlikely. For the pennies, we've quantified this. Compared with the sole configuration with all heads, it is a hundred billion billion billion times more likely that random shaking will yield a configuration with 50 heads and 50 tails. Shake that high-entropy configuration again, and getting a lower-entropy configuration such as all heads is not forbidden, but because of the highly skewed odds, in practice it doesn't happen.

For an everyday physical system made from far more than a hundred constituents, the odds against entropy decreasing become all the more daunting. As bread bakes it releases billions upon billions of molecules. Configurations in which those molecules spread throughout your home are spectacularly more numerous than those in which they collectively stream back toward the oven. Through their random jostling and bumping, the molecules **could** retrace their steps, find their way back to the loaf, fully undo the cooking process, and leave you with a mound of cold raw dough. But the odds of that happening are closer to zero than the likelihood of splattering paint on a canvas and replicating the **Mona Lisa.** Even so, the point is that were such an entropy-reversing process to happen, it would not contravene the laws of physics. While spectacularly unlikely, the laws of physics **do** allow entropy to go down.

Don't get me wrong. I'm not bringing this up to suggest we might one day uncook bread or witness a car uncrash or see a document unburn. Instead, I'm stressing an important point of principle. I explained earlier that the laws of physics put future and past on equal footing. The laws thus ensure that physical processes that unfold in one temporal sequence can unfold in reverse. And since those very same laws govern everything, including the physical processes responsible for how entropy changes over time, it would indeed be curious, erroneous really, to find that those laws only allow entropy to increase. They don't. All the entropically increasing processes you've experienced day in and day out during your entire life—from the mundane of a shattering glass to the profound of bodily aging—can happen in reverse. Entropy can decrease. It's just ridiculously unlikely.

So where does this leave our quest to explain why the future is different from the past? Well, given a configuration today of less than maximum entropy, the second law shows that the future is overwhelmingly likely to be different because entropy is overwhelmingly likely to increase. Configurations of matter that have less than the maximum possible entropy are chomping at the bit to proceed to higher entropy. And with that observation, some exploring the difference between past and future rest easy, reckoning that their work is done.

But the work is not done. Just as importantly, we need to explain how it is that we find ourselves today in such a special, unlikely, surprising state of less than maximum entropy—a universe replete with orderly structures from planets and stars to peacocks and people. Had that not been the case, had today's configuration been the expected, ordinary, unsurprising state of maximum entropy, then with great odds the universe would continue to inhabit such a state, yielding a future no different from the past. Like a bag of pennies jostling through the enormous number of configurations with roughly 50 heads and 50 tails, the universe would relentlessly meander through the enormous landscape of its highest entropy configurations— widely dispersed particles streaming this way and that across space, a cosmic version of your uniformly steam-filled bathroom.[15] Today's state of less-than-maximum entropy is, luckily for us, far more interesting. It provides the opportunity for particles to join into structures and for macroscopic change to occur. And so we are led to ask: How did today's less-than-maximum entropy state come to be?

Dutifully following the second law, we conclude that today's state derives from yesterday's even lower entropy state. And that state, we envision, derives from the day-before-yesterday's still lower entropy state, and so on, yielding a trail of ever-decreasing entropy taking us ever farther back in time until we finally reach the big bang. A highly ordered, exceedingly low entropy starting point at the big bang is why today's universe is not entropically maxed out, allowing for an eventful future that differs from the past.

Can we go further and explain why the beginning of the universe was so ordered? We will come back to this question in the next chapter, where we'll explore cosmological theorizing. For now we note that our survival requires order, from our internal molecular organization supporting a wealth of life-sustaining functions, to the food sources that provide us with high-quality energy, to the crafted tools and habitats that are essential for our continued existence. Without an environment chock-full of low-entropy ordered structures, we humans would not be here to notice.

Heat and Entropy

I began this chapter with Bertrand Russell lamenting a universe subject to relentless decline. With the second law's declaration of rising entropy we have now caught a glimpse of what inspired his dark prophecy. Think of rising entropy as increasing disorder and you have the gist of it. But to fully appreciate the future challenges that will face life, mind, and matter—a theme we will amply explore in subsequent chapters—we need to establish a link between the modern description of the second law of thermodynamics as I have laid it out and the original formulation developed in the mid-1800s. In that earlier version, the second law codified what was obvious to anyone working with steam engines: the process of burning fuel to run a machine always produces heat and waste—degradation. However, as the earlier version made no mention of counting configurations of particles and made no use of probabilistic reasoning, it might seem a world away from the statistical statement of entropic growth that we've been developing. But there is a deep and direct connection between the two formulations, one that reveals why the steam engine's conversion of high-quality energy into low-quality heat is illustrative of a ubiquitous degradation taking place throughout the cosmos.

I'll explain the link in two steps. First, let's look at the relationship between entropy and heat. Then, in the next section, we will tie together heat and the statistical statement of the second law.

Grab hold of a sauté pan's hot handle and it feels like heat is flowing to your hand. But does anything actually flow? There was a time long ago when scientists thought the answer was yes. They envisioned a fluidlike substance, called "caloric," which would flow from hotter locations to cooler ones much like a river flows from upstream to downstream. In time, the more refined understanding of matter's ingredients provided a different description. When you grasp the pan's handle, its faster-moving molecules collide with the slower-moving molecules in your hand, on average causing the speed of those in your hand to go up and those in the handle to go down. You sense the increased speed of the molecules in your hand as warmth; the temperature of your hand has increased. Correspondingly, the slower speed of the molecules in the handle means its temperature has decreased. What flows, then, is not a substance. The molecules in the handle stay in the handle, and those in your hand stay in your hand. Instead, much as information flows from one person to the next in a game of telephone, molecular agitation flows from molecules in the handle to those in your hand when you grab it. And so, whereas matter itself does not flow from handle to hand, a quality of matter—average molecular speed—does. That is what we mean by the flow of heat.

The same description applies to entropy. As the temperature of your hand increases, its molecules bounce around more quickly, the range of their possible speeds widens—increasing the number of attainable configurations that pretty much look the same—and so the entropy of your hand increases too. Correspondingly, as the temperature of the handle decreases, its molecules move slower, the range of their possible speeds narrows—decreasing the number of attainable configurations that pretty much look the same—and so the entropy of the handle decreases.

Whoa. Entropy **decreases**?

Yes. But this has nothing to do with rare statistical flukes like dumping a bag of pennies and getting all heads, as described in the previous section. The entropy of the hot handle will decrease every time you grab it. The simple yet vital point the sauté pan illustrates is that the second law's dictum of entropy increase refers to the **total** entropy of a complete physical system, which necessarily includes everything with which the system interacts. Since your hand interacts with the pan's handle, you can't apply the second law to the handle on its own. You must include both the handle and your hand (and, to be more precise, the entire pan, the stove, the surrounding air, and so on). And a careful accounting shows that the increase in the entropy of your hand outstrips the decrease in the entropy of the handle, ensuring that the total entropy does indeed go up.

So, much as with heat, there is a sense in which entropy can flow. For the pan, it flows from the handle to your hand. The handle becomes a little more ordered and your hand becomes a little less ordered. Again, the flow is not in the form of a tangible substance that was initially in the handle and has now moved to your hand. Rather, the entropy flow denotes an interaction between the molecules in the handle and those in your hand that affects the properties of each. In this case, it changes their average speeds—their respective temperatures—and that, in turn, affects the entropy they each contain.

As the description makes manifest, the flow of heat and the flow of entropy are intimately connected. To absorb heat is to absorb energy that is carried by random molecular motion. That energy, in turn, drives the receiving molecules to move more quickly or spread more widely, thus contributing to an increase in entropy. The conclusion, then, is that to shift entropy from here to there, heat needs to flow from here to there. And when heat flows from here to there, entropy shifts from here to there. In short, entropy rides the wave of flowing heat.

With this understanding of the interrelationship between heat and entropy, let's now revisit the second law.

Heat and the Second Law of Thermodynamics

Explaining why we experience events unfurling in one direction but not the reverse brought us to Boltzmann and his statistical version of the second law: entropy is overwhelmingly likely to increase toward the future, making reverse-run sequences (in which entropy would decrease) fantastically

improbable. How does this relate to the earlier formulation of the second law, inspired by the steam
engine, which was phrased in terms of the relentless
production of waste heat by physical systems?

The connection is that the two starting points—
reversibility and steam engines—are tightly linked.
The reason is that the steam engine relies on a cyclical process: a piston is thrust forward by expanding
steam and is then reset to its original position, where
it awaits the next thrust. The steam, too, reverts
to its original volume, temperature, and pressure,
as must all of the engine's vital parts, readying the
engine to heat back up and thrust the piston once
again. While none of this requires the ridiculously
improbable unfolding in which every molecule
finds its way back to exactly the same spot or acquires exactly the same speed as it had at the start
of the previous cycle, it does require that the overall
arrangement—the engine's macrostate—returns to
the same form to initiate each subsequent cycle.

What does that imply for entropy? Well, since
entropy is a count of the microscopic configurations that present as the same macrostate, if the
macrostate of the steam engine is reset at the start of
each new cycle, then its entropy must be reset too.
Which means that the entropy the steam engine
acquires during a given cycle (as it absorbs heat
from the burning fuel, as it generates additional
heat through friction of its moving parts, and so
on) must all be expelled to the environment by

the time the cycle concludes. How does the steam engine accomplish this? Well, we've seen that to transfer entropy you must transfer heat. Thus, for the steam engine to reset itself for the next cycle, **it must release heat into the environment.** That's the historical statement of the second law of thermodynamics, the inevitable expunging of waste heat into the environment—the very degradation that so weighed on Bertrand Russell—now derived from the statistical version of the second law.[16]

This is the destination I've been heading for, so feel free to jump to the next section. But if you have the patience, there is one detail I'd be remiss not to mention. You might wonder, if the steam engine absorbs heat from the burning fuel (thus absorbing entropy) only to release heat to the environment (thus releasing entropy), how does it have any remaining energy to accomplish useful tasks, like powering a locomotive? The answer is that the steam engine releases less heat than it absorbs and yet is still able to fully purge the entropy it has built up. Here's how it goes:

The steam engine absorbs heat and entropy from the burning fuel and releases heat and entropy to the cooler environment. The temperature difference between the fuel and the environment is what's vital. To see why, imagine turning on two identical space heaters, one in a room that's frigid and the other in a room that's hot. In the frigid room, the cold molecules of air are jolted by the space heater, causing them to move faster and disperse widely, and so their entropy increases significantly. In the hot room, the air molecules are already moving fast and flitting widely, and so the space heater only slightly increases their entropy. (It is kind of like turning up the beat at a wild New Year's party and barely noticing that the revelers dance a touch more quickly, but turn up the beat at the Thiksay Monastery, enticing monks to break from their meditative practice and start krumping, and you would readily see the change.) So even though the two space heaters are identical, the entropy they transfer to their surroundings is different: while each generates the same amount of heat, the space heater in the cooler environment transfers more entropy. A cooler environment thus amplifies a given amount of heat received into a larger entropic increase. With that realization, we see that the steam engine can discharge all of the entropy it acquires from the hotter fuel by expelling only part of that heat to the cooler environment. The remaining heat is then available to drive the steam to expand, pushing the piston and accomplishing useful work.

That's the explanation, but don't let the details cloud the larger conclusion: over time, physical systems will, with fantastic likelihood, evolve from configurations of lower entropy toward configurations of higher entropy. If a system, like a steam engine, seeks to maintain its structural integrity, it must stave off the natural drive toward increased entropy by transferring the entropy it builds up to its surroundings. To do so, the engine must release waste heat to the environment.

The Entropic Two-Step

If you carefully think through the steps we've followed, you'll see that although the steam engine has been peppered throughout, our conclusions transcend this eighteenth-century starting point. The essence of our analysis is a close accounting of entropy, and that accounting can be carried out in any context. This is a key realization, because the shifting of entropy from the steam engine to its surroundings through the release of heat is but one version of an utterly ubiquitous process we will encounter as we follow the unfolding of the cosmos. I call it the **entropic two-step,** by which I mean any process in which the entropy of a system decreases because it shifts a more than compensating increase in entropy to the environment. The two-step ensures that even though entropy may decrease here it will increase there, securing the net entropic increase we expect based on the second law.

The entropic two-step lies at the heart of how a universe heading toward ever-greater disorder can nevertheless yield and support ordered structures like stars, planets, and people. A theme we will encounter repeatedly is that when energy flows through a system—like the energy from burning coal flowing through the steam, driving work, and then exiting to the surrounding environment—it carries away entropy and can thus sustain or even produce order in its wake.

It is this entropic dance that will choreograph the rise of life and mind, as well as most everything that minds deem to matter.

You Are a Steam Engine

With the importance of resetting the entropy each time a steam engine goes through a cycle, you might wonder what would happen if the entropy reset were to fail. That's tantamount to the steam engine not expelling adequate waste heat, and so with each cycle the engine would get hotter until it would overheat and break down. If a steam engine were to suffer such a fate it might prove inconvenient but, assuming there were no injuries, would likely not drive anyone into an existential crisis. Yet the very same physics is central to whether life and mind can persist indefinitely far into the future. The reason is that what holds for the steam engine holds for you.

It is likely that you don't consider yourself to be a steam engine or perhaps even a physical contraption. I, too, only rarely use those terms to describe myself. But think about it: your life involves processes no less cyclical than those of the steam engine. Day after day, your body burns the food you eat and the air you breathe to provide energy for your internal workings and your external activities. Even the very act of thinking—molecular motion taking place in your brain—is powered by these energy-conversion processes. And so, much like the steam engine, you could not survive without resetting your entropy by purging excess waste heat to the environment. Indeed, that's what you do. That's what we all do. All the time. It's why, for example, the military's infrared goggles designed to "see" the heat we all continually expel do a good job of helping soldiers spot enemy combatants at night.

We can now appreciate more fully Russell's mind-set when imagining the far future. We are all waging a relentless battle to resist the persistent accumulation of waste, the unstoppable rise of entropy. For us to survive, the environment must absorb and carry away all the waste, all the entropy, we generate. Which raises the question, Does the environment—by which we now mean the observable universe—provide a bottomless pit for absorbing such waste? Can life dance the entropic two-step indefinitely? Or might there come a time when the universe is, in effect, stuffed and so is unable to absorb the waste heat generated by the very activities that define us, bringing an end to life and mind? In the lachrymose phrasing of Russell, is it true that "all the labors of the ages, all the devotion, all the inspiration, all the noonday brightness of human genius, are destined to extinction in the vast death of the solar system, and that the whole temple of Man's achievement must inevitably be buried beneath the debris of a universe in ruins"?[17]

These are among the central questions we will explore in coming chapters. But we've jumped a little ahead of ourselves. Before we discuss life and mind, let's understand how entropy and the second law play out in the formation of environments necessary for life and mind to take hold.

For that, we head back to the big bang.

3

ORIGINS AND ENTROPY

From Creation to Structure

When mathematics allows scientists to peer back to within a fraction of a second of what may well have been the beginning of the universe, the proximity to traditionally religious terrain suggests to some that there's a deep alliance or a deep connection or a deep conflict straining to be revealed. It's why I receive inquiries about my views on a creator almost as frequently as those asking about science. Indeed, questions often straddle the two. We will have ample time to consider such matters in later chapters, but here we will explore one point of contact, raised at the end of the previous chapter, essential to our larger story: If the second law of thermodynamics burdens the universe with a relentless increase in disorder, how can nature so readily produce exquisitely configured, highly ordered structures, from atoms and molecules, to stars and galaxies, to life and mind? If the universe began with an explosive bang, how could that fiery unfolding have given rise to all the organization—from the swirling arms of the Milky Way, to earth's stunning landscapes, to the intricate connections and corrugated folds of the human brain, to the art, music, poetry, literature, and science such brains produce?

One response, relied upon through the ages to address embryonic versions of such concerns, is that order is hewn from the chaos by a supreme intelligence. Human experience aligns with this anthropomorphically inspired turn. After all, much of the order we daily encounter in modern civilization **is** the handiwork of intelligence. But a proper exegesis of the second law renders an intelligent designer unnecessary. As surprising as it is remarkable, regions containing concentrated energy and order (stars being the archetypal example) are a natural consequence of the universe diligently toeing the second law's line and becoming ever more **disordered.** Indeed, such pockets of order prove to be catalysts that facilitate the universe, over the long run, to reach its entropic potential. Along the way, and as part of this entropic progression, they also facilitate the emergence of life.

To explore the dance between order and disorder

playing out across cosmological history, we begin at the beginning.

Sketching the Big Bang

In the mid-1920s, Jesuit priest Georges Lemaître used Einstein's newly minted description of gravity—the general theory of relativity—to develop the radical idea of a cosmos that began with a bang and has been expanding ever since. Lemaître was no armchair physicist. He received his doctorate from the Massachusetts Institute of Technology and was among the first scientists to apply the equations of general relativity to the cosmos as a whole. Einstein's intuition, which had successfully guided him through an exquisite decade of discoveries into the nature of space, time, and matter, was that objects **in** the universe have a beginning, middle, and an end, but the universe itself always was and would always be. When Lemaître's analysis of Einstein's equations suggested otherwise, Einstein dismissed him out of hand, telling the young researcher, "Your calculations are correct but your physics is abominable."[1] Einstein was emphasizing that you can be adept at manipulating equations and yet lack the good scientific taste to decide which of those mathematical manipulations reflect reality.

A few years later, Einstein performed one of the most famous scientific turnabouts. Detailed observations by the astronomer Edwin Hubble, working at the Mount Wilson Observatory, revealed that distant galaxies are all on the move. They're all rushing away. And the pattern of their exodus—the farther the galaxy, the higher the speed—agreed with the mathematical output of general relativity's equations. With data now supporting Lemaître's abominable physics, Einstein embraced wholeheartedly the conception of a universe that had a beginning.[2]

In the century since Lemaître's innovative calculations, the cosmological theorizing he initiated, together with independent work by the Russian physicist Alexander Friedmann, has been substantially developed and a corpus of observational evidence from ground and space-borne telescopes has been amassed. Here is the modern cosmological account that has emerged: Some fourteen billion years ago, the entire observable universe—all that we can see using the most powerful telescopes imaginable—was compressed into a stupendously hot, incredibly dense nugget, which then rapidly expanded. Cooling as it swelled, particles gradually slowed their frenzied motion and aggregated into clumps, which over time formed stars, planets, all manner of gaseous and rocky debris scattered across space—and us.

In two sentences, that's the story. Let's refine it. Let's examine how, without intent or design, without forethought or judgment, without planning or deliberation, the cosmos yields meticulously ordered configurations of particles from atoms to stars to life. Let's understand how the emergence of such ordered structures squares with the second law's decree of relentlessly increasing disorder. Let's witness the entropic two-step now performed on the cosmological stage.

To that end, we will need to understand various cosmological details more fully. First up: What drove the primordial nugget to start expanding in the first place? Or, in looser language, what ignited the big bang?

Repulsive Gravity

Antonyms abound because experience is full of opposites. Physics, too, has its share: order and disorder, matter and antimatter, positive and negative. But since the time of Newton, the force of gravity appeared to stand apart from this common pattern. Unlike the electromagnetic force, which can push or pull, gravity seemed to be solely an attractive force. According to Newton, gravity exerts a pull between objects, whether particles or planets, that draws them together, but never the reverse. Absent a principle that requires symmetry in all of nature's workings, most who thought deeply about gravity viewed its one-way character as an intrinsic quality that simply had to be accepted. Einstein changed this. According to the general theory of relativity, the gravitational force **can** be repulsive. Newton did not anticipate repulsive gravity, and neither you nor I have ever experienced it. But repulsive gravity does just what the name suggests. Instead of pulling inward, it pushes outward. According to Einstein's equations, big clumpy things like stars and planets exert the usual attractive version of gravity, but there are exotic situations in which the gravitational force can drive things apart.

While the capacity of the gravitational force to be repulsive was known to Einstein, as well as a number of subsequent scientists who worked on the general theory of relativity, its most profound application took more than half a century to be discovered. As a young postdoctoral fellow contemplating the big bang, Alan Guth realized that repulsive gravity might address a confounding cosmic mystery. Observations reveal that space is expanding. Einstein's equations concur. But the equations remained mute on the question of what force, billions of years ago, set the expansion off and running. Guth's detailed mathematical analyses, culminating with a late-night calculational frenzy in December 1979, coaxed the equations to speak.

Guth realized that if a region of space was filled with a particular kind of substance, something I like to call "cosmic fuel," and if the energy contained in the cosmic fuel was spread evenly throughout the region—not clumped like a star or planet—then the resulting gravitational force would indeed be repulsive. More precisely, Guth's calculations revealed that if a tiny region, perhaps as small as a billionth of a billionth of a billionth of a meter across, was suffused with a certain type of energy field (called the **inflaton field,** with the missing "i" being an intentional if quirky naming convention), and if the energy was distributed uniformly, like steam whose density is the same throughout a sauna, the repulsive gravitational push would be so forceful that the speck of space would inflate explosively, almost instantaneously stretching to as large as the observable universe, if not far larger. Repulsive gravity would thus power a bang. And a big bang at that.[3]

In the early 1980s, Soviet physicist Andrei Linde and the American duo Paul Steinhardt and Andreas Albrecht took Guth's handoff and ran with the concept, developing the first fully viable versions of **inflationary cosmology.** In the decades since, these early works have inspired thousands of pages of intricate mathematical calculations and a great many detailed computer simulations, filling journals worldwide with explanations and predictions based on the assumption of an inflationary past. Many of these predictions have now been confirmed by painstakingly precise astronomical measurements. While I won't lead you on a full tour of the observational case for inflationary cosmology, something amply covered in many articles and books, I'll describe one success that many physicists consider the most compelling of all. It is also the feature we will need for the next step in the cosmic unfolding: the formation of stars and galaxies.

The Afterglow

As the early universe rapidly stretched, its scorching heat spread over an ever-widening expanse, diminishing in intensity and cooling steadily.[4] Physicists as far back as the 1940s, long before the inflationary theory was developed, realized that the primordial heat, reduced by spatial expansion to a gentle glow, should still permeate the universe. Dubbed the "afterglow of creation" (or, in technical jargon, the "cosmic microwave background radiation"), this remarkable cosmological remnant was first detected in the 1960s by Bell Lab researchers Arno Penzias and Robert Wilson, whose advanced telecommunication antenna unwittingly tapped into a diffuse radiation permeating space, a mere 2.7 degrees above absolute zero. If you were around in the 1960s, you might have tapped into the radiation too. Part of the static on an old-style television tuned to a channel that had concluded its broadcast for the evening would have been due to this vestige of the big bang.

Inflationary cosmology refines the prediction of

an afterglow by taking into account quantum mechanics, the laws developed in the early decades of the twentieth century to describe physical processes playing out in the microworld. Since we're focused on the entire universe, something big, you might think the preoccupation of quantum physics with all things small would make it irrelevant. And if it weren't for inflationary cosmology, your intuition would be on the mark. But much as stretching a piece of spandex reveals the intricate pattern of its stitches, stretching space through a burst of inflationary expansion reveals quantum features usually cordoned off in the microworld. In essence, inflationary expansion reaches into the microworld and stretches quantum features clear across the sky.

The quantum effect of most relevance is the very one that established an irrefutable break from the classical tradition: the **quantum mechanical uncertainty principle.** Discovered in 1927 by German physicist Werner Heisenberg, the uncertainty principle demonstrated that there are features of the world—like the position and the speed of a particle—that a classical physicist in the mold of Isaac Newton would adamantly claim can be specified with complete certainty but that a quantum physicist realizes are burdened by a quantum fuzziness that makes them uncertain. It's as if the classical tradition viewed the world through pristine, polished spectacles that brought all physical features into perfectly sharp focus, while the spectacles donned by the quantum perspective are inherently foggy. In the large, everyday world of common experience, the quantum fog is too thin to impact our vision, so the classical and quantum perspectives are barely distinguishable. But the smaller you probe, the foggier the quantum lenses become and the fuzzier the view.

The metaphor might suggest that all we need do is clean the quantum lenses. But the uncertainty principle established that no matter how fastidious we are and regardless of the advanced equipment we use, there will always be a minimal amount of fogginess that cannot be wiped away. In fact, my phrasing betrays the bias of human experience. It is only by comparison with the demonstrably incorrect classical view—the view we humans discovered first because it's both simpler and extraordinarily accurate on the scales accessible to human senses—that quantum reality **seems** hazy. It is actually the classical perspective that provides an approximate and hence imprecise view of the true quantum reality.

I don't know why reality is governed by quantum laws. Nobody does. A century of experiments has confirmed a mountain of quantum mechanical predictions, and that's why scientists embrace the theory. Even so, for most of us quantum mechanics remains utterly foreign because its hallmark features emerge over distances so tiny that we just don't experience them in everyday life. If we did, common intuition would be shaped directly by quantum processes and quantum physics would be second nature. Much as you know the implications of Newtonian physics in your bones—you can quickly grab a falling glass, instantly intuiting its Newtonian trajectory—you'd know quantum physics in your bones too. But lacking such quantum intuition, we rely on experiment and mathematics to mold our understanding by portraying aspects of reality we can't directly experience.

The most widely discussed example, already mentioned, concerns the behavior of particles, where we learn to modify the sharp trajectories inherent in classical physics by overlaying incessant jittery motion from quantum uncertainty. As a particle transits from here to there, a classical physicist might draw its trajectory with a pointed quill, while a quantum physicist would run her finger along the wet ink, smearing out the path.[5] But quantum mechanics has relevance far beyond the motion of individual particles, and for cosmology the quantum uncertainty principle has a decisive influence on the inflaton field that fuels the rapid expansion of space. Although I described the inflaton's value as being uniform, taking on the same value at all locations within the inflating patch of space, quantum uncertainty fuzzes this out. Uncertainty overlays quantum jitters on the classical uniformity, resulting in the field's value, and hence its energy, being a tiny bit higher here and a tiny bit lower there.

When inflationary expansion rapidly stretches these minute quantum energy variations, they spread across space making the temperature a touch hotter over here and a touch cooler over there. Not by much. Mathematical analyses, first carried out by physicists in the 1980s, showed that the temperatures of hot and cold spots would differ by as little as one part in a hundred thousand. But the mathematical analyses also suggested that the tiny temperature variations would be visible if you knew how to look for them. The calculations revealed that the stretched-out quantum jitters result in a distinct pattern of temperature variations across space, a cosmological fingerprint available for astronomical forensics. Indeed, since the early 1990s, a sequence of telescopes deployed above the distortions caused by earth's atmosphere have confirmed the predicted pattern of temperature variations with ever-greater precision.

Take a moment to let this sink in. Physicists describe the earliest moments of the universe using Einstein's equations, updated to include Guth's hypothetical energy field filling space, subject to the quantum uncertainty we learned from Heisenberg. Mathematical analyses of the inflationary burst then reveal that it should have left an indelible imprint, a fossil of creation in the form of a specific pattern of minute temperature variations across the night sky. Sophisticated space-based thermometers built nearly fourteen billion years later by a species just coming of scientific age here in the Milky Way have now detected precisely that pattern.

It is a spectacular success, demonstrating once again the uncanny capacity of mathematics to encapsulate nature's patterns. But it would be too strong to conclude that the observations prove that a burst of inflationary expansion happened. When focusing on cosmological events that took place billions of years ago, at an energy scale likely millions of billions of times what we can probe in the laboratory, the best we can do is piece together observations and calculations to build confidence in our explanations. If an inflationary burst were the only way to understand the cosmological data then our confidence would head closer to certainty, but over the years imaginative scientists have developed alternative approaches (we will encounter one of these in chapter 10). All told, my view, shared by many researchers, is that while we need to be open to novel ideas that challenge dominant perspectives, the case for inflationary cosmology developed over the past forty years is formidable.[6] And so as our journey heads onward we will, for the most part, follow the inflationary trail.

With that assessment, let's now consider how an inflationary beginning interfaces with the second law's drive toward greater disorder.

The Big Bang and the Second Law

Notwithstanding centuries of scientific progress, we are no closer to answering the question raised by Gottfried Leibniz—"Why is there something rather than nothing?"—than we were when the German philosopher first expressed this lean distillation of the mystery of existence. Not that people haven't proposed creative ideas and provocative theories. But in asking a question of ultimate origin, we are seeking an answer that requires no antecedent, an answer that does not shift the question one step further back, an answer that is immune to the follow-on questions "Why were things **this** way instead of **that**?" or "Why **these** laws instead of **those**?" No explanation yet proposed has achieved this or even come close.

The inflationary framework surely hasn't. Inflation requires a list of ingredients that includes space, time, the cosmic fuel driving the expansion (the inflaton field), as well as the whole technical apparatus of quantum mechanics and general relativity, which themselves rest upon mathematics from multivariable calculus and linear algebra to differential geometry. There is no known principle that singles out these particular physical laws, articulated using these particular mathematical constructs, as the inevitable starting point for explaining the universe. Instead, we physicists use observation and experiment, together with a hard-to-describe intuitive mathematical sensibility, to guide us toward particular physical laws. We then analyze the laws mathematically to determine which environmental conditions in the earliest moments of the universe, if any, would have sparked the rapid expansion of space. Upon finding, happily, that there are such conditions, we **postulate** that they held near the big bang and we use the equations to determine what subsequently would have happened.

This is the best we can currently do. And it's nothing to sneeze at. The fact that we can use mathematics to describe what we think took place nearly fourteen billion years ago, and from that successfully predict what powerful telescopes should now see, well, it is breathtaking. Sure, profound questions abound, like what or who created space and time, and what or who imposed the guiding grip of mathematics, and what or who is responsible for there being anything at all, but even with all that left unanswered we've gained powerful insight into the cosmic unfolding.

My intent here is to use that insight to grasp how a universe with ever-increasing entropy, destined for ever-greater disorder, creates a wealth of order along the way. With that as our target, let's start with the most basic observation, alluded to in the previous chapter. If entropy has been steadily increasing since the big bang, then the entropy back at the bang must have been much lower than it is today.[7]

What are we to make of this?

Well, by now you've grown accustomed to shrugging your shoulders when encountering a high-entropy configuration—be it coins arranged in a random mixture of heads and tails, steam that uniformly permeates your bathroom, or aromas spread throughout your house. High-entropy configurations are expected, common, run-of-the-mill. But when encountering a low-entropy configuration, you realize that your reaction should be different. A low-entropy configuration is special. It is unusual. It calls out for an explanation for how such an ordered state of affairs came to be.

When applied to the early universe, this reasoning has generated its share of scientific and philosophical hand-wringing. By what force or process did the early universe acquire low entropy? A hundred pennies with all heads has low entropy and yet admits an immediate explanation—instead of dumping the coins on the table, someone carefully arranged them. But what or who arranged the special low-entropy configuration of the early universe? Without a complete theory of cosmic origins, science can't provide an answer. In fact, although it's a question that has kept me up for many a night (literally), science can't yet determine whether it's an issue worthy of any angst at all. Lacking an understanding of why there is something rather than nothing is tantamount to lacking the means to judge how exotic or ordinary that something actually is. To assess whether the detailed conditions of the early universe call for a shrug or a wide-eyed double take requires delineating the process by which those conditions were set.

One scenario that cosmologists have considered imagines that the early universe was a frenzied and chaotic environment, and as a result the value of the inflaton field across space would have fluctuated wildly, somewhat like the surface of boiling water. To generate repulsive gravity and set off the bang, we need a small region of space in which the inflaton's value was uniform (or very nearly so, taking into account quantum jitters). But finding such a uniform region amid the chaotic undulations would be like boiling a vat of water and finding a region on its agitated surface that had suddenly flattened. You have never seen that happen. Not because it's impossible but because it's extraordinarily unlikely. For a region of the vat's randomly bubbling water to pass through the same height at the same moment yielding a flat, orderly, uniform, low-entropy configuration would require an astounding coincidence. Similarly, for the wildly undulating inflaton field to have acquired a uniform value within a small region of space, thus igniting inflationary expansion, would have required an astounding coincidence too. And without an explanation for how this special, orderly, low-entropy, uniform configuration came to be, physicists are deeply uneasy.[8]

Seeking relief from the discomfort, some researchers rely on a simple observation: if you wait long enough even the most unlikely of things will happen. Shake a hundred pennies enough times and eventually they will land all heads. You'd be wise not to hold your breath waiting for this outcome, but it will happen. Similarly, we can argue that in a chaotic environment in which the inflaton's value fluctuates wildly, sooner or later—by sheer chance—there'll be a tiny region in which the random variations that drive the field's value up here or down there will align, resulting in the field having the same value throughout. This requires a statistical fluke, resulting in greater order and hence lower entropy, but on occasion it **will** happen. Not often. But according to this perspective, don't sweat it. Since all of these machinations would have taken place during a period of prehistory, before the rapid expansion of space we call the big bang, there was no one hanging around, arms crossed and shoe tapping, waiting for inflationary expansion to ignite. So let the inflationary preshow run as long as it takes. It is only when the statistical fluke of a uniform inflaton patch happens to happen that things finally change: the big bang is sparked, space inflates, and the cosmological performance begins.

While none of this addresses the most

fundamental questions of origin (the origin of space, or time, or fields, or mathematics, and so on), it shows how a chaotic environment can produce the special, ordered, low-entropy conditions inflation requires. When a tiny speck of space finally makes the statistically unlikely leap to low entropy, repulsive gravity jumps into action and propels it into a rapidly expanding universe—the big bang.

This is not the only proposal for how inflationary expansion may have gotten off the ground. Andrei Linde, one of inflationary cosmology's pioneers, has quipped that for every three researchers there are at least nine opinions on the matter.[9] So we must leave to future research, theoretical and observational, a more definitive answer for how a small region of space became uniformly filled with an inflaton field, thus setting off a burst of spatial expansion. For now, we will simply assume that one way or another, the early universe transitioned into this low-entropy, highly ordered configuration, sparking the bang and allowing us to declare that the rest is history.

Starting from this trailhead, we now set off on our trek, exploring how orderly structures like stars and galaxies form within a universe hurtling toward an ever-more disordered future.

The Origin of Matter and the Birth of Stars

Within a billionth of a billionth of a billionth of a second after the big bang, repulsive gravity stretched a tiny region of space enormously, perhaps far larger than the most distant reaches accessible to the most advanced telescopes possible.[10] Space remained filled with the inflaton field, but within another tiny fraction of a second that changed too. Like the energy in the surface of an expanding soap bubble, the energy in an expanding inflaton-filled region of space is precarious. It's unstable. Much as the soap bubble will eventually pop, transforming its energy into a mist of soapy water droplets, the inflaton field eventually "popped" too—it disintegrated, transforming its energy into a mist of particles.

We don't know the precise identity of these particles, but we can say with confidence that they were not the ordinary constituents of matter you learned about in junior high school. Yet, with the passage of just a few more minutes, a cascade of rapid particle reactions took place all across space—heavy particles disintegrating into sprays of lighter ones; particles with strong affinities joining together into tight conglomerates—transforming the primordial bath into a population of protons, neutrons, and electrons, the stuff of familiar matter (and, likely too, a supply of other more exotic particles, such as dark matter, attested to by a long history of astronomical observations[11]). Within a short time after the bang, the universe was thus filled with a hot, nearly uniform mist of particles, some familiar, others less so, wafting through a swelling spatial expanse.

I've qualified "uniform" with "nearly" because quantum jitters of the inflaton field not only yield temperature variations in the big bang's after-glow, but also ensure that when the inflaton disinte-grates, the density of the resulting particles will vary slightly across space—being a touch higher here, a touch lower there, and so on. These variations are crucial for what happens next: the all-important drive toward clumpy things like stars and galaxies. A region that's slightly denser than its neighbors exerts a slightly greater gravitational pull and so sucks in a slightly greater contingent of the sur-rounding particles. The region thus becomes denser still, and so exerts an even greater gravitational pull, sucking in yet more material. It's a gravitationally driven snowball effect that yields larger and larger clumps of matter. Wait long enough, on the order of hundreds of millions of years, and the gravita-tional snowballing yields particle agglomerations so massive, so compressed, and so hot that they ignite nuclear processes, giving birth to stars. Quantum uncertainty, magnified by inflationary stretching and concentrated by gravitational snowballing, results in the points of light dotting the night sky.

The question, then, is this: How does the star-

forming process, in which gravity coaxes a disordered, nearly uniform bath of particles to
form ordered astrophysical structures, square with
the second law's decree of increasing disorder? The
answer requires that we examine, with a little more
care, the pathways toward higher entropy.

Hurdles on the Path Toward Disorder

As bread bakes in your oven, the particles released
spread outward, occupying an increasingly large
volume, and so their entropy grows. But if you
are in a distant bedroom, you won't immediately
enjoy the bread's freshly baked aroma. It takes
time for the aroma to spread through your house.
You have to wait for the aroma molecules to
migrate outward and occupy the higher-entropy
arrangements that are available. This is typical.
Physical systems generally can't jump directly
to the maximum entropy configuration. Instead, as
the system's particles meander randomly, entropy
gradually increases toward the maximum possible.

Along the pathway toward higher entropy there
can also be hurdles that impede progress. Seal the
oven or close the kitchen door and you make it
harder for the aroma to spread, thus slowing the
rise in entropy. Such hurdles are due to human
intervention, but there are other situations in which
entropic hurdles arise from the laws governing

physical interactions themselves. An example with which I'm intimately familiar from a childhood incident also involves an oven.

One day during fourth grade I came home from school and decided to heat some leftover pizza I found in the refrigerator. I turned the oven to four hundred degrees, slid the pizza onto the middle rack, and waited. After about ten minutes I checked on its progress and was surprised that the pizza was just as cold as when I unwrapped it. It then dawned on me that although I'd turned on the gas, I'd forgotten to light the oven. (Our modest oven, typical of the day, did not have a built-in pilot light, so each use required it to be lit.) Following a procedure I'd witnessed my parents undertake hundreds of times, I leaned into the oven and struck a match, intending to poke it into the oven's small pilot hole. By this point substantial gas had accumulated in the oven's interior, and so when I lit the match it exploded. A wall of flames raced toward me. I tightly closed my eyes as the fire blew by, singeing off my eyebrows, eyelashes, and leaving my face and ears with second- and third-degree burns. The immediate life lesson, emphasized by my parents and reinforced by the months of painful healing, focused on proper use of kitchen appliances. (I eventually got back in the saddle and now do most of the cooking—although I do experience momentary unease when my kids, preparing their own meals, turn on the oven.) But

the larger scientific point is that there can be road-
blocks along the journey to higher entropy that can
be surmounted only with the help of a catalyst.
Here's what I mean.

Natural gas (which is mostly methane, a union of
carbon and hydrogen) can peacefully coexist with
the oxygen in air; the molecules of each gas can
uneventfully comingle. However, as the molecules
spread and intersperse, a distinct and far higher
entropy configuration beckons. But that configuration cannot be reached by simply allowing molecules to continue fanning out. The higher-entropy
configuration requires a chemical reaction. Don't
sweat the details, but let me briefly spell it out.
One molecule of natural gas can combine with two
molecules of oxygen resulting in one molecule of
carbon dioxide, two of water, and, of prime importance, a burst of energy. At the level of molecules,
this is what it means for natural gas to burn. The
chemical reaction releases energy pent up in the
tight bonds holding the gas molecules together,
kind of like what happens when a collection of taut
rubber bands snap. In the case of my oven escapade,
that searingly energetic burst—highly agitated and
fast-moving molecules—scorched my face. All of
which tells us that by releasing energy stored in
orderly chemical bonds and transforming it into the
chaotic motion of rapidly moving molecules, such
chemical reactions yield a sharp increase in entropy.

Though the details are specific to one child's

regrettable mishap, the episode demonstrates a widely applicable physical principle. There can be speed bumps in the entropic road: left on their own, natural gas and oxygen won't combine, they won't burn, and they won't reach the higher-entropy configuration that's available. These chemical constituents are able to clear the entropic hurdle only with the help of a catalyst that can jump-start the reaction. For me, the catalyst was a burning match. The little flame struck by my fourth-grade self set off a domino effect. The flame's energy broke the bonds holding some of the natural gas molecules together, allowing newly freed carbon and hydrogen atoms to combine with ambient atoms of oxygen, which released additional energy that severed more natural gas bonds, driving the process onward. And onward. The explosion was the cascade of energy generated by the rapid re-arranging of chemical bonds.

Note that chemical bonds rely on the electromagnetic force. Positively charged protons attract negatively charged electrons ("unlike electric charges attract"), clasping atomic constituents into molecular unions. Which means that the entropic leap from the calm intermingling of gas molecules to the explosive burning generated by the breaking and forging of chemical bonds is driven by the electromagnetic force. Such is the case for many of the entropy-increasing processes we experience in everyday life.

Although less familiar here on earth, in episodes that repeatedly play out in the cosmos, the evolution toward higher entropy is frequently driven by nature's other forces: the gravitational force and the nuclear forces (the strong nuclear force holds atomic nuclei together, while the weak nuclear force generates radioactive decay). And much as we've now seen in the case of the electromagnetic force, the path toward higher entropy blazed by gravity and the nuclear forces need not be smooth either. There can be hurdles, and there often are. The way the universe surmounts these hurdles—the cosmic analog of my striking a match—is a subtle business. But it's a business we should all care about deeply. Among the transient structures that form as gravity and the nuclear forces guide the universe toward higher entropy are stars and planets, and here on earth, life. For all their majesty, these orderly arrangements are nature's workhorses, harnessing gravity and the nuclear forces to drive the cosmos toward realizing its entropy potential.

Let's focus first on gravity.

Gravity, Order, and the Second Law

Gravity is the weakest of nature's forces, a fact made evident by the simplest demonstration. Pick up a coin. The muscles in your arm beat out the gravitational pull of the entire earth. Whether you consider yourself soft or strapping, victory over the gravitational pull of a planet highlights gravity's intrinsic weakness. The only reason we're even aware of gravity is that it's a cumulative force: every bit of the earth pulls on every bit of a coin, and on every bit of this book, and on every bit of you, and since there's a whole lot of earth, these pulls add up to downward forces we can feel. But the gravitational attraction between two smaller things, like two electrons, is a million billion billion billion billion times weaker than their electromagnetic repulsion.

The intrinsic weakness of gravity is why we didn't even mention it during our earlier exploration of entropy. Were we to include the effects of gravity in everyday situations like the spreading of steam in your bathroom or the drifting of aromas through your house, our discussion of entropy would hardly change. Sure, gravity gently tugs the molecules downward, causing the steam's density to be slightly larger closer to the bathroom floor, but the effect is so small that for a qualitative understanding it just doesn't matter. However, if we shift our attention from the everyday and consider astronomical processes involving a great deal more matter, we encounter a profoundly important interplay between entropy and the gravitational force.

Admittedly, the ideas I'll now explain are somewhat challenging, so feel free to skip to the next section for a summary if at any point the discussion gets too thick for your taste. But the payoff for sticking with me is worthwhile: an understanding of how gravity spontaneously sculpts order from an ever-more-disordered cosmos.

Imagine a cosmic version of the bread-baking scenario. Instead of your house, imagine an enormous box, much larger than the sun, floating in otherwise empty space. And instead of aromas seeping from your oven, imagine that in the middle of the box we start with a ball of gas (to be definite, imagine it's hydrogen, the simplest element on the periodic table) whose molecules are oozing outward. From our experience with the bread's aroma drifting throughout your house, we expect that the gas will evolve toward higher entropy via the molecules dispersing until they uniformly fill the box. But now let's change things a bit. Unlike the case of baking bread, let's add so many molecules into the ball of gas that gravity **does** matter: the gravitational pull experienced by any given molecule, due to the combined gravitational pull exerted by each of the gargantuan number of other gas molecules, significantly affects the molecule's motion. How does this impact our conclusion?

Well, put yourself in the shoes of a gas molecule leading the outward migration. As you stream away from the central cluster, you feel a gravitational pull exerted by all the other molecules tugging you back. That force slows you down. Slower speed means lower temperature. And so as the gas cloud increases its overall volume by expanding outward, the temperature toward the frontier decreases. Hold that in mind, and now jump with me to the perspective of a molecule located nearer to the bulk of the cloud. Being closer, you feel a much stronger gravitational pull compared to your previous experience on a distant frontier. In fact, with enough molecules, the combined gravitational pull will be sufficiently strong to prevent you from migrating outward at all. Instead, you will be pulled inward. You'll thus fall toward the center of the gas cluster, picking up speed as you go. Faster speed means higher temperature, and so as gravity causes the core of the gas cloud to shrink inward, decreasing its volume, its temperature goes up.

Compared with our expectation from baking bread—that the gas would, over time, become evenly distributed throughout the box and attain a uniform temperature—we see that when gravity matters the unfolding is completely different. Gravity results in some molecules being pulled into a hotter, denser core, while others drift outward into a cooler, more diffuse shell that surrounds it.

Modest though these observations may appear, we've now uncovered one of the most influential guiding hands of order in the universe. Let me elaborate.

You've never clutched hold of your morning coffee and found it hotter than when you poured it. That's because heat flows only from higher temperature to lower temperature, and so your hot coffee transfers some of its heat to the cooler environment, causing the coffee's temperature to decrease.[12] For our large cloud of gas, heat also flows from the hot central core to the cooler surrounding shell. Now, I can't fault you for thinking that this flow of heat will cool the core and warm the shell, bringing their temperatures closer together, much as heat transferred from your coffee to the air brings your hot mug closer to room temperature. But—and this is remarkable and remarkably important—when gravity is directing the show, the conclusion is reversed. **As heat flows out from the core, the core gets hotter and the shell gets cooler.**

This is surely counterintuitive, but understanding it is just a matter of connecting dots we've already marked. As the surrounding shell absorbs heat from the core, the additional energy drives the cloud to swell even further. The outward moving molecules, once again, strain against the inward pull of gravity, and hence are slowed even further.[13] The net effect is that the temperature of the expanding shell goes down, not up. Conversely, as the core relinquishes heat, the decrease in energy causes it to contract even further. The inward-moving molecules, flowing in the same direction as the inward pull of gravity, pick up speed as they fall, and so the temperature of the contracting core goes up, not down.

If your coffee behaved this way, you'd be well advised to drink it quickly. The longer you'd wait, the more heat it would relinquish to the surrounding air and the hotter it would become. For coffee, that's absurd. But for a gas cloud large enough for gravity to play a dominant role, this is what happens.

Dwell on this conclusion for a moment and you'll realize that we've encountered a self-amplifying process, much like what happens with credit-card debt—the more you owe, the more interest you're assessed and the greater your debt becomes, driving the cycle to spiral. For a gas cloud, as the core shrinks and its temperature rises, it will relinquish yet more heat to the cooler surroundings, causing the core to shrink yet further and its temperature to rise yet higher. At the same time, the heat absorbed by the shell causes it to expand yet further and its temperature to fall yet lower. The ever-widening gap in temperature between the core and the shell causes heat to flow yet more vigorously and drives the cycle to spiral onward.

Barring intervention or change of circumstance, such self-amplifying cycles continue unabated. For mounting credit-card debt, you intervene by sending a payment or declaring bankruptcy. For the compressed core that's getting ever hotter, nature intervenes with a new physical process: **nuclear fusion.** When a collection of atoms gets sufficiently hot and dense, they slam together with such force that they can meld more deeply than they do in chemical processes like the burning of natural gas. Whereas chemical burning is a reaction that involves the electrons that surround atoms, nuclear fusion is a reaction that joins nuclei at the center of atoms. Through such deep melding, nuclear fusion generates copious quantities of energy manifested as rapidly moving particles. And it is such rapid thermal motion that generates an outward pressure capable of balancing the inward force of gravity. Nuclear fusion in the core thus halts the contraction. The result is a concentrated, stable, and sustained source of heat and light.

A star is born.

To appreciate how the formation process tallies on the entropy scoreboard, let's add up the contributions. Both the core of the gas cloud, which becomes the star, as well as the shell of the gas that surrounds it, are subject to two competing entropic effects. For the core, temperature goes up, acting to increase entropy, and volume goes down, acting to decrease entropy. Only detailed calculation[14] can determine the winner, with the result being that the decrease exceeds the increase, so the core's net entropy goes down. The formation of large gravitational clumps, like stars, is indeed a move toward greater order. For the surrounding shell, volume goes up, acting to increase entropy, and temperature goes down, acting to decrease entropy. Again, detailed calculation is required to determine the winner, with the result being that the increase exceeds the decrease, so the shell's net entropy goes up. Just as important, the calculations establish that the entropy increase of the shell exceeds the entropy decrease of the core, ensuring that the entire process results in an overall increase of entropy, earning a well-deserved nod of approval from the second law.

The chain of events, highly idealized and simplified to be sure, shows how a star—a pocket of low entropy, a pocket of order—can be produced spontaneously even though no engineer directs the action and even though the second law of thermodynamics, with its dictum that total entropy increases, remains in full force. Compared with a steam engine, the cosmic setting is more exotic, but what we've found is another instance of the entropic two-step. Much as a steam engine and its surrounding environment engage in a thermodynamic dance—the steam engine releases waste heat, causing its entropy to decrease, while the environment soaks up the heat, causing its entropy to increase—a gas cloud that's large enough for gravity to matter engages in an analogous pas de deux. As the core of such a gas cloud contracts under the pull of gravity, its entropy decreases, but in the process it releases heat that causes the entropy of the surroundings to increase. A local region of order is created within an environment that undergoes a more than compensating surge in disorder.

The new feature of the gravitational version of the entropic two-step is that it's self-sustaining. As the gas cloud contracts and emits heat, its temperature rises, causing yet more heat to flow outward and driving the two-step to continue stepping. By contrast, when the steam engine performs work and emits heat, its temperature drops. Without burning more fuel to heat the steam back up, the engine peters out. That's why the steam engine requires a clever intelligence to design, build, and power it, while the region of order created by a contracting cloud of gas—a star—is sculpted and powered by the mindless force of gravity.

Fusion, Order, and the Second Law

Let's take stock.

When gravity's influence is minimal, the second law drives a system toward homogeneity. Things spread out, energy diffuses, entropy increases. And if that's all there was, the story of the universe, beginning to end, would be bland. But when there's enough matter for gravity's influence to be significant, the second law undertakes a rapid U-turn, and drives the system away from homogeneity. Matter clumps here and spreads out there. Energy concentrates here and diffuses there. Entropy decreases here and increases there. The manner in which the second law's directive is carried out thus depends sensitively on the force of gravity. When there's enough gravity—enough sufficiently concentrated stuff—ordered structures can form. With that, the story of the unfolding universe becomes far richer.

As described, the starring role in this process is played by the force of gravity. By comparison, the nuclear force, responsible for fusion, seems decidedly secondary. Its job appears limited to an intervention: fusion results in the outward pressure that halts the inward collapse driven by gravity. Indeed, an offhand summary scientists commonly rehearse is to say that gravity is the ultimate source of all structure in the cosmos, offering nary a nod to the role of the nuclear force. But a more generous appraisal is that there's an equitable partnership between gravity and the nuclear force as they work in tandem to advance the second law's narrative.

The point is that the nuclear force dances the entropic two-step too. When atomic nuclei fuse—as in the sun, where hydrogen nuclei fuse into helium billions and billions of times each second—the result is a more complex, more intricately organized, lower-entropy atomic cluster. In the process, some of the mass of the original nuclei is converted into energy (as prescribed by $\mathbf{E = mc^2}$), mostly in the form of a burst of photons that heats the star's interior and powers the release of light from the star's surface. And it is through such fiery starlight, which is itself a torrent of outward streaming photons, that the star transfers copious quantities of entropy to the environment. Indeed, much as we found with the steam engine, and with the

contracting gas cloud, the increase in environmental entropy more than compensates for the
decrease in entropy from fusing nuclei, ensuring
that the net entropy goes up and the integrity of
the second law is once again secured.

Just as natural gas and oxygen need a catalyst
(such as my striking a match) to initiate chemical
burning, atomic nuclei need a catalyst to spark
nuclear fusion. For stars, that catalyst is none other
than the force of gravity, crushing matter in the
core until it becomes sufficiently hot and dense for
fusion to ignite. Once fusion begins, it can power
a star for billions of years, relentlessly synthesizing
complex atomic nuclei as it extracts an otherwise
inaccessible trove of entropy that it sprays outward
through heat and light. And as we will discuss in
the next chapter, these products—complex atoms
and a steady bath of streaming light—are essential
for the formation of even richer and more intricate
structures, including you and me. Thus, although
gravity is the vital force in the formation of a star
and in maintaining a stable stellar environment, for
billions of years it's the nuclear force that's on the
front line, spearheading the entropic charge. From
this perspective, gravity's role shifts from leading
protagonist to indispensable partner in a long duet.

The upshot, anthropomorphized, is that the
universe cleverly leverages the gravitational and
nuclear forces to wrest a cache of untapped entropy
that's locked up inside of its material constituents.

Without gravity, particles that are uniformly dispersed, like an aroma that has filled your house, have attained the highest entropy available. But with gravity, particles that are squeezed into massive and dense balls supported by nuclear fusion drive the entropy tally yet higher.

Catalyzed by gravity and executed by the nuclear force, this version of the entropic two-step is danced by matter clear across the universe. It's a process that has dominated the cosmic choreography since shortly after the big bang, resulting in vast numbers of stars—orderly astronomical structures whose heat and light, in at least one instance, enabled the emergence of life. That development, as we will explore in the next chapter, involves a counterpart to entropy—evolution—that is capable of shaping the most exquisitely complex structures in the universe.

4

INFORMATION AND VITALITY

From Structure to Life

D ear Professor Schrödinger," began the unassuming 1953 letter from biologist Francis Crick to Erwin Schrödinger, one of the founding fathers of quantum mechanics and the 1933 Nobel laureate in physics. "Watson and I were once discussing how we came to enter the field of molecular biology, and we discovered that we had both been influenced by your little book, 'What is Life?'" Crick followed the reference to Schrödinger's book with an exhilaration he could barely contain: "We thought you might be interested in the enclosed reprints—you will see that it looks as though your term 'aperiodic crystal' is going to be a very apt one."[1]

The Watson to whom Crick refers is, of course,

James Watson, coauthor with Crick of the "enclosed reprints," which, still hot off the press, included a scientific paper destined to be one of the most celebrated of the twentieth century. In published form this manuscript would take up less than a single journal page, yet that proved adequate for laying out the double helix geometry of DNA and garnering Crick and Watson, together with Maurice Wilkins from King's College, the 1962 Nobel Prize.[2] Remarkably, Wilkins too credited Schrödinger's book with sparking his passion for determining the molecular basis for heredity; in Wilkins's words, "it set me in motion."[3]

Schrödinger wrote **What Is Life?** in 1944 based on a series of public lectures he had given the previous year at the Dublin Institute for Advanced Studies. In announcing the lectures, Schrödinger noted that his topic was challenging and that "the lectures could not be termed popular," a laudable commitment to a thorough exploration of the topic even at the potential expense of a diminished audience.[4] Despite that, for three consecutive Fridays in February 1943, with World War II raging on the continent, an audience of more than four hundred—including the Irish prime minister, various dignitaries, and wealthy socialites—crammed a lecture theater perched atop the grey stone Fitzgerald Building on the Trinity College campus to hear the Vienna-born physicist grapple with the science of life.[5]

Schrödinger's self-described charge was to make headway on one primary question: "How can the events **in space and time** which take place within the spatial boundary of a living organism be accounted for by physics and chemistry?" Or, to loosely paraphrase: Rocks and rabbits are different. But how? And why? Each is an enormous collection of protons, neutrons, and electrons, and all these particles—whether confined to rock or rabbit—are governed by the very same laws of physics. So what takes place within the body of a rabbit that renders its collection of particles so profoundly different from the collection of particles constituting a rock?

It's the kind of question a physicist would ask. More often than not, physicists are reductionists and so tend to look beneath complex phenomena for explanations that rely on properties and interactions of simpler constituents. Whereas biologists often define life by its core activities— life imbibes raw materials for powering self-sustaining functions, eliminates waste generated by the process, and in the most successful instances reproduces—Schrödinger sought an answer to "What is life?" that would draw on life's fundamental physical underpinnings.

The lure of reductionism is strong. If we could identify what animates a collection of particles, what molecular magic sparks the fires of life, we would take a significant step toward understanding life's origin and the ubiquity, or not, of life in the cosmos. More than a half century later, not-withstanding monumental strides in physics and especially molecular biology, we are still pursuing variations of Schrödinger's question. While there has been impressive progress in decomposing life (and matter more generally) into its constituent parts, researchers still face the formidable task of laying out how life emerges when collections of these constituents are arranged in particular config-urations. Such synthesis is an essential component of the reductionist program. After all, the more finely you examine something that's alive, the more challenging it is to see that it's living. Concentrate on a single molecule of water, an atom of hydrogen, or an individual electron, and you will find that none bear any mark delineating whether they are a constituent of something living or dead, of something animate or inanimate. Life is recogniz-able from the collective behavior, the large-scale organization, the overarching coordination of an enormous number of particulate constituents—even a single cell contains more than a trillion atoms. Seeking insight into life by homing in on fundamental particles is akin to experiencing a Beethoven symphony instrument by instrument, note by single note.

Schrödinger himself emphasized a version of this very point in his first lecture. If a body or a brain could be impaired by the errant movement of a single atom or a handful of atoms, the survival prospects of that body or brain would be dim. To avoid such sensitivity, Schrödinger pointed out, bodies and brains are made of large collections of atoms that can maintain their overall highly coordinated functioning even as the individual atoms randomly jitter about. So Schrödinger's goal was not to reveal life hovering within a single atom but to build upon the understanding of atoms to construct a physicist's explanation of how a large collection might assemble into something that lives. In his view, this was an expansive quest that would likely require science to broaden its base of conceptual structures. Indeed, in an epilogue to **What Is Life?** touching on consciousness, Schrödinger raised some eyebrows (and lost his first publisher) when he invoked the Hindu Upanishads to suggest that we are all part of an "omnipresent, all-comprehending eternal self," and the freedom of will we each exert reflects our divine powers.[6]

While my take on free will differs from Schrödinger's (as we will see in chapter 5), I do share his affinity for a wide explanatory landscape. Deep mysteries call for clarity delivered through a collection of nested stories. Whether reductionist or emergent, whether mathematical or figurative, whether scientific or poetic, we piece together the richest understanding by approaching questions from a range of different perspectives.

Nested Stories

During the past few centuries, physics has refined its own collection of nested stories organized by the distances over which each story is relevant. It's central to an approach we physicists relentlessly drill into our students. To understand how a baseball momentarily deformed by the blazing swing of Mike Trout's bat springs back to its spherical shape, you need to analyze the ball's molecular structure. That's where innumerable microphysical forces push back on the deformation and launch the ball on its way. But this molecular perspective is useless for understanding the ball's trajectory. The voluminous data required to track the motion of trillions of trillions of molecules as the ball spins and soars over the left-field fence would be utterly incomprehensible. When it comes to the trajectory, you need to zoom out from the molecular weeds and examine the ball's motion as a whole. You need to tell a related but distinct higher-level story.

The example illustrates a simple but widely relevant realization: the questions we ask determine the stories that provide the most useful answers. It's a narrative structure that capitalizes on one of nature's most fortuitous qualities. At each scale the universe is coherent. Newton had no knowledge of quarks and electrons, and yet if you gave him a baseball's speed and direction as it left Mike Trout's bat, he'd calculate its trajectory in his sleep. As physics has progressed since Newton's time, we've been able to probe finer layers of structure, and this has significantly filled out our understanding. But the description at each step makes sense on its own. If it didn't—if, for example, understanding the motion of a baseball required understanding the quantum behavior of its particles—it's hard to see how we would have ever made progress. Divide and conquer has long been the rallying call of physics, a strategy that has resulted in rousing triumphs.

An equally important charge is to synthesize the individual stories into a seamless narrative. For the physics of particles and fields, such synthesis was brought to its most refined form by Ken Wilson, earning him the 1982 Nobel Prize.[7] Wilson developed a mathematical procedure for analyzing physical systems over a range of different distances—from scales far smaller, say, than those probed by the Large Hadron Collider to the far larger atomic distances that have been accessible for well over a century—and then systematically connecting the stories, clarifying how each hands off the narrative burden to the next as the scale migrates beyond its particular domain. The method, called the **renormalization group,** lies at the core of modern physics. It shows how the language, conceptual framework, and equations used to analyze physics on one distance scale need to shift as we change focus to a different scale. By using it to develop a nested collection of distinct descriptions and delineating how each informs those it borders, physicists have extracted detailed predictions that have been confirmed through a great many experiments and observations.

While Wilson's technique is tailored for the mathematical tools of the modern high-energy particle physicist (quantum mechanics and its generalization, quantum field theory), the overarching realization is broadly applicable. There are many ways of understanding the world. In the traditional organization of the sciences, physics deals with elementary particles and their various unions, chemistry with atoms and molecules, and biology with life. That categorization, still with us today but far more prominent when I was a student, provides a reasonable if coarse demarcation of the sciences by scale. In more recent times, however, the deeper researchers have probed, the more they've realized that grasping the crossovers between disciplines is essential. The sciences are not separate. And when focus shifts from life to intelligent life, yet other overlapping disciplines—language, literature, philosophy, history, art, myth, religion, psychology, and so on—become central to the chronicle. Even the staunch reductionist realizes that as fatuous as it would be to explain a baseball's trajectory in terms of molecular motion, it would only be more so to invoke such a microscopic perspective in explaining what a batter was feeling as the pitcher went through his windup, the crowd roared, and the fastball approached. Instead, higher-level stories told in the language of human reflection provide far greater insight. Nevertheless—and this is key—these better-suited human-level stories must be compatible with the reductionist account. We are physical creatures subject to physical law. And so there's little to be gained by physicists clamoring that theirs is the most fundamental explanatory framework or from humanists scoffing at the hubris of unbridled reductionism. A refined understanding is gleaned by integrating each discipline's story into a finely textured narrative.[8]

In this chapter we commit to a reductionist stance, recognizing that later chapters will explore life and mind from a complementary humanist sensibility. Here, we will discuss the origin of the atomic and molecular ingredients necessary for life, the origin of one particular environment—the earth and sun—in which those ingredients have comingled in just the right way for life to arise and flourish, and we will explore the deep unity of life on earth by examining some of the astounding microphysical structures and processes common to all living things.[9] Although we won't answer the question of life's origin (still a mystery), we will see that all life on earth can be traced to a common single-celled ancestral species, sharply delineating what a science of life's origin will ultimately need to explain. This will lead us to examine life from the broadly applicable thermodynamic perspective developed in previous chapters, making clear that living things share a deep kinship not just with one another but with stars and steam engines too: life is one more means the universe employs to release the entropy potential locked within matter.

My aim is not to be encyclopedic but to provide just enough detail so that you sense nature's rhythms, the resonant patterns playing out from the big bang to life on earth.

The Origin of the Elements

Grind up anything previously alive, pry apart its complex molecular machinery, and you'll find an abundance of the same six types of atoms: carbon, hydrogen, oxygen, nitrogen, phosphorus, and sulfur, a collection of elements students sometimes remember with the acronym SPONCH (not to be confused with the Mexican marshmallow cookie of the same name). Where do these life-supporting atomic ingredients come from? The answer that has emerged represents one of the great success stories of modern cosmology.

The recipe for building any atom, however complex, is direct. Join the right number of protons with the right number of neutrons, jam them together in a tight ball (the nucleus), surround them with electrons equal in number to that of the protons, and set the electrons in particular orbits dictated by quantum physics. That's it. The challenge is that, unlike Lego pieces, the atomic constituents don't just snap together. They strongly push and pull one another, making the assembly of nuclei a difficult task. Protons, in particular, all have the same positive electric charge, and so it takes enormous pressure and temperature for them to ram through their mutual electromagnetic repulsion and get close enough for the strong nuclear force to dominate, locking them in a powerful subatomic embrace.

The ferocious conditions in the immediate aftermath of the big bang were more extreme than anything encountered anywhere any time since, and so would seem an environment ripe for surmounting electromagnetic repulsion and assembling atomic nuclei. Within a phenomenally dense and energetic brew of colliding protons and neutrons, you might suppose that agglomerations would naturally form, synthesizing the periodic table one atomic species after another. Indeed, that's what George Gamow (a Soviet physicist whose first attempt to defect, in 1932, involved paddling a kayak stocked mostly with coffee and chocolate across the Black Sea) and his graduate student Ralph Alpher suggested in the late 1940s.

They were partially right. One catch, which they realized, is that in the earliest moments the temperature of the universe was too high. Space was flooded with extraordinarily energetic photons that would have blasted apart any incipient unions of protons and neutrons. But, as they also realized, just about a minute and a half later—a long time when considering the whirlwind speed at which the early universe developed—the situation changed. By then, the temperature dropped sufficiently for typical photon energies to no longer overwhelm the strong nuclear force, finally allowing unions of protons and neutrons to persist.

The second catch, which became clear later on, is that building up complex atoms is an intricate process that requires time. It requires a highly specific series of steps in which prescribed numbers of protons and neutrons are melded together into various lumps, which then need to fortuitously encounter particular complementary lumps, fuse with them too, and so on. Like a gourmet's recipe, the order in which the ingredients are combined is essential. And what makes the process particularly tricky is that some intermediate lumps are unstable, meaning that after they form they tend to disintegrate quickly, disrupting the culinary preparations and slowing atomic synthesis. This hindrance is a big deal because the steadily falling temperature and density as the early universe rapidly expands implies that the window of opportunity for fusion quickly closes. By roughly ten minutes after creation, the temperature and density drop below the threshold required for nuclear processes.[10]

When these considerations are made quantitative, as initiated by Alpher in his PhD dissertation and refined by many researchers since, we find that in the immediate aftermath of the big bang only the first few atomic species would have been synthesized. The mathematics allows us to work out their relative abundances: about 75 percent hydrogen (one proton), 25 percent helium (two protons, two neutrons), and trace amounts of deuterium (a heavy form of hydrogen, with one proton and one neutron), helium-3 (a light form of helium with two protons and one neutron), and lithium (three protons, four neutrons).[11] Detailed astronomical observations of atomic abundances have confirmed that these ratios are spot-on, a triumph of mathematics and physics in illuminating the detailed processes that happened within minutes of the big bang.

What about more complex atoms, like those essential to life? Suggestions for their origin go back to the 1920s. British astronomer Sir Arthur Eddington (who when asked what it was like to be among only three people who understood Einstein's general relativity, famously responded, "I'm trying to think who the third person is") hit on the right idea: the scorching interior of stars might provide cosmic Crock-Pots for slow-cooking more complex atomic species. The proposal passed through the hands of many brilliant physicists, including those of Nobel laureate Hans Bethe (my first faculty office was next to his, and I could set my watch by his utterly reliable four p.m. exuberant sneeze) and, perhaps most consequentially, those of Fred Hoyle (who in a 1949 BBC Radio program dismissively referred to the universe being created in "one big bang," unwittingly coining one of science's most pithy monikers[12]), which turned the suggestion into a mature and predictive physical mechanism.

Compared with the breakneck pace of change in the immediate aftermath of the big bang, stars provide stable environments that can persist for millions if not billions of years. The instability of particular intermediate lumps slows the fusion pipeline in stars too, but when you've got time to kill you can still get the job done. So unlike the situation with the big bang, after hydrogen fuses to helium nuclear synthesis in stars is far from over. Stars that are sufficiently massive will continue to crush nuclei together, forcing them to fuse into the more complex atoms of the periodic table, while producing substantial heat and light in the process. For example, a star that's twenty times the mass of the sun will spend its first eight million years fusing hydrogen into helium, then devote its next million years to fusing helium into carbon and oxygen. From there, with its core temperature getting ever higher, the conveyor belt continually revs up: it takes about a thousand years for the star to burn its storehouse of carbon, fusing it into sodium and neon; over the next six months, further fusion produces magnesium; within a month more sulfur and silicon; and then in a mere ten days fusion burns the remaining atoms, producing iron.[13]

We pause at iron, for good reason. Of all atomic species, iron's protons and neutrons are bound together most tightly. This matters. If you try to build yet heavier atomic species by cramming in additional protons and neutrons, you'll find that the iron nuclei have little interest in participating. The nuclear bear hug gripping together iron's twenty-six protons and thirty neutrons has already squeezed out and released as much energy as is physically possible. To add protons and neutrons would require a net input—not output—of energy. As a result, when we reach iron, stellar fusion's orderly production of larger and more complex atoms, with the accompanying release of heat and light, grinds to a halt. Like ash that's fallen to the hearth of your fireplace, iron can't be burned further.

What then of all the atomic species with yet larger nuclei, including utilitarian elements like copper, mercury, and nickel; sentimental favorites like silver, gold, and platinum; and exotic heavyweights like radium, uranium, and plutonium?

Scientists have identified two sources for these elements. When a star's core is mostly iron, fusion reactions no longer generate the outward pushing energy and pressure necessary to counteract the inward pull of gravity. The star begins to collapse. If the star is massive enough, this collapse accelerates into an implosion so powerful that the core temperature rockets; the imploding material bounces off the core and triggers a spectacular shock wave that surges outward. And as the shock wave rumbles from the core toward the star's surface, it compresses the nuclei it encounters with such fury that a slew of larger nuclear agglomerations form. In the maelstrom of chaotic particle motion, all of the periodic table's heavier elements can be synthesized, and when the shock wave finally reaches the star's surface, it blasts the rich atomic smorgasbord into space.

A second source of heavy elements is the violent collisions between neutron stars, celestial bodies produced in the death throes of stars whose mass is roughly ten to thirty times that of the sun. That neutron stars are mostly made of neutrons—chameleonic particles that can transform into protons—bodes well for building atomic nuclei, as we have a profusion of the right raw materials. One obstacle, though, is that to form atomic nuclei the neutrons need to free themselves from the star's powerful gravitational grip. That's where a collision between neutron stars comes in handy. The impact can throw off plumes of neutrons, which, having no electric charge and thus experiencing no electromagnetic repulsion, more easily coalesce into groups. After some of these neutrons then flip the chameleonic switch and become protons (releasing electrons and anti-neutrinos in the process), we acquire a supply of complex atomic nuclei. In 2017, neutron-star collisions migrated from theoretical plaything to observational fact when scientists detected the gravitational waves such collisions generate (which followed on the heels of the very first gravitational waves detected, which were produced by the collision of two black holes). A flurry of analyses have determined that neutron-star collisions produce heavier elements more efficiently and abundantly than supernova explosions, and so it may be that the majority of the universe's heavy elements were produced through these astrophysical smashups.

Fused in stars and ejected in supernova explosions, or jettisoned by stellar collisions and amalgamated in particle plumes, an assortment of atomic species float through space, where they swirl together and coalesce into large clouds of gas, which over yet more time clump anew into stars and planets, and ultimately into us. Such is the origin of the ingredients constituting anything and everything you have ever encountered.

The Origin of the Solar System

At just over four and a half billion years old, the sun is a cosmic newcomer. It was not among the universe's first generation of stars. We saw in chapter 3 that those stellar trailblazers originated from quantum variations in the density of matter and energy that were stretched across space by inflationary expansion. Computer simulations of these processes reveal that the first stars ignited about one hundred million years after the big bang, with an entrance on the cosmic stage that was anything but dainty. The first stars were likely mammoth, hundreds or perhaps even thousands of times the mass of the sun, burning with such intensity that they quickly died out. The heaviest ended their lives in a gravitational implosion so emphatic that they collapsed all the way down to black holes, extreme configurations of matter that will be a prime focus later in our journey. Less massive early stars ended their lives with a fiery supernova explosion that, beyond seeding space with complex atoms, initiated the next round of stellar formation. Much as a supernova shock wave ripping through a star forcefully fuses its atomic constituents, a shock wave thundering through space compresses the clouds of molecular ingredients it encounters. And because compressed regions are denser, they exert a greater gravitational pull on their surroundings, drawing in yet more particulate constituents and setting off a new round of gravitational snowballing en route to the next generation of stars.

Based on the composition of the sun—the quantities of various heavy elements it now contains, determined by spectroscopic measurements—solar physicists believe the sun is a grandchild of the universe's first stars, a third-generation arrival. But there is much uncertainty regarding where the sun originally formed. One candidate that has been investigated is a region known as Messier 67, about three thousand light-years away, which contains a cluster of stars whose chemical compositions appear similar to the sun's, suggesting a close family resemblance. The challenge, still unresolved, is to explain how the sun and the planets of the solar system (or the protoplanetary disk from which the planets would subsequently form) would have been ejected from that distant stellar nursery and migrated here. Some studies of the potential trajectories conclude that there's virtually no chance that Messier 67 is the sun's birthplace, while others, invoking various modified assumptions, have yielded more encouraging results.[14]

What we can say with more confidence is that some 4.7 billion years ago a supernova shock wave likely plowed through a cloud containing hydrogen, helium, and small quantities of more complex atoms, compressing part of the cloud, which, now being denser than its surroundings, exerted a stronger gravitational pull and thus began to draw material inward. Over the next few hundred thousand years, this region of the gas cloud continued to contract, rotating slowly at first and then more rapidly, like a graceful skater pulling in her arms while spinning. And much as the spinning skater experiences an outward pull (which splays out any loose fringe on the skater's costume), so did the spinning cloud, which spread and flattened its outer regions into a rotating disk, which surrounded a smaller spherical region at the core. During the following fifty to one hundred million years, the gas cloud then performed a slow and steady rendition of the gravitational entropic two-step discussed in chapter 3: The gravitational force squeezed the spherical core, which grew ever hotter and denser, while the surrounding material cooled and thinned. The entropy of the core decreased; the entropy of the surroundings offset that by a more than compensating increase. Ultimately, the core's temperature and density crossed the threshold for igniting nuclear fusion.

The sun was born.

During the next few million years, the detritus left over from the sun's formation, amounting to just a few tenths of a percent of the original swirling disk, coalesced through numerous instances of gravitational snowballing into the solar system's planets. Lighter and more volatile substances—like hydrogen and helium as well as methane, ammonia, and water—which would be disrupted by the sun's intense radiation, accumulated more abundantly in the solar system's cooler outer regions, forming the gas giants, Jupiter, Saturn, Uranus, and Neptune. Heavier and more robust constituents, like iron, nickel, and aluminum, which better withstood the hotter environment closer to the sun, consolidated into the smaller rocky inner planets, Mercury, Venus, Earth, and Mars. Being much less massive than the sun, planets are able to support their modest weight via their own atoms' intrinsic resistance to compression. Core temperatures and pressures within planets rise but nowhere near the levels necessary for nuclear fusion to ignite, resulting in the comparatively temperate environments for which life—surely our form and possibly all life in the universe—owes significant gratitude.

Young Earth

Earth's first half billion years are referred to as the Hadean period, invoking the Greek god of the underworld to connote an infernal era of raging volcanoes, gushing molten rock, and thick noxious fumes of sulfur and cyanide. But some scientists now suspect that as a standard-bearer for young earth, Poseidon may well be the god of choice. The still-debated sea change rests on evidence no more substantial than flecks of dust. Although we lack rock samples from that early era, researchers have identified ancient translucent specks—called **zircon crystals**—that formed when the early earth's molten lava cooled and solidified. Zircon crystals are proving pivotal to understanding earth's early development because not only are they virtually indestructible, surviving billions of years of geological battering, but they also act as miniature time capsules. When they form, zircon crystals snare molecular samples of the environment, which we can time-stamp through standard radioactive dating. By closely analyzing impurities in the zircon crystals, we sample conditions of archaic earth.

One find in Western Australia turned up zircon crystals dated to 4.4 billion years ago, just a couple of hundred million years after the earth and the solar system formed. By analyzing their detailed composition, researchers have suggested that ancient conditions may have been far more agreeable than previously thought. Early earth may have been a relatively calm water world, with small landmasses dotting a surface mostly covered by ocean.[15]

That's not to say that earth's history didn't have its moments of flaming drama. Roughly fifty to one hundred million years after its birth, earth likely collided with a Mars-sized planet called Theia, which would have vaporized the earth's crust, obliterated Theia, and blown a cloud of dust and gas thousands of kilometers into space. In time, that cloud would have clumped up gravitationally to form the moon, one of the larger planetary satellites in the solar system and a nightly reminder of that violent encounter. Another reminder is provided by the seasons. We experience hot summers and cold winters because earth's tilted axis affects the angle of incoming sunlight, with summer being a period of direct rays and winter being a period of oblique ones. The smashup with Theia is the likely cause of earth's cant. And though less sensational than a planetary collision, both the earth and the moon endured periods of significant pummelings by smaller meteors. The moon's lack of eroding winds and its static crust have preserved the scars but earth's thrashing, less visible now, was just as severe. Some early impacts may have partially or even fully vaporized all water on earth's surface. Despite that, the zircon archives provide evidence that within a few hundred million years of its formation, earth may have cooled sufficiently for atmospheric steam to rain down, fill the oceans, and yield a terrain not all that dissimilar from the earth we now know. At least, that's one conclusion reached by reading the crystals.

The duration required for earth to simmer down and sport an abundance of water—whether hundreds of millions of years or far longer—is intensely debated because it speaks directly to the question of when in our geological history life first arose. While it's too strong to say that where there's liquid water there's life, we can say with some confidence that in the absence of liquid water there's an absence of life, at least the kind of life with which we are familiar.

Let's see why.

Life, Quantum Physics, and Water

Water ranks among nature's most familiar yet consequential substances. Its molecular makeup, H_2O, has become for chemistry what Einstein's $\mathbf{E = mc^2}$ is to physics, the subject's most famous formula. By fleshing out that formula, we gain insight into water's distinctive properties and develop some of the key ideas in Schrödinger's program of understanding life at the level of physics and chemistry.

By the mid-1920s, many of the world's leading physicists could sense that the accepted order was on the verge of radical upheaval. Newtonian ideas, whose predictions for the motion of orbiting planets and flying rocks had for centuries set the gold standard of accuracy, were failing miserably when applied to tiny particles like electrons. As unruly data bubbled up from the microworld, the calm seas of Newtonian understanding became turbulent. Physicists quickly found themselves struggling just to stay afloat. Werner Heisenberg's lament, muttered as he aimlessly walked through an empty park in Copenhagen after a grueling night of intense calculations with Niels Bohr, summed up the situation well: "Can nature possibly be as absurd as it seemed to us in these atomic experiments?"[16] The answer, a resounding yes, came in 1926 from an unassuming German physicist, Max Born, who broke the conceptual logjam by introducing a radically new quantum paradigm. He argued that an electron (or any particle) can only be described in terms of the **probability** that it will be found at any given location. In one stroke, the familiar Newtonian world in which objects always have definite positions gave way to a quantum reality in which a particle might be here or there or somewhere else entirely. And far from a failing, the uncertainty inherent in a probabilistic schema revealed an intrinsic feature of quantum reality long overlooked by the deeply insightful yet demonstrably coarse Newtonian framework. Newton based his equations on the world he could see. A couple of hundred years later, we learned that there is an unexpected reality beyond the reach of our frail human perceptions.

Born's proposal came with mathematical precision.[17] He explained that an equation Schrödinger had published a few months earlier could be used to predict the quantum probabilities. This was news to Schrödinger, and everyone else too. But as scientists followed Born's directive, they found that the mathematics worked. Spectacularly so. Data that had previously been subsumed under ad hoc rules of thumb or had resisted explanation entirely could finally be understood through systematic mathematical analyses.

When applied to atoms, the quantum perspective jettisons the old "solar-system model," which pictured electrons in orbit around the nucleus much as planets orbit the sun. In its place, quantum mechanics envisions an electron as a fuzzy cloud surrounding the nucleus whose density at any given location indicates the probability that the electron will be found there. An electron is unlikely to be found where its probability cloud is thin, likely to be found where its probability cloud is thick.

Schrödinger's equation makes this description mathematically explicit, determining the shape and density profile of an electron's probability cloud as well as stipulating—and for our current discussion, this is key—precisely how many of the atom's electrons each such cloud can accommodate.[18] The details quickly become technical, but to grasp the essential features, think of an atom's nucleus as a central stage and its electrons as an audience that watches the action from seats on surrounding tiers, arranged for theater in the round. In this "quantum theater," Schrödinger's math applied to atoms dictates how the electron audience fills in the seats.

Much as you'd expect from your experience climbing stairs in a real theater, the higher the tier the more energy an electron needs to reach it. So when an atom is as calm as it can possibly be, in its lowest energy configuration, its electrons constitute the most orderly of audiences, populating a higher tier only if the lower tiers are fully occupied. With the atom possessing minimal energy, no electron climbs any higher than it absolutely has to. How many electrons can a given tier hold?

Schrödinger's math provides the answer, a universal fire code that applies to all quantum theaters: at most two electrons are allowed on tier one, eight electrons on tier two, eighteen on tier three, and so forth, as specified by the equation. Should an atom's energy be pumped up, say, by having been zapped by a powerful laser, some of its electrons may be sufficiently agitated to jump up to a higher tier, but this exuberance will be short-lived. Such excited electrons quickly fall back to their original tier, emitting energy (carried away by photons) and returning the atom to its calmest configuration.[19]

The math also reveals one further peculiarity, a kind of atomic OCD that's a primary driver of chemical reactions throughout the cosmos. Atoms have an aversion to tiers that are only partially filled. Tiers that are empty? Fine. Tiers that are full? Fine. But partial occupancy? That drives atoms up the wall. Some atoms are lucky, being endowed with just the right number of electrons to achieve full occupancy on their own. Helium contains two electrons, to balance the electric charge of its two protons, and they happily fill the first tier. Neon has ten electrons, to balance the electric charge of its ten protons, and they just as happily fill its first tier, which accommodates two, and its second tier, which accommodates the remaining eight. But for most atoms, the number of electrons needed to balance the number of protons does not fill a complete set of tiers.[20]

So what do they do?

They barter with other atomic species. If you're an atom with an upper tier that needs two more electrons and I'm an atom with an upper tier occupied by two electrons, then if I donate two electrons to you, we will each scratch the other's occupancy itch: the donation results in us each having fully complete tiers. Notice too that by accepting my electrons you will acquire a net negative charge, and by donating my electrons I will acquire a net positive charge—and since opposite charges attract, you and I will embrace to form an electrically neutral molecule. Alternatively, if you and I, for example, both need one more electron to fill out our upper tiers, there is a different type of deal we can strike: we can each donate one electron to a communal pool that we share, again scratching each other's occupancy itch, and—through the bond of our shared electrons—again combining into an electrically neutral molecule. These processes, which fill out electron tiers by joining atoms together, are what we mean by chemical reactions. They provide the template for such reactions here on earth, within living systems, and throughout the universe.

Water provides an important case in point. Oxygen contains eight electrons, two on tier one and six on tier two. Oxygen thus strives for two more electrons, seeking to fill out its second tier to the maximum occupancy of eight. One readily available source is hydrogen. Every hydrogen atom has a single electron, hanging solo and twiddling its thumbs on tier one. If a hydrogen atom has the opportunity to fill this tier with one more electron it happily will. So hydrogen and oxygen agree to share a communal pair of electrons, fully satisfying hydrogen and bringing oxygen one electron closer to orbital bliss. Include a second hydrogen atom that similarly shares a pair of communal electrons with oxygen, and it is rapture all around. The sharing of these electrons binds the oxygen atom to the two hydrogen atoms, giving rise to a molecule of water, H_2O.

The geometry of this union has far-reaching implications. The interatomic pushes and pulls shape all water molecules into a wide V, with oxygen at the vertex and each hydrogen perched on one of the letter's upper tips. Although H_2O has no net electrical charge, because oxygen is so manic about filling its orbital tiers, it hoards the shared electrons, resulting in a distribution of charge across the molecule that is lopsided. The vertex of the molecule, oxygen's home, has a net negative charge, while the two upper tips, where the hydrogens dwell, have a net positive charge.

The distribution of electrical charge across a water molecule might seem like an esoteric detail. But it's not. It proves essential to the emergence of life. Because of water's skewed charge distribution, it can dissolve nearly everything. The negatively charged oxygen vertex grabs hold of anything with even a slight positive charge; the positively charged hydrogen tips grab hold of anything with even a slight negative charge. In tandem, the two ends of a water molecule act like charged claws that pull apart most anything that's submerged for a sufficient time.

Table salt is the most familiar example. Composed of an atom of sodium bonded to an atom of chlorine, a molecule of table salt has a slight positive charge near the sodium (which donates an electron to the chlorine) and a slight negative charge near the chlorine (which accepts an electron from the sodium). Drop salt into water, and the oxygen side of H_2O (negatively charged) grabs hold of the sodium (positively charged), while the hydrogen side of H_2O (positively charged) grabs hold of the chlorine (negatively charged), ripping salt molecules apart and dissolving them into solution. And what's true for salt is true for a great many other substances too. The details vary, but water's asymmetric charge arrangement makes it an uncanny solvent. Wash your hands, even without soap, and water's electrical polarity will be hard at work, dissolving foreign matter and carrying it away.

Well beyond its utility in personal hygiene, water's capacity to grab hold of and ingest substances is indispensable to life. Cell interiors are miniature chemistry labs whose workings require the rapid movement of a vast collection of ingredients:

nutrients in, waste out, comingling of chemicals to synthesize substances required for cellular function, and so on. Water makes this possible. Water, constituting some 70 percent of a cell's mass, is life's ferrying fluid. Nobel laureate Albert Szent-Györgyi summarized it eloquently: "Water is life's matter and matrix, mother and medium. There is no life without water. Life could leave the ocean when it learned to grow a skin, a bag in which to take the water with it. We are still living in water, having the water now inside."[21] As poetry, this is a graceful ode to water and life. As science, there is as yet no argument to establish the statement's universal validity, but we know of no form of life that challenges the necessity of water.

The Unity of Life

Having surveyed the synthesis of simple and complex atoms, the origin of the sun and earth, the nature of chemical reactions and the necessity of water, we are now equipped to turn to life itself. While it might seem natural to begin with the genesis of life, that topic, still unsettled, is better approached after exploring the quintessential molecular qualities of life itself. And for someone like me, having spent the past thirty years pursuing a unified theory of nature's fundamental forces, such an exploration reveals a stunning biological unity.

We don't know the exact number of distinct species on earth, microbes to manatees, but studies have provided estimates ranging from a low in the millions to a high in the trillions. Whatever the exact number, it's huge. The wealth of different species, however, belies the singular nature of life's inner workings.

Examine living tissue closely enough and you'll encounter life's "quanta"—cells—the tissue's smallest units we'd identify as being alive. Regardless of their source, cells share so many features that the untrained eye examining individual specimens would be hard-pressed to distinguish mouse from mastiff, turtle from tarantula, housefly from human. That's remarkable. Surely our cells must show an obvious and significant distinguishing imprint. Yet they don't. The reason, established during the past few decades, is that all complex multicellular life descended from the same single-celled ancestral species. Cells are similar because their lineages radiate from the same starting point.[22]

That's a telling realization. With its copious incarnations, life might have had many distinct origins. Tracing the lineage of the sea mollusk all the way back might have revealed one starting point, while doing the same for wombats or orchids might have revealed others. But the evidence strongly suggests that in seeking life's origin, the lineages converge to a common ancestor. Two ubiquitous qualities of life make the case yet more convincing. Each illustrates the deep commonalities shared by all that lives. The first, and more familiar, concerns **information:** how cells encode and utilize the information that directs life-sustaining functions. The second, equally important but less widely celebrated, concerns **energy:** how cells harness, store, and deploy the energy required for carrying out life-sustaining functions. In both we will see that clear across the spectacular breadth of life on earth the detailed processes are identical.

The Unity of Life's Information

One way we recognize that a rabbit is alive is by seeing it move. A rock can move too, of course. A strong river current can push it downstream or a volcanic eruption can launch it skyward. The difference is that the rock's motion can be fully understood, even predicted, based on the external forces that act upon it. Tell me enough about the current or the eruption and I can do a reasonably good job of determining what will happen. Predicting the motion of the rabbit is harder. Activity within what Schrödinger called the "spatial boundary" of the rabbit—its internal activity—is a decisive factor in its locomotion. The rabbit twitches its nose, turns its head, pounds its legs, and all this makes it appear to have a will of its own. Whether the rabbit or any life form (including us) actually has such an autonomous will is a question that has been debated for centuries, and one we'll take up in the next chapter, so let's not get bogged down with it here. For now, we can all agree that whereas activity within the rock is of virtually no consequence to the motion we observe, the rabbit's coordinated, complex, and self-directed movements clue us in that it is alive.

It's not a foolproof diagnostic. Automated systems can execute motion of a broadly similar sort, and as technological progress continues, the ability to emulate life will become sharper still. But that serves only to underscore the larger point: motion of the kind we're considering arises from an interplay between information and execution, between what we might call software and hardware. For an automated system, the description is literal. Drones, self-driving cars, Roombas, and so on are governed by software that takes environmental data as input and as output determines a response executed by onboard hardware, from wings to rotors to wheels. For a rabbit, the description is metaphor. Nevertheless, the software-hardware paradigm is a particularly useful way of thinking about life too. The rabbit accumulates sensory data from the environment, runs it through a "neural computer" (its brain), which sends information-laden signals down nerve pathways—eat clover patch, hop over fallen twigs, and so on—generating physical actions. The motion of the rabbit arises from the internal processing and transmission of a complex set of instructions that flows through its physical structure: biological software driving biological hardware. Such processes are wholly absent for a rock.

If we dive deep into a single cell of the rabbit we encounter a similar set of ideas playing out on a smaller scale. The vast majority of a cell's functions are executed by proteins, large molecules that catalyze and regulate chemical reactions, transport essential substances, and control detailed properties like cellular shape and movement. Proteins are built from combinations of twenty smaller subunits, **amino acids,** similar to the way English words arise from various combinations of twenty-six letters. And much as sensible words require letters to be arranged in specific orders, usable proteins require amino acids to be linked in specific sequences. If such assembly were left to blind chance, the likelihood that the requisite amino acids would happen to bump into one another in just the right way to build a particular protein would be next to nothing. The sheer number of ways that twenty distinct amino acids can be linked in a long chain makes this evident: for a chain with one hundred and fifty amino acids (a small protein), there are about 10^{195} different arrangements, far larger than the number of particles in the observable universe. Much as the proverbial team of monkeys typing random letters for decades will fail to spell out more than "To be or not to be," random chance will fail to create the specific proteins required for life.

Instead, the synthesis of complex proteins requires a set of instructions that spell out a step-by-step process—hook this amino acid to that one, then tack on this one, followed by that one, and so on. That is, protein synthesis requires cellular software. And within every cell such instructions exist. They are encoded by DNA, the life-supporting chemical whose geometrical architecture was discovered by Watson and Crick.

Every molecule of DNA is configured in the famous spiral of the double helix, a long twisting ladder whose rungs consist of pairs of struts, shorter molecules called bases, usually denoted A, T, G, and C (the technical names won't matter for us, but these stand for adenine, thymine, guanine, and cytosine). Members of a given species mostly share the same sequence of letters. For humans, the DNA sequence runs about three billion letters long, with your sequence differing from that of Albert Einstein or Marie Curie or William Shakespeare or anyone else by less than about a quarter of a percent, roughly one letter out of every string of five hundred.[23] But while basking in the glow of possessing a genome so similar to that of any of history's most revered luminaries (or infamous villains), note that your DNA sequence also has a 99 percent overlap with any given chimpanzee's.[24] Minor genetic differences can have major impact.

In constructing the rungs of the DNA ladder, the bases pair off according to a rigid rule: an A strut on one rail of the ladder attaches to a T strut on the other rail, a G strut on one rail attaches to a C strut on the other. The sequence of bases on one side of the ladder thus uniquely determines the sequence on the other. And it is within the sequence of letters that we find, among other vital cellular information, instructions that specify which amino acids will be linked to which, directing the synthesis of a species-specific collection of proteins essential to that form of life.

All life codes the instructions for building proteins in the same way.[25]

In one perhaps overly detailed paragraph, here's the manual for how it works, the molecular Morse code hardwired into all life. Groups of three consecutive letters on a given rail of DNA denote one particular amino acid from the collection of twenty.[26] For example, the sequence CTA denotes the amino acid leucine; the sequence GCT denotes another, alanine; the sequence GTT denotes valine; and so on. If you were examining the rungs attached to one rail of a DNA segment and read off the nine-letter sequence CTAGCTGTT, that would instruct you to attach leucine (the first three letters, CTA) to alanine (the second three letters, GCT), which you would then attach to valine (the last three letters, GTT). A protein built from, say, a thousand linked amino acids would be coded by a specific sequence of three thousand letters (the starting location and the ending location of any such sequence is also coded by particular three-letter sequences, much as a capitalized letter and a period denote the start and end of this sentence). Such a sequence constitutes a **gene,** the instructional blueprint for assembling a protein.[27]

I've laid out the details for two reasons. First, seeing the code makes the concept of cellular software explicit. Given a segment of DNA, we can read off the instructions which direct the cell's inner workings, a sophisticated coordination wholly absent in inanimate matter. Second, seeing the code demonstrates what biologists mean when they call it universal. Every molecule of DNA, whether from seaweed or Sophocles, encodes the information needed to build proteins in the same way.

That is the unity of life's information.

The Unity of Life's Energy

Just as a steam engine needs a steady supply of energy to repeatedly push its piston, life requires a steady supply of energy to carry out essential functions from growth and repair to movement and reproduction. For the steam engine, we extract energy from the environment. We burn coal, wood, or some other fuel, and the heat generated is consumed by the engine's inner mechanism, driving the steam to expand. Living beings also extract energy from the environment. Animals extract energy from food, plants from sunlight. But unlike the steam engine, life doesn't generally use such energy on the spot. The processes of life, being more complex than the expansion or contraction of steam, require a more refined system for the delivery and distribution of energy. Life needs energy from the fuel it burns to be stored and doled out on a regular and reliable basis as cellular constituents require it.

All life meets the challenge of energy extraction and distribution in the same way.[28]

The universal solution life has come up with, a complex sequence of processes taking place right now inside you and me and, as far as we know, all else that lives, ranks among nature's most astonishing accomplishments. Life extracts energy from the environment through a type of slow chemical burning and stores that energy by charging up biological batteries built into all cells. These cellular battery packs then provide a steady source of electricity that cells use to synthesize molecules tailor-made for transporting and delivering energy to every cellular component.

That may sound heavy. It is heavy. It is also vital. So let's briefly unpack it. If you don't grasp every detail, that's fine. Even a cursory tour reveals the wonders of how life powers its inner workings.

The chemical burning central to life's processing of energy is called a **redox reaction.** Not the

most inviting name, but the archetypal example—
a burning log—clarifies the nomenclature. As a log
burns, carbon and hydrogen in the wood relinquish
electrons to oxygen in the air (remember, oxygen
yearns for electrons), bonding them into molecules
of water and carbon dioxide, and releasing energy
in the process (the very reason fire is hot). When
oxygen grabs electrons, we say that it has been
reduced (you can think of this as a reduction in
oxygen's yearning for electrons). When carbon
or hydrogen relinquishes electrons to oxygen, we
say that it has been **oxidized.** Together, we have a
reduction-oxidation reaction, or redox for short.

Scientists now use the term "redox" more broadly,
referring to a collection of reactions in which electrons are passed between chemical constituents,
regardless of whether oxygen is involved. Still, a
flaming log provides a widely relevant template
for describing chemical burning. Ravenous atoms,
burdened by partially filled tiers, grab electrons
from atomic donors with such a powerful clasp that
significant pent-up energy is released in the process.

In living cells—let's focus on animals to be
definite—similar redox reactions take place but,
importantly, the electrons stripped from atoms
that you ingested at breakfast are not transferred
directly to oxygen. If they were, the energy released
would create something akin to a cellular fire, an
outcome life has learned the benefit of avoiding.

Instead, electrons donated by food pass through a series of intermediate redox reactions, rest stops on a trek that ultimately ends with oxygen but that allows smaller amounts of energy to be released at each step. Like a ball in the bleachers cascading down a stadium's steps, electrons jump from one molecular receptor to another, with each receptor more electron crazed than the previous, ensuring that each jump results in the release of energy. Oxygen, the most electron-crazed receptor of all, waits for the electron at the bottom of the stairs, and when it finally arrives, the oxygen hugs the electron tight, squeezing out the marginal energy it can still provide, thus concluding the energy extraction process.

The process for plants is largely the same. The main difference is the source of the electrons. For animals, they come from food. For plants, they come from water. Sunlight striking chlorophyll in the green leaves of plants strips electrons from water molecules, pumps up their energy, and sets them off on a similar energy-extracting redox cascade. And so the energy supporting all the actions of all living things can be traced to one and the same process, jumping electrons executing a series of cellular redox reactions. It's why Albert Szent-Györgyi, continuing his poetic reflections, mused, "Life is nothing but an electron looking for a place to rest."

From the perspective of physics, it's worth emphasizing how surprising this all is. Energy is the coin that pays for all comings and goings throughout the cosmos, a coin minted in a wide range of currencies and earned through an even wider range of callings. One currency is nuclear energy, generated by fission and fusion among a wealth of atomic species; electromagnetic energy is another, generated by pushes and pulls among a wealth of charged particles; gravitational energy is another still, generated by interactions among a wealth of massive bodies. And yet of all the innumerable processes, life on planet earth leverages one and only one energy mechanism: a specific sequence of electromagnetic chemical reactions in which electrons engage in a downward-directed sequence of jumps, starting with food or water and ending with the clutching embrace of oxygen.

How and why did this energy extraction process become life's go-to mechanism? No one knows. But the universality, like that of the genetic code, speaks again, and strongly so, to the unity of life. Why do all living things power themselves in the same way? The immediate answer is that all life must have descended from a common ancestor, a single-celled species that researchers believe likely existed around four billion years ago.

Biology and Batteries

Evidence for the unity of life grows even more convincing as we follow the subsequent journey of the energy released by electrons hopping from one redox reaction to another. That energy is used to charge up biological batteries that are built into each and every cell. In turn, the biological batteries power the synthesis of molecules particularly adept at transporting and delivering energy wherever and whenever it is needed throughout a cell. It is an elaborate process. But across life, it is the same process.

In broad outline here is how it goes. As an electron jumps into the outstretched molecular arms of a given redox receptor, the receiving molecule twitches, causing it to shift its orientation relative to other molecules closely packed around it, much like a gear ratcheting one step forward. When the fickle electron subsequently jumps to the next redox receptor, the first molecule clicks back to its original orientation, while the new molecular recipient experiences the twitch. As the electron executes further jumps, the pattern continues. Molecules receiving an electron twitch, ratcheting their orientations forward; molecules losing an electron twitch too, ratcheting their orientations back.

The sequence of electron hops and resulting molecular twitches accomplishes a subtle but significant task. As the molecules ratchet back and forth, they push against a group of protons, forcing them through a surrounding membrane, where they accumulate in a thin compartment, which amounts to an overcrowded holding cell. Or, in more prosaic language, a proton battery.

In an ordinary battery, chemical reactions force electrons to accumulate on one side of the battery (the anode), where the mutual repulsion of these like-charged particles means they're primed to flee at the first opportunity. When you complete an electrical circuit by pushing an "on" button or flipping a switch, you free the pent-up electrons, allowing them to flow out of the anode, pass through a device—bulb, laptop, phone—and finally return to the battery's other side (the cathode). Commonplace though batteries are, they are utterly ingenious. They store energy in a crowded collection of electrons standing at the ready to relinquish that energy on a moment's notice to power devices of our choosing.

In a living cell we encounter an analogous situation, with pent-up protons replacing pent-up electrons. But it's a distinction that hardly makes a difference. Protons, like electrons, all carry the same electric charge, and so they also repel another. When cellular redox reactions pack protons closely together, they too stand at the ready waiting for the chance to rush away from their enforced companions. Cellular redox reactions thus charge up biological proton-based batteries. In fact, because the protons are all clustered on one side of an extremely thin membrane (just a few dozen atoms wide), the electric field (the membrane voltage divided by the membrane thickness) can be enormous, upwards of tens of millions of volts per meter. A cellular bio battery is no slouch.

What, then, do cells do with these mini power stations? Here's where things get yet more astounding. Attached to the membrane are a great many nanoscale-sized turbines. When the packed protons are allowed to flow back across specific sections of the membrane, they cause the tiny turbines to rotate, much as flowing gusts of air cause windmills to rotate. In centuries past, such wind-powered turning motion was used to crush wheat or other grains into flour. The cellular windmills undertake an analogous grinding project but instead of pulverizing structure the process builds it. As they turn, the molecular turbines repeatedly cram together two particular input molecules (ADP, adenosine diphosphate plus a phosphate group), synthesizing one particular output molecule (ATP, adenosine triphosphate). Forced together by the turbine, the constituents of each resulting ATP molecule are in a tense arrangement: mutually repelling charged constituents are clasped together by chemical bonds, and so, much like a compressed spring, they strain to be released. That's extraordinarily useful. Molecules of ATP can travel throughout a cell, releasing that stored energy when needed by snapping the chemical bonds and allowing the constituent particles to relax into a lower energy, more comfortable state. It is that very energy, released by the dissociation of ATP molecules, that powers cellular functions.

The tireless activity of these cellular power stations becomes clear when you consider a few numbers. The functions that keep a typical cell alive for just a single second require the energy stored in about **ten million** ATP molecules. Your body contains tens of trillions of cells, which means that every second you consume on the order of one hundred million trillion (10^{20}) ATP molecules. Each time an ATP is used, it splits up into the raw materials (ADP and a phosphate), which the proton battery-powered turbines then cram back together into freshly minted, fully rejuvenated ATP molecules. These ATP molecules then hit the road again, delivering energy throughout the cell. To meet your body's energy demands, your cellular turbines are thus astoundingly productive. Even if you're an extremely fast reader, as you scan through this very sentence your body is synthesizing some five hundred million trillion molecules of ATP. And just now, another three hundred million trillion more.

Summary

Putting the details to the side, the conclusion is that as energetic electrons from food (or electrons energized by sunlight in plants) cascade down a flight of chemical stairs, the energy released at each step charges up biological batteries that reside in all cells. The energy stored in the batteries is then used to synthesize molecules that do for power what UPS trucks do for packages: the molecules reliably deliver packets of energy wherever they are called for within the cell. This is the universal mechanism that powers all life. This is the singular energy pathway that underlies **every** action we take and **every** thought we have.

As with our brief foray into DNA, the main point hovers above the particulars: the intricate and seemingly baroque collection of processes that power cells is universal across all life. That unity, together with the unity of DNA's coding of cellular instructions, provides overwhelming evidence that all life emerged from a common ancestor.

Much as Einstein sought a unified theory of nature's forces, and much as physicists today dream of an even grander synthesis embracing all matter and perhaps space and time too, there is something thoroughly seductive in identifying a common core within a vast range of seemingly distinct phenomena. That the deep inner workings of all life—from my two dogs resting quietly on the carpet, to the chaotic swirl of insects attracted by the lamp near my window, to the chorus of frogs rising up from the nearby pond, to the coyotes I now hear howling in the distance—rely on the same molecular processes, well, it is spectacular. So set aside the details, take a break before concluding the chapter, and allow that wondrous realization to sink in fully.

Evolution Before Evolution

Vital realizations not only provide unforeseen clarity, they also energize us to dig deeper. How did the common ancestor of all complex life come to be? Deeper still, how did life begin? Scientists have yet to determine the origin of life, but our discussion has made clear that the question is a three-parter. How did the genetic component of life—the capacity to store, utilize, and replicate information—come to be? How did the metabolic component of life—the capacity to extract, store, and utilize chemical energy—come to be? How did the packaging of genetic and metabolic molecular machinery into self-contained sacks—cells—come to be? The story of life's origin requires definitive answers to these questions, but even without a complete understanding we can turn to an explanatory framework—Darwinian evolution—that will almost certainly be an integral part of that future narrative.

When I first learned about Darwinian evolution, my biology teacher presented the theory as if it were the clever solution to a brain teaser that, once understood, should elicit a gentle slap to the forehead and the exclamation "Why didn't I think of that?" The puzzle is to explain the origin of the rich, varied, and bountiful array of species inhabiting planet earth. Darwin's solution comes down to two connected ideas: First, when organisms reproduce, progeny are generally similar but not identical to their parents. Or, as Darwin put it, reproduction yields descent with modification. Second, in a world with finite resources, there's competition for survival. Those biological modifications that enhance success in the competition increase the likelihood that the bearer will survive long enough to reproduce and thus pass on their survival-enhancing traits to future generations. Over time, different combinations of successful modifications slowly accumulate, driving an initial population to branch into groups that form distinct species.[29]

Simple and intuitive, Darwinian evolution almost seems self-evident. Yet however compelling its explanatory framework, were Darwinian evolution not supported by data it would have failed to achieve scientific consensus. Logic is not enough. Confidence in Darwinian evolution rests on the overwhelming support it has received from scientists who have traced gradual changes in the structure of organisms and delineated the adaptive advantages many of the changes conferred. If such transformations were absent, or if they occurred without any evident pattern, or if they bore no relation to the bearer's capacity to survive or reproduce, schoolkids would not be learning Darwinian evolution.

Darwin did not specify the biological basis for descent with modification. How do living beings bequeath traits to their offspring? And how do some of those traits descend in modified form? In Darwin's day, the answers were not known. Sure, everyone realized that little Mary looked like mom and dad, but an understanding of the molecular mechanism for passing on traits was still many discoveries away. That Darwin could develop the theory of evolution in the absence of such details speaks to the generality and power of the ideas. They transcend nitty-gritty details. It wasn't until nearly a century later, in 1953, that the illumination of DNA's structure made the path toward a molecular basis for heredity visible. With genteel restraint, Watson and Crick concluded their paper with an understatement ranking among the world's most famous: "It has not escaped our notice that the specific pairing we have postulated immediately suggests a possible copying mechanism for the genetic material."

Watson and Crick revealed the process by which life duplicates the very molecules that store the cell's internal instructions, allowing copies of the instructions to be passed on to progeny. As we have seen, the information that directs cellular function is encoded in the sequence of bases strung along the rails of DNA's twisted ladder. When a cell prepares to reproduce, to divide in two, the DNA ladder splits down the middle, yielding two rails, each comprising a sequence of bases. Because the sequences are complementary (an A on one rail ensures there's a T in the corresponding position on the second rail; a C on one rail ensures there's a G in the corresponding position on the second rail), each rail provides a template for building a copy of the other. By attaching the partner bases to those on each of the separated rails, the cell creates two complete copies of the original DNA strand. When the cell subsequently divides, each daughter cell receives one of the duplicate copies, passing genetic information from one generation to the next—the copying mechanism that had not escaped the notice of Watson and Crick.

As described, the copying process would yield identical strands of DNA. So how might new or modified traits arise in daughter cells? Errors. No process is 100 percent perfect. Although rare, mistakes will crop up, sometimes by chance and other times inflicted by environmental influences such as energetic photons—ultraviolet or X-ray radiation—that can corrupt the copying process. The DNA sequence a daughter cell inherits can thus differ from the one contributed by its parent. Oftentimes, such modifications are of little consequence, like a single typo on page 413 of **War and Peace.** But some modifications can impact a cell's functioning, for good or ill. The former, by enhancing fitness, stand a better chance of being passed on to subsequent generations and thus spreading through the population.

Sexual reproduction adds complexity because genetic material is not simply duplicated but is instead formed by melding contributions from the male and female parents. But while such reproduction represented a momentous step in the history of life on earth—one whose origin is still debated—the Darwinian principles apply all the same. The blending and copying of genetic material yield variations in inherited traits, and the ones most likely to persist across generations are those that enhance the carrier's prospects of survival and reproduction.

Essential to evolution is that in the descent from parent to progeny, modifications to DNA are typically few in number. This stability protects genetic improvements built up over previous generations, ensuring that they are not rapidly degraded or wiped out. To give a feel for just how rare such changes are, copying errors creep in at the rate of roughly one per every one hundred million DNA base pairs. That's like a medieval scribe getting a single letter wrong per every thirty copies of the Bible. And even that tiny rate is an overestimate, because 99 percent of the misprints are repaired by chemical proofreading mechanisms operating within each cell, reducing the net error rate to about one per every ten billion base pairs.

Even such minimal genetic modification, when accumulated over a great many generations, can give rise to massive physical and physiological development. This is not obvious. Some who encounter the wonder of the eye, the capacities of the brain, or the complexity of the cellular energy mechanisms will conclude that these systems could not have evolved without a guiding intelligence. And that conclusion would be justified if evolutionary development took place over familiar timescales. It didn't. Life has evolved for **billions** of years. That's **thousands** of **millions** of years. If each year were represented by a sheet of printer paper, then a billion years would correspond to a stack nearly a hundred kilometers high. Think of those pages as constituting a flip-book whose thickness is more than ten times the height of Mount Everest. Even if the drawing on each page differs only slightly from the one before it, the drawings at the beginning and end of the stack can easily be as different as a chimp is from an amoeba.

That is not to suggest that evolutionary change follows a carefully designed plan that gradually and efficiently progresses, page by page, from simple to complex organisms. Instead, evolution by natural selection is better described as innovation by trial and error. The innovations arise from random combinations and mutations of genetic material. The trials pit one innovation against another in the arena of survival. The errors, by definition, are innovations that lose. It is an approach to innovation that would bankrupt most businesses. Trying out one random possibility and then another, hoping against hope that sooner or later one of them lights up the market—well, try pitching that strategy to your board of directors. But nature has a surplus of a resource that for business is scarce: time. Nature is not in a hurry and does not need to meet a bottom line. The cost of innovating by small random changes is a cost nature can bear.[30]

An essential factor, too, is that there wasn't a single, isolated evolutionary flip-book. Every cell division in every organism occupying every nook and cranny of the planet contributed to the Darwinian narrative. Some of these story lines fizzled (genetic modifications that were detrimental). Most added nothing new to the ongoing plot (genetic material passed on without change). But some provided unexpected twists (genetic modifications that were adaptively useful) that would develop into their own evolutionary flip-books. Many of these, in fact, would support interdependent plots and subplots, so the evolutionary narrative in one flip-book would be influenced by that in others. The richness of life on earth thus reflects the enormous duration of the evolutionary chronicles, certainly, but also the enormous number of chronicles nature has written.

Like any healthy field of research, Darwinian evolution has been debated and refined over the decades. At what rate do species evolve? Does that speed vary widely over time? Are there long periods of stasis followed by short periods of more rapid change? Or is change always gradual? How should we think about traits that might decrease an organism's survival prospects while increasing the likelihood that it will reproduce? What is the full slate of mechanisms by which genes can change from generation to generation? How should we respond to gaps in the evolutionary record? Some of these issues have led to impassioned scientific brawls but—and this is key—none have cast any doubt on evolution itself. Details of any explanatory framework can and should and will be finessed over time, but the foundation of Darwinian theory is rock-solid.

Which raises a question: Might the Darwinian framework have relevance to a wider arena than life? After all, the essential ingredients—replication, variation, and competition—are not limited to living things. Printers replicate pages. Optical distortions yield variations in the copies. The printer's wireless receiver competes for limited bandwidth.

Let's imagine, then, a context closer to life than office printers but one decidedly inanimate: molecules that have acquired the ability to replicate. DNA is a prime example, so keep it in mind. But the replication of DNA—the splitting of its twisted ladder and the subsequent rebuilding of each component rail into two fully fledged DNA daughter molecules—relies on an army of cellular proteins, and so requires the processes of life to already be in place.

Imagine instead a molecule that can replicate on its own, long before any life anywhere has emerged. We don't need to commit to a definite replication mechanism, but just so you have a concrete mental image, perhaps when floating in a rich chemical stew this type of molecule acts like a molecular magnet, strongly attracting the very constituents that compose it and providing a template to assemble them into a molecular impersonator. Imagine, too, that the replication process, like all processes in the real world, is imperfect. Much of the time a newly synthesized molecule is identical to the original, but sometimes it's not. Over the course of a great many molecular generations, we thus build up an ecosystem inhabited by a spectrum of molecules that are variations on the original.

In any environment there are always limited raw materials, limited resources. So as our ecosystem of molecules continues to replicate, those that replicate most efficiently and accurately—fast, cheap, but far from out of control—will prevail. Such molecules garner the title most "fit" and over time will dominate the molecular population. Each subsequent mutation arising from imperfect replication offers yet further modifications to the molecular fitness. And as with all things alive, so with all things that aren't: those modifications that enhance molecular fitness will triumph over those that don't. The greater fecundity of molecules that are more fit swings the demographics toward those very molecules.

What I've described is a molecular version of evolution—**molecular Darwinism.** It shows how groups of jostling particles guided solely by the laws of physics can become ever more adept at reproduction—something we ordinarily associate with life. When we seek life's origin, this suggests that molecular Darwinism may have been an essential mechanism during the era leading up to the emergence of the first life. A version of that suggestion, far from consensus but one that has gained a significant following, relies on a special, multitalented molecule: RNA.

Toward the Origins of Life

Back in the 1960s, a number of prominent researchers, including Francis Crick, chemist Leslie Orgel, and biologist Carl Woese, drew attention to a close cousin of DNA, called RNA (ribonucleic acid), which some four billion years ago may have jump-started a phase of molecular Darwinism that was the precursor to life.

RNA is an extraordinarily versatile molecule that is an essential component of all living systems. You can think of it as a shorter, one-sided version of DNA, comprising a single rail along which a sequence of bases is attached. Among its various cellular roles, RNA is a chemical mediator that takes imprints of various small sections of an "unzipped" strand of DNA, similar to the way a dentist can take a mold of your teeth when you separate your upper and lower jaws, and transports the information to other parts of the cell, where it directs the synthesis of specific proteins. Like DNA, molecules of RNA thus embody cellular information and so are a component of a cell's software. But there's an important difference between RNA and DNA: whereas DNA is content to be a cell's oracle, a fount of wisdom directing cellular activity, RNA is willing to get its hands dirty with the manual labor of chemical processes. Indeed, the cell's ribosomes—miniature factories that snap together amino acids to yield proteins—have a particular variety of RNA (ribosomal RNA) at their core.

RNA is thus both software and hardware. It can direct as well as catalyze chemical reactions. And among such reactions are some that promote the replication of RNA itself. While the molecular machinery that makes copies of DNA uses an elaborate collection of chemical cogs and wheels, RNA itself can promote the synthesis of the base pairs necessary for its own replication. Consider the implication. Molecules of RNA, blending software and hardware, have the potential to sidestep the chicken and egg conundrum: How do you assemble molecular hardware without first having the molecular software, the instructions to carry out the assembling? How do you synthesize molecular software without first having the molecular hardware, the infrastructure to carry out the synthesizing? Embodying both functions, RNA melds chicken and egg, and thus has the capacity to propel an era of molecular Darwinism forward.

Such is the RNA World proposal. It imagines that before there was life there was a world suffused with RNA molecules, which through molecular Darwinism evolved over an almost unfathomable number of generations into the chemical structures that constituted the first cells. While details are tentative, scientists have sketched what this phase of molecular evolution may have been like. In the 1950s, Nobel laureate Harold Urey and his graduate student Stanley Miller mixed gases (hydrogen, ammonia, methane, water vapor) that they believed constituted earth's early atmosphere, zapped the gaseous cocktails with electric currents to simulate strikes of lightning, and famously announced that the resulting brown sludge contained amino acids, the building blocks of proteins. Although subsequent research showed that the initial gas mixtures Miller and Urey studied did not accurately reflect the chemical makeup of earth's early atmosphere, similar experiments carried out with other gaseous cocktails that did (including a mixture Miller and Urey themselves had concocted to model the toxic fumes from active volcanoes, which, curiously, sat unanalyzed for more than a half century[31]) were just as successful in generating amino acids. Moreover, amino acids have now been detected in interstellar clouds, in comets, and in meteorites. So, plausibly, a chemical stew on young earth may have blended replicating RNA molecules with a plentiful assortment of amino acids.

Imagine, then, that as RNA molecules continued to replicate, a chance mutation facilitated something novel: the mutant RNA coaxed some of the amino acids in the environmental stew to hook up into chains yielding the first rudimentary proteins (a crude version of the kinds of processes that now take place in ribosomes). If, by chance, some of these basic proteins happened to increase the efficiency of RNA replication—after all, catalyzing reactions is, in part, what proteins do—they would be richly rewarded: the proteins would usher the mutant form of RNA to dominance, and the newly plentiful supply of mutant RNA would help synthesize more of the proteins. In tandem, they would constitute a self-reinforcing chemical loop that would propel the chance molecular aberrations to become the norm. Over time, the continued molecular machinations might hit upon another chemical novelty, a double-railed ladder—a rudimentary form of DNA—that proved to be a more stable and more efficient structure for molecular replication, and thus gradually usurped the replication processes and relegated RNA to a supporting role. The chance formation of molecular bags—cell walls—would increase fitness further by concentrating chemicals in sequestered regions and offering protection from environmental disruption. Spreading throughout the chemical population, the structures necessary for the first rudimentary cells would assemble.[32]

Life would be born.

The RNA World is but one of numerous proposals. It's an example that places a premium on the genetic component of life: molecules that embody information and through replication pass that information on to subsequent generations. Should the proposal prove correct, we would still need to address the origin of RNA itself; perhaps an even earlier stage of molecular evolution might have generated RNA from yet simpler chemical constituents. Other proposals place more weight on the metabolic component of life: molecules that catalyze reactions. Instead of a replicating molecule that can act as a protein, these scenarios begin with protein molecules that can replicate.

Yet other proposals envision two wholly distinct developments, one that leads to molecules that replicate and another that leads to molecules that catalyze chemical reactions, and only later do these processes fuse into cells that can carry out the basic functions of reproduction and metabolism.

Proposals also abound for where the chemical antecedents to life first formed. Some researchers conclude that Darwin's offhand suggestion of a "warm little pond" is not particularly promising because for hundreds of millions of years rocky debris rained down on earth, rendering the surface less than hospitable.[33] Even so, biologist David Deamer has suggested that essential to the origin of life is an environment that cycles between wet and dry, like land at the edge of a pond or lake. His team's research has demonstrated that such wet and dry cycles can propel lipids to form membranes—cell walls—within which molecular snippets can be coaxed to connect into longer chains, akin to RNA and DNA.[34] Chemist Graham Cairns-Smith has proposed that the crystals constituting clay beds—structures that grow by continually locking atoms into an orderly, repeating pattern—may have constituted an early system of replication that was a precursor to such behavior in more complex organic molecules en route to life.[35] Another compelling contender, suggested and developed by geochemist Mike Russell and biologist Bill Martin, are cracks in the ocean floor that spew out warm, mineral-rich plumes generated by the interaction of seawater with the rock constituting earth's mantle.[36] These so-called alkaline hydrothermal vents precipitate limestone chimneys rising up from the seabed— some grow to a height of more than fifty meters, taller than the Statue of Liberty—laden with nooks and crannies through which an energetic flood of chemicals continually streams. The proposal envisions that within the many eddies that form within the towers, molecular Darwinism performs its chemical wizardry, yielding replicators that over time ratchet up in complexity and sophistication, ultimately spawning life on earth.

The details occupy forefront research. To date, laboratory attempts to recreate these processes are intriguing but inconclusive. We have yet to create life from scratch. I have little doubt that one day, perhaps not far off, we will. In the meantime, an overarching scientific narrative for life's origin is emerging. Once molecules acquire the capacity to replicate, chance errors and mutations will feed molecular Darwinism, driving chemical concoctions along the all-important vector of increased fitness. Playing out over hundreds of millions of years, the process has the capacity to build the chemical architecture of life.

The Physics of Information

By this point you may have concluded that life's molecules must have aced their studies of organic chemistry. Otherwise, how in the world would they know what they are supposed to? How does DNA know to split down the middle and attach complementary bases to the ones it has exposed, creating a duplicate molecule? How does RNA know to make copies of sections of DNA, transport that information to the appropriate cellular structures where yet other distinct but related molecules know how to read the genetic code and link up appropriate sequences of amino acids into functioning proteins?

Of course, the molecules don't know anything. Their behavior is governed by the blind, mindless, unschooled laws of physics. But the question remains: How do they consistently and reliably carry out a stunningly intricate series of complex chemical processes? It's a question that harks back to my paraphrasing of Schrödinger's primary query in **What Is Life?**: The jostling and careening of molecules within a rock are governed by the laws of physics. The jostling and careening of molecules within a rabbit are also governed by the laws of physics. How do they differ? We have now seen that the rabbit's particles are guided by an additional influence—the rabbit's internal archive of information, its cellular software. Importantly, critically, vitally: This information does not supersede the laws of physics. Nothing does. Instead, much as a water slide doesn't supersede the laws of gravity but through its shape guides riders along a specific trajectory they would otherwise not follow, the rabbit's cellular software is carried by chemical arrangements that through their shape, structure, and constituents guide various molecules along trajectories that they, too, would otherwise not follow.

How do such molecular guides work? Because of the detailed arrangement of its constituent atoms, a given molecule might attract this amino acid, repel that one, and be thoroughly indifferent toward others. Or, like matched Lego pieces, a given molecule might snap together with only one specific other molecule. All of this is physics. When atoms and molecules push or pull or snap together, it is the electromagnetic force in action. The point, then, is that information in a cell is not abstract. It is not a free-floating set of instructions that molecules need to study, memorize, and execute. Instead, the information is encoded in the molecular arrangements themselves, arrangements that coax other molecules to bump or join or interact in a manner that carries out cellular processes like growth, repair, or reproduction. Even though the molecules inhabiting a cell lack intent or purpose, and even though they are thoroughly oblivious, their physical structure allows them to accomplish highly specialized tasks.

In this sense, the processes of life are molecular meanderings fully described by physical law that simultaneously tell a higher-level, information-based story. For the rock, there is no higher-level story. When you use the laws of physics to describe the bumping and jostling of the rock's molecules, you're done. But when you use the very same laws of physics to describe the bumping and jostling of rabbit molecules, you are not done. Not by a long shot. Overlaid on the reductionist story is a whole additional story that tells of the rabbit's unique internal molecular arrangements that choreograph an exquisite spectrum of organized molecular motions. And it is these molecular motions which carry out higher-level processes within the rabbit's cells.

Indeed, for the rabbit, and for us, too, such biological information is also organized on larger scales, guiding processes that act not just within individual cells but across collections of cells, yielding the hallmark quality of coordinated complexity. When you reach for a cup of coffee, the motion of every atom constituting every molecule in your hand, arm, body, and brain is fully governed by the laws of physics. Again, with gusto: Life does not and cannot contravene physical law. Nothing can. But the fact that a huge number of your molecules can act in concert, coordinating their overall motion to cause your arm to reach out across a table and your hand to clutch a mug, reflects the wealth of biological information, embodied in atomic and molecular arrangements, directing a profusion of complex molecular processes.

Life is physics orchestrated.

Thermodynamics and Life

Evolution, per Darwin, guides the development of structures from molecules to single cells to complex multicellular organisms. Entropy, per Boltzmann, charts the unfolding of physical systems, from wafting aromas to clanking heat engines to burning stars. Life is subject to both of these guiding influences: Life arose and was refined via evolution. Life, like all physical systems, abides by the dictates of entropy. In the final couple of chapters of **What Is Life?**, Schrödinger explored the seeming tension between the two. When matter coalesces into life, it sustains order over long periods of time. And as life reproduces, it generates additional collections of molecules that are also arranged in orderly structures. Where in all of this is entropy, disorder, and the second law of thermodynamics?

In his answer, Schrödinger explained that organisms resist the rise to higher entropy by "feeding upon negative entropy,"[37] a phrasing that through the decades has generated minor confusion and persnickety criticism. But it's clear that while he expressed it in somewhat different language, Schrödinger's answer is the very one we have been developing: the entropic two-step. Living things are not isolated, and so any accounting of the second law must incorporate their environment. Take me. For more than a half century I've successfully kept my entropy from shooting through the roof. I've done this by taking in orderly structures (mostly vegetables, nuts, and grains), slowly burning them (through redox reactions, electrons from the food cascade down the stadium stairs and ultimately combine with oxygen I have inhaled), using the energy released to power various metabolic activities, and dispensing entropy to the environment through waste and heat. Overall, the two-step has allowed my entropy to seemingly thumb its nose at the second law while the environment has diligently had my back, taking up the entropic slack. The process of burning, storage, and release of energy to power cellular functions is more elaborate than the corresponding process that powers steam engines, but entropically speaking the essential physics is the same.

Beyond Schrödinger's choice of language, a less fussy concern is the origin of the high-quality, low-entropy nourishment. Heading from animals down the food chain we encounter plants, which feed directly on sunlight. Their energy cycle provides another instance of the entropic two-step. Incoming solar photons absorbed by plant cells kick electrons into higher energy states, which cellular machinery then harnesses (via a series of redox reactions that guide the electrons down the stadium stairs) to power various cellular functions. Photons from the sun are thus the low-entropy, quality nourishment that plants absorb, exploit for the processes of life, and then release in a higher-entropy, degraded form as waste (for each photon received from the sun, the earth sends a less orderly collection of a couple dozen energetically depleted and widely dispersed infrared photons back into space).[38]

Following the trail toward the low-entropy source yet further, we seek the origin of the sun, which dovetails with the gravitational story from chapter 3: gravity squeezes gas clouds into stars, lowering internal entropy and, through heat released, raising the entropy of the surrounding environment. Ultimately, nuclear reactions ignite, stars light up, and photons are sent streaming outward. When that star is the sun, those photons that reach earth are the low-entropy source of energy that powers plant metabolism, making clear why researchers often say that the gravitational force sustains life. While true, by now you know that I like to share the credit more equitably, lauding gravity for causing matter to clump and securing stable stellar environments, but also extolling nuclear fusion for the relentless production of a steady stream of high-quality photons over millions and billions of years.

The nuclear force, in tandem with gravity, is a fount of life-giving low-entropy fuel.

A General Theory of Life?

In his 1943 lectures, Schrödinger emphasized that the torrent of scientific developments had been so intense that "it has become next to impossible for a single mind fully to command more than a small specialized portion."[39] Consequently, he encouraged thinkers to extend the reach of their expertise by exploring realms outside their traditional intellectual stomping ground. With **What Is Life?** he unabashedly brought the training, intuition, and sensibility of a physicist to bear on the puzzles of biology.

In the decades since, as knowledge has become increasingly specialized, a growing cohort of researchers has continued to sound Schrödinger's interdisciplinary call. Many have responded. Researchers with training across fields including high-energy physics, statistical mechanics, computer science, information theory, quantum chemistry, molecular biology, and astrobiology, among many others, have developed new and insightful ways of probing the nature of life. I'll close this chapter by focusing on one such development that extends our thermodynamic theme and, if the program succeeds, may one day help answer some of science's most profound questions: Could life be such a long-shot possibility that it arose only once in a universe containing hundreds of billions of galaxies, each with hundreds of billions of stars, many of which have orbiting planets? Or is life the natural outcome, perhaps even the inevitable outcome, of certain basic and relatively common environmental conditions, suggesting a cosmos teeming with life?

To approach questions of such broad sweep we need principles with comparable sweep. By now, we've seen ample evidence of the expansive applicability of thermodynamics, a physical theory Einstein described as the only one for which he could confidently declare "it will never be overthrown."[40] Perhaps in analyzing the nature of life—its origin and evolution—we can push the thermodynamic perspective yet further.

Over the past few decades, scientists have done just that. The research discipline that has emerged (called **nonequilibrium thermodynamics**) systematically analyzes the kinds of situations we have now encountered repeatedly: high-quality energy coursing through a system, powering the entropic two-step and thus allowing the system to resist the pull toward internal disorder that would otherwise hold sway. Belgian physical chemist Ilya Prigogine, who was awarded the 1977 Nobel Prize for his pioneering work in the field, developed the mathematics for analyzing configurations of matter that, when subject to a continual source of energy, can spontaneously become ordered—what Prigogine called "order out of chaos." If you had a good high school physics class, you may have encountered a simple yet impressive example, Bénard cells. Heat a flat dish containing a puddle of viscous oil. At first not much happens. But as you gradually increase the energy streaming through the liquid, random molecular motions conspire to yield visible order. Looking down on the oil, you will see it tessellate into a collection of small hexagonal chambers. Looking from the side, you will see the liquid flowing in a stable and regular pattern, rising from the bottom of each hexagonal chamber, reaching the top, and then looping back to the chamber's bottom.

From the standpoint of the second law of thermo-dynamics, such spontaneous order is wholly unex-pected. It arises because the liquid's molecules are subject to a particular environmental influence: they are continually heated by the flame. And this persistent injection of energy has significant impact. In any system there will occasionally be spontaneous fluctuations that momentarily form a small, localized, orderly pattern. Usually such tiny fluctuations quickly disperse back into a dis-ordered form. But Prigogine's analysis showed that when molecules are in certain special patterns they become exceptionally adept at absorbing energy, and this dictates a different fate. If the physical system is receiving a steady flow of concentrated energy from the environment, the special molecular patterns can use the energy to sustain or even enhance their orderly form, while dumping a degraded form of that energy (less accessible, more spread out) back into the environment. The orderly patterns are said to dissipate the energy and hence are called **dissipative structures.** Total entropy, including environmental, increases, but by steadily pumping energy into a system we can drive and maintain order via a sustained entropic two-step.

Prigogine's description parallels the physical explanation, going back to Schrödinger, for how organisms stave off entropic degradation. Not that Bénard cells are alive, but living beings are dissipative structures, too, absorbing energy from the environment, using it to sustain or enhance their orderly form, and releasing a degraded form of that energy back to the environment. Prigogine's results provided a mathematically precise articulation of his slogan "order from chaos"; many subsequent researchers speculated that the math might be developed further, perhaps yielding insight on how the orderly molecules necessary for life emerged from the chaos of random molecular motions taking place on early earth.

Of the many contributions to this program,

recent work by Jeremy England (extending earlier results developed by researchers including Christopher Jarzynski and Gavin Crooks) is particularly exciting.[41] Through clever mathematical manipulations, England has teased out the implications of the second law of thermodynamics when it is applied to systems powered by an external source of energy. To get a feel for his result, imagine you are on a playground swing. As every kid knows intuitively, you need to pump your legs (and angle your body) at the right rate to get the swing going and maintain a smooth, rhythmic motion. And that rate, according to basic physics, depends on the distance between the seat and the swing's pivot. If you pump your legs at the wrong rate, the rhythmic mismatch prevents the swing from efficiently absorbing the energy you are providing, and so you won't swing high. Imagine, however, that this particular swing has an unusual feature: as you pump your legs, the length of the swing changes, adjusting the period of its motion to agree with that of your legs. This "adaptation" allows the swing to rapidly get into the groove, take in the energy you offer, and quickly reach a satisfying height on each cycle. Subsequently, the energy of your pumping action is absorbed by the swing, but it doesn't drive the swing any higher. Instead, the energy you input keeps the swing's motion steady by working against countervailing frictional forces and, in the process, producing waste (heat, sound, and so on) that is dissipated back to the environment (assuming you're not a daredevil like my daughter, who awaits the swing's high point to fly from the seat, soar, and then dissipate energy by tumbling on the ground).

England's mathematical analysis revealed that in the molecular domain, particles that are being "pushed" by an external source of energy can have an experience analogous to your playground escapade. An initially disordered collection of particles can adapt their configuration to "get in the groove"—to form an arrangement that more efficiently absorbs energy from the environment, uses it to maintain or enhance orderly internal motion or structure, and then dissipates a degraded form of that energy back to the environment.

England calls the process **dissipative adaptation.** Potentially, it provides a universal mechanism for coaxing certain molecular systems to get up and dance the entropic two-step. And as that's what living things do for a living—they take in high-quality energy, use it, and then return low-quality energy in the form of heat and other wastes—perhaps dissipative adaptation was essential to the origin of life.[42] England notes that replication itself is a potent tool of dissipative adaptation: if a small collection of particles has become adept at absorbing, using, and dispensing energy, then two such collections are better still, as are four or eight, and so on. Molecules that can replicate might then be an **expected** output of dissipative adaptation. And once replicating molecules appear on the scene, molecular Darwinism can kick in, and the drive to life begins.

These ideas are in their early stages, yet I can't help but think they would have made Schrödinger happy. Using fundamental physical principles, we have developed an understanding of the big bang, the formation of stars and planets, the synthesis of complex atoms, and now we are determining how those atoms might arrange into replicating molecules well adapted for extracting energy from the environment to build and sustain orderly forms. With the power of molecular Darwinism to select for ever-fitter molecular collections, we can envision how some might acquire the capacity to store and transmit information. An instruction manual passed from one molecular generation to the next, which preserves battle-tested fitness strategies, is a potent force for molecular dominance. Acting out over hundreds of millions of years, these processes may have gradually sculpted the first life.

Whether or not the details of these ideas survive future discoveries, the outline of life's story according to physics is taking shape. And if that story proves to be as general as recent work suggests, life might well be a common feature of the cosmos. Exciting as this would be, life is one thing and intelligent life quite another. Finding microbes on Mars or on Jupiter's moon Europa would be a monumental discovery. But as thinking, conversing, creative beings, we would still be alone.

What, then, is the path from life to consciousness?

PARTICLES AND CONSCIOUSNESS

From Life to Mind

Somewhere between the first prokaryotic cells four billion years ago and the human brain's ninety billion neurons entangled in a network of one hundred trillion synaptic connections, the ability emerged to think and feel, to love and hate, to fear and yearn, to sacrifice and revere, to imagine and create—newfound capacities that would ignite spectacular achievement as well as untold destruction. "Everything begins with consciousness and nothing is worth anything except through it,"[1] is how Albert Camus put it. Yet, until recent years, consciousness was an unwelcome word in the hard sciences. Sure, doddering researchers in the twilight of their careers might be forgiven for turning to the fringe topic of mind, but the goal of mainstream scientific research is an understanding of objective reality. And for many, and for a long time, consciousness didn't qualify. The voice chattering inside your head, well, it can be heard only inside your head.

It is an ironic stance. Descartes's **"Cogito, ergo sum"** summarizes our contact with reality. All else could be an illusion, but thinking is the one thing even the die-hard skeptic can be sure of. And notwithstanding Ambrose Bierce's "I think that I think, therefore I think that I am,"[2] if you are thinking, the case for existing is strong. For science to pay no mind to consciousness would be to turn from the very thing, the only thing, we each can count on. Indeed, for thousands of years many have denied the finality of death by hanging existential hope on consciousness. The body dies. That's apparent, obvious, undeniable. But our seemingly persistent inner voice, as well as the abundant thoughts, sensations, and emotions filling each of our subjective worlds, speaks to an ethereal presence that, some have imagined, stands outside the base facts of physical existence. Atman, anima, immortal soul—it has been given many names, but all connote the belief that the conscious self taps into something that outlasts the physical form, something that transcends traditional mechanistic science. Not only is mind our tether to reality, perhaps it is our tether to eternity.

Therein lies a more revealing clue for why the

hard sciences have long resisted all things consciousness. Science reacts to talk of realms beyond the reach of physical law with an exasperated grimace, a turning on its heels, and a swift return to the lab. Such scoffing represents a dominant scientific attitude but also highlights a critical gap in the scientific narrative. We have yet to articulate a robust scientific explanation of conscious experience. We lack a conclusive account of how consciousness manifests a private world of sights and sounds and sensations. We cannot yet respond, or at least not with full force, to assertions that consciousness stands outside conventional science. The gap is unlikely to be filled anytime soon. Most everyone who has thought about thinking realizes that cracking consciousness, explaining our inner worlds in purely scientific terms, poses one of our most formidable challenges.

Isaac Newton ignited modern science by finding patterns in the parts of reality accessible to human senses and codifying them in his laws of motion. In the centuries since, we've recognized that pressing on from Newton requires blazing three distinct trails: We need to understand reality on scales far smaller than Newton considered, a path that has taken us to quantum physics, which has explained the behavior of fundamental particles and, among much else, the biochemical processes underlying life. We need to understand reality on scales far larger than Newton considered, a path that has taken us to general relativity, which has explained gravity and, among much else, the formation of stars and planets essential for the emergence of life. And for the third frontier, most labyrinthine of all, we need to understand reality on scales far more complex than Newton considered, a path we anticipate leading to an explanation of how large collections of particles can coalesce to yield life and generate mind.

By training his intellectual might on highly simplified problems—ignoring, for example, the churning internal structures of the sun and planets and treating each instead as a solid ball—Newton did the right thing. The art of science, of which Newton was the master, lies in making judicious simplifications that render problems tractable while retaining enough of their essence to ensure that the conclusions drawn are relevant. The challenge is that simplifications effective for one class of problems can be less so for others. Model the planets as solid balls and you can work out their trajectories with ease and precision. Model your head as a solid ball and the insights into the nature of mind will be less enlightening. But to jettison unproductive approximations and lay bare the inner workings of a system containing as many particles as the brain—a laudable goal—would require mastering a level of complexity fantastically beyond the reach of today's most sophisticated mathematical and computational methods.

What's changed in recent years is newfound access to observable and measurable features of brain activity that, at the very least, access processes that reliably accompany conscious experience. When researchers can use functional magnetic resonance imaging to meticulously track blood flow supporting neural activity, or insert deep brain probes to detect electrical impulses firing along individual neurons, or use electroencephalograms to monitor electromagnetic waves rippling across the brain, and when the data reveal clear patterns that mirror both observed behavior and reports of inner experience, the case for approaching consciousness as a physical phenomenon strengthens substantially. Indeed, encouraged by these impressive advances, daring researchers have deemed the time ripe to develop a scientific basis for conscious experience.

Consciousness and Storytelling

Some years ago, during a good-natured but heated exchange on the role of mathematics in describing the universe, I emphatically told a late-night television host that he was nothing but a bag of particles governed by the laws of physics. Not as a joke, although without missing a beat he turned it into one. ("Hey, that's a great pickup line.") And not as a jibe, for in this regard, whatever holds true for him applies equally to me. Instead, the remark sprang from my deep-seated reductionist commitment, which holds the view that by fully grasping the behavior of the universe's fundamental ingredients we tell a rigorous and self-contained story of reality. We don't have a finished draft of this story in hand since a great many problems at the forefront of research remain unsolved, some of which we'll encounter shortly. Nonetheless, I can envision a future when scientists will be able to provide a mathematically complete articulation of the fundamental microphysical processes underlying anything that happens, anywhere and anywhen.

There is something comforting in this prospect, something that gracefully resonates with a twenty-five-hundred-year-old sentiment of Democritus, "Sweet is sweet, bitter is bitter, hot is hot, cold is cold, color is color; but in truth there are only atoms and the void."[3] The point being that everything emerges from the same collection of ingredients governed by the same physical principles. And those principles, as attested to by a few hundred years of observation, experimentation, and theorizing, will likely be expressed by a handful of symbols arranged in a small collection of mathematical equations. **That** is an elegant universe.[4]

As powerful as such a description would be, it would remain but one among many stories we tell. We have the capacity to shift focus, to reset resolution, to engage with the world in a wide variety of ways. While a complete reductionist description would provide a scientific bedrock, other descriptions of reality, other stories, provide insights that many deem more relevant because they are closer to experience. Telling some of these stories, as we've already seen, requires new concepts and language. Entropy helps us tell the story of randomness and organization within large collections of particles, whether they're wafting from your oven or coalescing into stars. Evolution helps us tell the story of chance and selection as collections of molecules—living or not—replicate, mutate, and gradually become better adapted to their environment.

A story many deem more relevant still focuses on consciousness. To embrace thoughts, emotions, and memories is to embrace the core of human experience. It is also a story that requires a perspective qualitatively different from any we have taken so far. Entropy, evolution, and life can all be studied "out there." We can fully tell their stories as third-person accounts. We are witnesses to these stories and, if we are sufficiently diligent, our account can be exhaustive. These stories are inscribed in open books.

A story that encompasses consciousness is different. A story that penetrates into the inner sensations of sight or sound, of elation or grief, of comfort or pain, of ease or anxiety, is a story that relies on a first-person account. It is a story informed by an inner voice of awareness speaking from a personal script each one of us seemingly authors. Not only do I experience a subjective world, but I have a palpable sense that from within that world I control my actions. No doubt, when it comes to your actions you have a similar sense. Laws of physics be damned; I think, therefore I control. Understanding the universe at the level of consciousness requires a story that can grapple with an utterly personal and seemingly autonomous subjective reality.

To illuminate conscious awareness we thus encounter two distinct but related challenges. Can matter, on its own, produce the sensations infusing conscious awareness? Can our conscious sense of autonomy be nothing more than the laws of physics acting themselves out on the matter constituting brain and body? To these questions, Descartes answered with a definitive no. In his view, the manifest difference between matter and mind reflects a deep division. The universe has physical stuff. The universe has mind stuff. Physical stuff can affect mind stuff and mind stuff can affect physical stuff. But the two kinds of stuff are different. In modern language, atoms and molecules are not the stuff of thought.

Descartes's stance is alluring. I can attest that tables and chairs, cats and dogs, grass and trees are different from the thoughts inside my head, and I suspect you would confirm a similar sentiment. Why would the particles that constitute the tangible elements of external reality and the physical laws that govern them have any relevance for explaining my inner world of conscious experience? Perhaps, then, we should expect an understanding of consciousness to not merely be a higher-level story, to not merely be a story that shifts its gaze from outward to inward, but to be a fundamentally different kind of story, one that requires a conceptual revolution on par with those of quantum physics and relativity.

I'm all for intellectual revolutions. There is nothing more exciting than a discovery that turns the accepted worldview on its head. And in what follows, we will discuss upheavals that some consciousness researchers envision to be heading our way. But for reasons that will become clear, I suspect that consciousness is less mysterious than it feels. Resonating with my late-night TV exclamation and, more importantly, with a segment of researchers who've devoted their professional lives to these questions, I anticipate that we will one day explain consciousness with nothing more than a conventional understanding of the particles constituting matter and the physical laws that govern them. That would yield its own variety of revolution, establishing a virtually unlimited hegemony for physical law, reaching arbitrarily far into the outer world of objective reality and arbitrarily deep into the inner world of subjective experience.

In the Shadows

Not all brain function commands the reverence accorded consciousness. Much neurological activity is orchestrated beneath the surface of conscious awareness. As you watch a sunset, your brain rapidly processes the data carried by trillions of photons striking photoreceptors in your retinas each second, diligently interpolating the image to account for your blind spots (where, in each eye, your optic nerve connects to the retina, carrying data to your brain's lateral geniculate nucleus and on to the visual cortex), continually compensating for the shifting of your eyes and movement of your head, correcting for photons blocked or scattered by ocular irregularities, flipping each image right side up, fusing the parts of each image common to both eyes, and so on, and yet as you quietly contemplate the sun's final rays, you are completely unaware of all that is happening just behind your eyes. A similar description holds as you read these words. The architecture of awareness allows you to focus on the conceptual ideas the words symbolize, relegating massive visual and linguistic data processing to brain functions that go unnoticed. More innate still, day in and day out, you walk, you talk, your heart beats, your blood flows, your stomach digests, your muscles flex, and on and on, and it all happens without the need for you to pay the slightest attention.

That the brain is awash with influential processes escaping introspection is a premise with a long history, one that has been expressed in myriad forms. Vedic texts written three thousand years ago invoke a notion of the unconscious, and references continue across the centuries as penetrating thinkers have surmised flavors of mental qualities unavailable to the palate of conscious awareness: Saint Augustine ("The mind is not large enough to contain itself: but where can that part of it be which it does not contain?"[5]), Thomas Aquinas ("The mind does not see itself through its essence"[6]), William Shakespeare ("Go to your bosom, / Knock there, and ask your heart what it doth know"[7]), Gottfried Leibniz ("Music is the hidden arithmetical exercise of a mind unconscious that it is calculating"[8]). Intriguing too are processes that **seem** to reside below the radar and yet generate echoes accessible to conscious processing. Stories abound, for example, of the unconscious mind solving problems and delivering the solutions unbidden. One of the most colorful comes from German pharmacologist Otto Loewi, who during the night before Easter Sunday 1921 briefly awoke and scribbled down an idea that had just come to him in a dream. In the morning, Loewi had an overwhelming sense that the nocturnal note contained a vital insight, but however hard he tried he was unable to decipher it. The next night he had the same dream, but this time he immediately went to the lab and followed the dream's directive to carry out an experiment testing his long-standing hypothesis that chemical processes, not electrical, are central to cellular communication. By Monday, the dream-inspired experiment was done, and its success would ultimately lead to Loewi's winning the Nobel Prize.[9]

Popular culture tends to entwine the subterranean workings of the mind with the contributions of Sigmund Freud (notwithstanding a cadre of scientists that years earlier had pursued related ideas[10]) and the churning undercurrents of repressed memories, desires, conflicts, phobias, and complexes that he conceived as buffeting human behavior to and fro. The weighty difference in modern times is that speculations, hunches, and intuitions regarding the life of the mind now confront data that were previously unavailable. Researchers have developed clever ways to peek over the mind's shoulder and track brain activity lying beneath the level of conscious awareness.

Some of the most striking studies involve patients who have lost some degree of neurological function. A well-known case, involving a subject known as P.S. who sustained right cerebral damage, was documented in the late 1980s by Peter Halligan and John Marshall.[11] As anticipated with this type of impairment, P.S. would fail to report details on the far left side of any image she was shown. She claimed, for instance, that two dark green line drawings of a house were identical even though the left side of one of the houses was being consumed by a raging red fire. Yet, when asked which of the two houses she'd prefer to call home, P.S. consistently chose the house that was not burning. The researchers argued that although P.S. was unable to acquire conscious awareness of the blaze, the information had entered covertly and was influencing her decision from behind the scenes.

Healthy brains, too, reveal their own dependence on hidden influences. Psychologists have established that even if you are paying close attention, an image flashed on-screen for less than about forty milliseconds (and sandwiched between somewhat longer flashes of other images known as masks) will fail to enter your conscious awareness. Nevertheless, such subliminal images can influence conscious decisions. The famous claim of an uptick in soft drink consumption caused by subliminal frames of "Drink Coke" being flashed in movie theaters is an urban myth propagated in the late 1950s by a struggling market researcher.[12] But clever laboratory studies have provided compelling evidence for specific types of clandestine mental processes.[13] For example, imagine facing a screen on which numbers, each between 1 and 9, are flashed and your task is to rapidly classify each as either larger or smaller than 5. Your reaction times will be faster when a given number is preceded by a subliminal flash of a digit lying on the same side of 5 as the given number (for example, when a 4 is preceded by a subliminal 3). Conversely, your reaction times will be slower when a given number is preceded by a subliminal flash of a digit lying on the opposite side of 5 as the given number (for example, when a 4 is preceded by a subliminal 7).[14] Even though you are not consciously aware of the fleeting numerical cameos, they've whisked across your brain and impacted your response.

The upshot is that your brain surreptitiously coordinates a regulatory, a functional, and a data-mining marvel. Wondrous though these brain activities are, they do not constitute a conceptual mystery. The brain rapidly sends and receives signals along nerve fibers, allowing it to control biological processes and generate behavioral responses. To delineate the precise neural pathways and physiological details underlying such functions and behaviors, scientists face the daunting task of mapping out vast territories dense with complex biological circuitry at a level of precision well beyond what has so far been achieved. Still, everything we're learning suggests that however challenging, however vast the reserves of creativity and diligence required, there is every reason to believe that the familiar strategies of science will prevail.

And were it not for one pesky quality of mind, that would be that. But when we look beyond the mind's tasks and consider instead the mind's sensations—the inner experience we identify as the essence of being human—some researchers have reached a different and far less optimistic prognosis for the capacity of traditional science to provide insight. This takes us to what some call the "hard problem" of consciousness.

The Hard Problem

In a letter to Henry Oldenburg, one of the most prolific correspondents during the formative years of modern science, Isaac Newton noted, "To determine more absolutely, what Light is . . . and by what modes or actions it produceth in our minds the Phantasms of Colours, is not so easie. And I shall not mingle conjectures with certainties."[15] Newton was struggling to explain the most common of experiences: the inner sensation of one or another color. Consider a banana. It's no big deal, of course, to look at a banana and determine that it's yellow. If you have the right app, your phone can do it. But as far as we know, when your phone reports that the banana is yellow, the phone does not have an inner feeling of yellow. It does not have an inner sensation of yellow. It does not see yellow in its mind's eye. You do. So do I. As did Newton. His predicament was to understand how in the world we do this.

The predicament is relevant well beyond mental "phantasms" of yellow or blue or green. As I type these words, snacking on popcorn, music playing softly in the background, I feel a range of inner experiences: pressure on my fingertips, a salty aftertaste, the magnificent voices of Pentatonix, a mental monologue negotiating the next phrase in this sentence. Your inner world is taking in these words, perhaps hearing them spoken by your mind's inner voice, while perhaps also feeling distracted by that last piece of chocolate pie in the refrigerator. The point is that our minds host a range of inner sensations—thoughts, emotions, memories, images, desires, sounds, smells, and more—that are all part of what we mean by consciousness.[16] As with Newton and the banana, the challenge is to determine how our brains create and sustain these vibrant worlds of subjective experience.

To take in the full depth of the puzzle, imagine you are endowed with superhuman vision allowing you to peer into my brain and see every one of its roughly thousand trillion trillion particles—electrons, protons, and neutrons—bumping and jostling, attracting and repelling, flowing and scattering.[17] Unlike the large collections of drifting particles from baking bread or those coalescing into a star, the particles constituting a brain are arranged in a highly organized pattern. Even so, focus in on any one such particle and you'll find that it interacts with others via the very same forces described by the very same mathematics whether that particle is floating in your kitchen, in the corona of the North Star, or inside my prefrontal cortex. And within that mathematical description, affirmed by decades of data from particle colliders and powerful telescopes, there is nothing that even hints at the inner experiences those particles somehow generate. How can a collection of mindless, thoughtless, emotionless particles come together and yield inner sensations of color or sound, of elation or wonder, of confusion or surprise? Particles can have mass, electric charge, and a handful of other similar features (nuclear charges, which are more exotic versions of electric charge), but all these qualities seem completely disconnected from anything remotely like subjective experience. How then does a whirl of particles inside a head—which is all that a brain is—create impressions, sensations, and feelings?

Philosopher Thomas Nagel gave an iconic and particularly evocative account of the explanatory gap.[18] What's it like, he asked, to be a bat? Picture it: Aloft on a bed of air as you soar across a dark landscape, you cry out with an incessant patter of clicks, generating echoes from trees, rocks, and insects, which allow you to map the environment. From the reflected sound you realize a mosquito is up ahead and darting to the right, so you swoop in and enjoy a tiny morsel. Since our mode of engagement with the world is profoundly different, there is just so far our imagination can take us into the bat's inner world. Even if we had a complete accounting of all the underlying fundamental physics, chemistry, and biology that make a bat a bat, our description would still seem unable to get at the bat's subjective "first-person" experience. However detailed our material understanding, the inner world of the bat seems beyond reach.

What's true for the bat is true for each of us. You are a swarm of interacting particles. So am I. And while I understand how your particles can result in your report of having seen the color yellow— the particles in your vocal tract, mouth, and lips need only choreograph their motions to yield that external behavior—I have a much harder time understanding how the particles provide you with the subjective inner experience of yellow. While I understand how your particles can cause you to smile or frown—again, the particles just need to appropriately choreograph their motions—I am at a loss to understand how the particles yield an inner sensation of happiness or sadness. Indeed, although I have direct access to my own inner world, I am similarly at a loss to understand how that world emerges from the motion and interaction of my own particles.

I would be stymied too, of course, in trying to explain many other things in staunchly reductionist terms, from Pacific typhoons to raging volcanoes. But the challenge presented by these happenings, and a world chock-full of examples like them, is solely that of describing the complex dynamics of a fantastically large number of particles. If we could surmount that technical hurdle, we would be done.[19] And that's because there is no inner sensation for "what it is like to be" a typhoon or a volcano. Typhoons and volcanoes, as far as we know, don't have subjective worlds of inner experience. We aren't missing first-person accounts. But for anything conscious, that is precisely what our objective third-party description lacks.

In 1994 David Chalmers, a young Australian philosopher, hair flowing past his shoulders, took the stage at the annual consciousness conference in Tucson and described this deficit as the "hard problem" of consciousness. Not that the "easy" problem—understanding the mechanics of brain processes and their role in imprinting memories, responding to stimuli, and molding behavior—is easy. It's just that we can envision what the shape of a solution to those sorts of problems would look like; we can articulate an in-principle approach at the level of particles or more complex structures like cells and nerves, which seems coherent. The challenge to envision such a solution for consciousness motivated Chalmers's assessment. He argued that not only are we lacking a bridge from mindless particles to mindful experience, if we try to build one using a reductionist blueprint—making use of the particles and laws that constitute the fundamental basis of science as we know it—we will fail.

The assessment struck a chord—consonant for some, dissonant for others—which has been echoing across consciousness research ever since.

Something About Mary

It is easy to be flippant about the hard problem. In the past, my own response may have seemed so. When asked, I would often say that conscious experience is merely what it feels like when a certain kind of information processing takes place in the brain. But because the core issue is to explain how there can be a "what it feels like" at all, the response too quickly dismisses the hard problem as not being hard and not even being a problem. More charitably, it is a response that sides with a widely held view that thinks too much is made of thought. While some hard-problem aficionados argue that to understand consciousness we will need to introduce concepts from outside conventional science, others—so-called **physicalists**—anticipate that cleverly construed and creatively applied, traditional scientific methods, solely invoking physical properties of matter, will be up to the task. The physicalist perspective does indeed summarize my own long-held view.

Yet over the years as I have thought about the

question of consciousness more carefully, I have had significant moments of doubt. The most startling came when I encountered an influential argument that philosopher Frank Jackson put forward a decade before the hard problem had been labeled hard.[20] Jackson tells a simple story that, gently dramatized, goes like this. Imagine that in the far future there is a brilliant girl, Mary, who is profoundly color-blind. Since birth, everything in her world has appeared solely in black and white. Her condition baffles the most renowned doctors, and so Mary decides that it will be up to her to figure it out. Driven by the dream of curing her deficit, Mary undertakes years of intensive study, observation, and experiment. And through it all, Mary becomes the greatest neuroscientist the world has ever known, reaching a goal that has long eluded humankind: she fully unravels every last detail about the structure, function, physiology, chemistry, biology, and physics of the brain. She masters absolutely everything there is to know about the brain's workings, both its global organization and its microphysical processes. She understands all the neural firings and particle cascades that happen when we marvel at a rich blue sky, enjoy a succulent plum, or lose ourselves in Brahms's Third Symphony.

With this achievement Mary is able to identify the cure for her visual impairment, and she undergoes the surgical procedure to correct it. Months later the doctors are ready to remove the bandages, and Mary prepares to take in the world anew. Standing in front of a bouquet of red roses, Mary slowly opens her eyes. Here's the question: From this first experience of the color red, will Mary learn anything new? By finally having the inner experience of color, will she acquire new understanding?

Playing this story out in your mind, it seems dead obvious that the very first time Mary experiences the inner sensation of red she'll be overwhelmed. Surprised? Yes. Thrilled? Of course. Touched? Deeply. It seems self-evident that this first direct experience of color will expand her understanding of human perception and the inner response it can generate. From this commonly held intuition, Jackson then encourages us to consider the implication. Mary had mastered everything there is to know about the physical workings of the brain. And yet, through this one encounter, she has apparently expanded that knowledge. She has gained knowledge of the conscious experience accompanying the brain's response to the color red. The conclusion? **Complete knowledge of the brain's physical workings leaves something out.** It fails to expose or explain subjective sensations. Had such physical knowledge been all-encompassing, Mary would have taken off the bandages and shrugged.

When I first read this account, I felt a sudden kinship with Mary, as if I had also undergone a corrective surgery that opened a previously obscured window on the nature of consciousness. My offhand confidence that physical processes in the brain **are** consciousness, that consciousness **is** the sensation of such processes, was suddenly shaken. Mary possessed all possible knowledge of all the brain's physical processes and yet from the scenario it seems clear that such understanding is incomplete. This suggests that when it comes to conscious experience, physical processes are part of the story but not the full story. When Jackson's paper first appeared, long before I encountered it, experts too were roused, and in the decades that followed Mary has sparked much response.

Philosopher Daniel Dennett asks us to really consider the implication of Mary's exhaustive knowledge of the physical facts. His point is that the concept of complete physical understanding is so utterly foreign that we grossly underestimate the explanatory power it would provide. With such an all-encompassing grasp, from the physics of light to the biochemistry of eyes to the neuroscience of the brain, Dennett argues that Mary **would** be able to discern the inner sensation of red long before experiencing it.[21] Remove the bandages and Mary may respond to the beauty of the red roses, but seeing their red color will simply confirm her expectations. Philosophers David Lewis[22] and Laurence Nemirow[23] take a different tack, arguing that Mary acquires a new ability—to identify, remember, and imagine the inner experience of red—but that does not constitute a new fact that stands outside her previous mastery. Upon removing the bandages, Mary may not shrug, but the "wow" she may utter speaks solely to her delight at a new way of cogitating on old knowledge. Even Jackson himself now argues against his original conclusion, having undergone a change of heart after years of contemplating Mary. We are so accustomed to learning things about the world through direct experience, like grasping how it feels to sense red by seeing red, that we tacitly assume these experiences provide the only means for acquiring such knowledge. According to Jackson, that's unjustified. While Mary's learning process would be unfamiliar, invoking deductive reasoning when more ordinary folk rely on direct experience, her complete command of the physical knowledge would allow her to determine what it is like to see red.[24]

Who is right? The original Jackson and the followers of his first foray? Or the later Jackson and all those who are convinced that upon seeing the roses Mary doesn't learn anything new?

The stakes are high. If consciousness can be explained by facts about the world's physical forces acting on its material constituents, our charge will be to determine how. If not, our charge will be more sweeping. We will need to determine the new concepts and processes that understanding consciousness requires, a journey that almost certainly will take us well beyond the current bounds of science.

Historically, we have navigated with confidence through the choppy waters of human intuition by identifying testable consequences of conflicting viewpoints. As yet, no one has proposed an experiment or an observation or a calculation that can definitively settle the question raised by Mary's story or, more ambitiously, reveal the source of inner experience. For the most part, the considerations we have for adjudicating among those perspectives that pass basic muster are plausibility and intuitive appeal, flexible measures that, as we will see, have allowed for a diverse collection of viewpoints.

A Tale of Two Tales

Strategies for explaining consciousness fan out across an impressive terrain of ideas. At the extremes are positions that either dismiss consciousness as an illusion (**eliminativism**) or declare that consciousness is the only quality of the world that is real (**idealism**). In between, we encounter a spectrum of proposals. Some operate within the confines of traditional scientific thought, others slip between the cracks of current scientific understanding, and others still augment the qualities we have long held to define reality at its most fundamental level. Two short tales provide these proposals with historical context.

Had you overheard discussions in biological circles during the eighteenth and nineteenth centuries, you would be familiar with **vitalism.** It was a concept addressing what one might have called the "hard problem" of life: Since the world's fundamental ingredients are inanimate, how can collections of such ingredients possibly be alive? Vitalism's answer, stark and direct, was that such collections cannot be alive. At least not on their own. Vitalism proposed that the missing ingredient is a nonphysical spark or life force that endows inanimate matter with the magic of life.

Had you moved in particular physics circles during the nineteenth century, you would have heard excited talk of electricity and magnetism as Michael Faraday and others delved ever more deeply into this increasingly intriguing realm. One perspective you would have encountered argued that these novel phenomena could be explained within the standard mechanistic approach of science handed down by Isaac Newton. Finding the clever combination of flowing fluids and miniature cogs and wheels responsible for the new phenomena might be a challenge, but the basis for understanding was already in hand. Because of the anticipated adequacy of conventional scientific reasoning, one might have called this the "easy problem" of electricity and magnetism.

History has revealed that the expectations described in each of these tales were misguided. With two centuries of hindsight, the near-mystical enigma that life once conjured has diminished. Although we still lack a complete understanding of life's origin, there is nearly universal scientific consensus that no magical spark is required. Particles configured into a hierarchy of structures—atoms, molecules, organelles, cells, tissues, and so on—are all that's necessary. The evidence strongly favors the existing framework of physics, chemistry, and biology as being fully sufficient for explaining life. The hard problem of life, while surely difficult, has been reclassified as easy.

For electricity and magnetism, data collected from careful experiments demanded that scientists go beyond the features of physical reality that were on the books prior to the 1800s. The existing understanding gave way to a wholly new physical quality of matter (electric charge) responding to a wholly new type of influence (space-filling electric and magnetic fields) described by a wholly new set of equations (twenty such equations in the initial formulation) developed by James Clerk Maxwell. Although solved, the "easy" problem of electricity and magnetism turned out to be hard.[25]

Many researchers envision that vitalism's tale will be recapitulated with consciousness: as we gain an ever-deeper understanding of the brain, the hard problem of consciousness will slowly evaporate.

Although currently mysterious, inner experience will gradually be seen as a direct consequence of the brain's physiological activities. What we are missing is a full command of the brain's inner workings, not a new variety of mind-stuff. One day, according to this physicalist perspective, folks will smile as they think back on how we once invested consciousness with such impassioned but unwarranted mystery.

Others envision that electromagnetism's tale provides the relevant model for consciousness. When your understanding of the world confronts puzzling facts, you naturally try to incorporate them within the existing scientific framework. But some facts may not fit existing templates. Some facts may reveal new qualities of reality. Consciousness, according to this camp, abounds with facts of just this sort. If this perspective proves right, understanding subjective experience will require a substantial reconfiguration of the intellectual playing field, with the potential of profound ramifications that may have impact well beyond questions of mind.

One of the most radical of such proposals comes from David Chalmers, Mr. Hard Problem himself.

Theories of Everything

Chalmers, convinced that conscious awareness cannot emerge from a swirl of mindless particles, encourages us to take the tale of electromagnetism to heart. Much as nineteenth-century physicists bravely faced the futility of cobbling together strained explanations of electromagnetic phenomena using the conventional science of the time, we need the same courage in recognizing that to demystify consciousness we must look beyond known physical qualities.

But how? One possibility, simple and bold, is that individual particles themselves are endowed with an innate attribute of consciousness—call it **proto-consciousness** to avoid imagery of elated electrons or cranky quarks—that cannot be described in terms of anything more fundamental. That is, our description of reality must widen to include an intrinsic and irreducible subjective quality that is infused in nature's elementary material ingredients. And it is this quality of matter that we have long overlooked, which is why we've so far failed to explain the physical basis of conscious experience. How can a swirl of mindless particles create mind? They can't. To create a conscious mind you need a swirl of mindful particles. By pooling their proto-conscious qualities, a large collection of particles can yield familiar conscious experience. The proposal, then, is that particles are endowed with a well-studied collection of physical properties (mass, electric charge, nuclear charges, and quantum mechanical spin) as well as the previously neglected quality of proto-consciousness. Reviving panpsychist beliefs, whose historical roots reach as far back as ancient Greece, Chalmers thus entertains the possibility that consciousness is relevant to anything and everything made of particles, whether a bat's brain or a baseball bat.

If you're wondering what proto-consciousness really is or how it's infused into a particle, your curiosity is laudable, but your questions are beyond what Chalmers or anyone else can answer. Despite that, it is helpful to see these questions in context. If you asked me similar questions about mass or electric charge, you would likely go away just as unsatisfied. I don't know what mass is. I don't know what electric charge is. What I do know is that mass produces and responds to a gravitational force, and electric charge produces and responds to an electromagnetic force. So while I can't tell you what these features of particles **are,** I can tell you what these features **do.** In the same vein, perhaps researchers will be unable to delineate what proto-consciousness is and yet be successful in developing a theory of what it does—how it produces and responds to consciousness. For gravitational and electromagnetic influences, any concern that substituting action and response for an intrinsic definition amounts to an intellectual sleight of hand is, for most researchers, alleviated by the spectacularly accurate predictions we can extract from our mathematical theories of these two forces. Perhaps we will one day have a mathematical theory of proto-consciousness that can make similarly successful predictions. For now, we don't.

However exotic this all sounds, Chalmers argues

that his approach sits squarely within the bounds of science, properly construed. For centuries, scientists have focused exclusively on the objective unfolding of reality, and with that as the target they developed equations that do a wonderful job of explaining experimental and observational data. But such data are fully available to a third-person review. Chalmers is suggesting that there are other data, the data of inner experience, and presumably other equations too, that capture pattern and regularity in the inner domain. Conventional science would thus explain external data while science's next era would explain internal data.

Said in a slightly different way, for many years there has been a movement afoot, often credited to physicist John Wheeler (known to the public for popularizing the term "black holes"), that envisions information as the most fundamental of all physical currencies. To describe the state of the world now, I provide information that specifies the configuration of all the dancing particles and undulating fields permeating space. The laws of physics take that information as input and yield as output information that delineates the state of the world later on. Physics, according to this framing, is in the business of information processing.

Using this language, Chalmers's proposal is that there are two sides to information: There is the objective, third-party-accessible quality of information—the information that has, for hundreds of years, been the province of conventional physics. There is also a subjective, first-person-accessible quality of information that physics has so far not considered. A complete theory of physics would need to embrace not just outer but also inner information and would need laws that describe the dynamic evolution of each type. The processing of inner information would provide the physical basis of conscious experience.

Einstein's dream of a unified theory of physics, one capable of describing all of nature's particles and forces within a single mathematical formalism, has been called the search for a theory of everything. That unfortunately bombastic description, often applied to my own field of string theory, explains why I am so often asked for my views on consciousness. After all, consciousness would seem to fit comfortably within a theory that can explain **everything.** Yet, as I have frequently told those who've asked, it is one thing to grasp the physics of elementary particles and quite another to parlay that into an understanding of the human mind. Building the scientific apparatus to connect the vastly different scales, both in size and complexity, ranks among our most difficult scientific challenges. However, should Chalmers be right, consciousness would enter the scientific account on the ground floor, at the level of fundamental equations and primitive constituents. Which means we might one day have an understanding that incorporates from the get-go the external and internal sides of information processing—objective physical processes and subjective conscious experiences. **That** would be a unified theory. I would continue to resist the locution "theory of everything"—I expect scientists would still have a hard time predicting what I'm going to have for breakfast tomorrow—but such understanding would be revolutionary.

Is this the right direction? I'd be thrilled if it were. We would be standing at the frontier of a whole new terrain of reality awaiting exploration. But as you have likely surmised, there is great skepticism that in its effort to find the source of consciousness, science will need to travel to lands this exotic. Carl Sagan's famous dictum that extraordinary claims require extraordinary evidence is an apt guide. There **is** overwhelming evidence of something extraordinary—our inner experiences—but far less convincing evidence that these experiences are beyond the explanatory reach of conventional science.

Understanding would deepen if we could identify the physical conditions required for generating subjective experiences, a task central to the theory of consciousness we now consider.

The Mind Integrates Information

That the brain is a crenellated, moist, information-processing collection of cells is uncontroversial.

Brain scans and invasive probes have established that distinct parts of the brain specialize in processing particular types of information—optical, auditory, olfactory, linguistic, and so on.[26] By itself, however, information processing does not capture the brain's distinctive qualities. A great many physical systems process information, from the abacus to the thermostat to the computer, and taking Wheeler's perspective to heart, there is a sense in which each and every physical system can be thought of as an information processor. So what distinguishes the variety of information processing that results in conscious awareness? This is a question guiding psychiatrist and neuroscientist Giulio Tononi, joined in the pursuit by neuroscientist Christof Koch. It has led to an approach called **integrated information theory.**[27]

To get a sense of the theory, imagine I present you with a brand-new red Ferrari. Regardless of whether you are a fan of high-end sports cars, the encounter stimulates your brain with a wealth of sensory data. Information expressing the car's visual, tactile, and olfactory qualities, as well as more abstract connotations from the car's power on the road to associations of luxury and wealth, immediately become entwined in a unified cognitive experience. It is an experience whose information content Tononi would characterize as **highly integrated.** Even focusing more narrowly on the car's color, note that your experience is decidedly not one of a colorless Ferrari that your mind subsequently paints red. Nor is it of an abstract red environment that your mind subsequently shapes into a Ferrari. Although shape information and color information activate different parts of the visual cortex, your conscious experience of the Ferrari's shape and color are inseparable. You experience them as one. This, according to Tononi, is an intrinsic quality of consciousness: the information threading through conscious experience is tightly stitched together.

A second intrinsic quality of consciousness is that the range of things you are capable of holding in your mind is enormous. From a dizzying array of sensory experiences, to stirrings of the imagination, to abstract planning and thinking and worrying and anticipating, you have a virtually limitless mental repertoire. Which means that when your mind is focused on any one particular conscious experience, like the red Ferrari, it is highly differentiated from the vast majority of other mental experiences you could be having. Tononi's proposal elevates these observations to a defining characterization: **conscious awareness is information that is highly integrated and highly differentiated.**

Most information lacks these qualities. Take a photograph of the red Ferrari and consider the resulting digital file. To keep things simple, don't worry about details like image compression, and instead imagine that the file is an array of numbers whose values record color and brightness information for each pixel in the image. These numbers are generated by photodiodes in your camera responding to the light reflecting off distinct locations on the car's surface. How integrated is the information? Because each photodiode's response is independent of the others—there's no communication or linkage between them—the information in the digital file is completely balkanized. You could store the datum for each pixel in a separate file and the total information content would remain unchanged. Which means there is no information integration at all. How differentiated is the information in the digital file? While there is a vast assortment of possible images a camera's digital file can store, the information content is constrained to a fixed array of independent numbers. That's it. A digital photographic file isn't set up to contemplate the ethics of capital punishment or struggle with the proof of Fermat's Last Theorem. In this sense, the information content is extremely limited, which means the camera is not a high scorer when it comes to information differentiation.

And so, as your brain constructs a mental representation, its information content rapidly becomes highly integrated and highly differentiated, but as the camera constructs a digital photograph, its information acquires neither of these features. That, according to Tononi, is why you have a conscious experience of the Ferrari but your digital camera does not.

With the goal of making these considerations quantitative, Tononi has proposed a formula that assigns a numerical value to the information contained in any given system, usually denoted ϕ, with larger values of ϕ indicating greater differentiation and deeper integration—and hence, as the theory goes, a higher level of conscious awareness. The approach thus presents a continuum from simple systems, with less information integration and differentiation that may experience rudimentary forms of consciousness, to more complex systems like you and me, with sufficient integration and differentiation to yield the familiar level of conscious awareness, to the possibility of yet other systems whose informational capacities—and conscious experience—could outpace our own.

As with Chalmers's approach, Tononi's theory has a panpsychist leaning. Nothing in the proposal is intrinsically tied to a particular physical structure. Your experience of conscious awareness resides in a biological brain, but according to Tononi and his math, a sufficiently high value of ϕ, whether contained in neural synapses or neutron stars, would be consciously aware. For some, like computer scientist Scott Aaronson, this leaves the proposal open to what he deems a devastating attack. Aaronson's calculations have shown that by cleverly linking together simple logic gates (the most basic of electronic switches), the resulting network can have values of ϕ as large as you like—on par with that of the human brain or even larger.[28] According to the theory, the network of switches should be conscious. And that's a conclusion Aaronson—and most people's intuition—considers absurd. Tononi's response? However strange and unfamiliar the conclusion, the network **will** be conscious.

Now, you might think, He can't **really** believe that. But consider your incredulity in context. How can it be that a three-pound clump of brain, when appropriately connected to a blood supply and network of nerves, has familiar conscious experience? **That** is the claim, based on all that science has so far revealed, that stretches credulity. Yet, because of your own inner world, it is a claim that you readily accept. If I then hand you something else, lacking body and brain, and suggest that it too is conscious, the stretch to accept this new claim may seem significant, but actually, it is comparatively modest. By embracing the nearly ludicrous suggestion that a gloppy grey knot of neurons has consciousness, you've already taken the big step. That's not an argument for Tononi's proposal, but it makes clear that familiarity can skew our sense of the absurd.

If this approach should prove correct, it will clarify the qualities that a system must have to yield a conscious experience. That would be substantial progress. Still, in its current form, integrated information theory would leave us wondering why consciousness **feels** the way it does. How does highly differentiated and highly integrated information yield inner awareness? According to Tononi, it just does. Or, more precisely, he suggests that this question may be the wrong one to ask. Our charge, in his view, is not to explain how conscious experience emerges from whirring particles but rather to determine the conditions required for a system to have such experiences. And that's what integrated information theory seeks to do. While I appreciate this perspective, my intuition, shaped by the spectacular successes of reductionist explanations, will remain unsatisfied until we connect physical processes involving familiar particulate ingredients to the sensations of mind.

One final proposal we will now take up pursues a different strategy. It is a physicalist account through and through, and provides one of the most illuminating approaches to the mystery of consciousness.

The Mind Models the Mind

Neuroscientist Michael Graziano's theory of consciousness begins with a couple of well-known qualities of brain functioning that we can all readily buy into.[29] To appreciate them, return to the Ferrari. Imagine you see the car's sleek red exterior, feel the smooth ergonomic shape of the door handles, smell the unmistakable new car aroma, and so on. Intuitively, we think of these as direct experiences of an external reality, but as we have known for centuries they are not. Modern science makes this explicit. Red light reflecting off the Ferrari's surface is an electric field that oscillates at roughly four hundred trillion times each second at right angles to a similarly oscillating magnetic field, all traveling toward you at three hundred million meters per second. That is the physics of red light, and that is the stimulus your eyes encounter.[30] Notice that there is no "red" in the physics description. Red happens when the electromagnetic field enters your eyes, tickles light-sensitive molecules in your retina, and generates an impulse carried to your brain's visual cortex, which specializes in visual information processing and interprets the signal. Red is a human construct that happens deep inside your head. And that new car smell? A similar story. The seats, carpet, and plastic wrap off-gas molecules that permeate the car's interior. There is no new car smell until those molecules waft into your nostrils, brush up against receptor neurons on your olfactory epithelium, and generate an impulse that fires along your olfactory nerve toward your olfactory bulb, which relays the processed signal to various neurological structures for interpretation. As with red, the only place the new car smell happens is within your brain.

And so, when the Ferrari grabs your attention, a collection of cognitive data processing wheels is set in motion. Red, fragrant, shiny, metallic, glass, wheels, engine, power, movement, velocity, and so on—a range of physical qualities and functional capacities are both conjured and bound by your brain into the version of the car you hold in your mind. So far, this sounds similar to integrated information theory, but Graziano's proposal takes these realizations in a different direction. His central thesis is that however heedful of detail you might be, your mental representations are always vastly simplified. Even describing the car as "red" is a shorthand for the many similar but distinct frequencies of light—the many shades of red— that reflect off different parts of the car's surface: electromagnetic waves, for instance, oscillating at 435, 172, 874, 363, 122 cycles per second from a spot on the driver's side door, 447, 892, 629, 261, 106 cycles per second from a spot on the hood, and so on.[31] Your mind would reel if it dealt with such an overabundance of detail. Instead, "red" is the mind's welcome, albeit schematic, simplification. So too for the vast collection of similar simplifications the mind constantly makes. For just about everything you ever encounter in the environment, a schematic representation is not only adequate but also frees up mental resources for other life-supporting purposes. Long ago, brains that may have become distracted by the billowing details of the physical world are brains that would have been swiftly eaten. Brains that survived are brains that avoided being consumed by details that lacked survival value. Replace the red Ferrari with a rumbling avalanche or a quaking earth, and you can see the survival advantage of having a quick and dirty mental representation that facilitates a rapid response.

When your attention is not directed at cars or avalanches or earthquakes, but is instead focused on animals or humans, you similarly create schematic mental representations. But beyond representations of their physical forms, you also create schematic mental representations of their minds. You try to assess what's going on inside their heads—whether a given animal or human is friend or foe, offers safety or danger, is seeking mutual opportunity or selfish gain. Clearly, there is significant survival value in quickly sizing up the nature of our encounters with other life. Researchers call this capacity, refined over generations by natural selection, our **theory of mind**[32] (we theorize, intuitively, that living things are endowed with minds that operate more or less like ours), or **the intentional stance**[33] (we attribute knowledge, beliefs, desires, and thus intentions to the animals and humans we encounter).

Graziano emphasizes that you routinely apply this very capacity to yourself: you continuously create a schematic mental representation of your own state of mind. If you are looking at the red Ferrari, not only do you create a schematic representation of the car, you also create a schematic representation of your Ferrari-focused attention. All the features you bind together to represent the Ferrari are augmented by an additional quality summarizing your own mental focus: the Ferrari is red, smooth, and shiny, **and** your attention is focused on the Ferrari being red, smooth, and shiny. That is how you keep track of your engagement with the world.

As with the representation of the Ferrari, and as with your representation of the attention of others, the representation of your own attention leaves out vast swaths of details. It ignores the underlying neuronal firings, the information processing and complex signal exchanges that generate your focus and instead sketches the attention itself, what in common language we normally call our "awareness." And this, according to Graziano, is the heart of why conscious experience seems to float unmoored in the mind. When the brain's penchant for simplified schematic representations is applied to itself, to its own attention, the resulting description ignores the very physical processes responsible for that attention. That is why thoughts and sensations seem ethereal, as if they come from nowhere, as if they hover in our heads. If your schematic representation of your body were to leave out your arms, the motion of your hands would seem ethereal too. And that is why conscious experience seems utterly distinct from the physical processes carried out by our particulate and cellular constituents. The hard problem seems hard—consciousness seems to transcend the physical—only because our schematic mental models suppress cognizance of the very brain mechanics that connect our thoughts and sensations to their physical underpinnings.

The allure of a physicalist theory like Graziano's (and others that have been proposed and developed[34]) is that consciousness, like life, would be reduced to conducive arrangements of lifeless, thoughtless, and emotionless constituents. Certainly, there is a vast neurological landscape stretching between us and such a promised land of reductionist understanding. But unlike the terra incognita envisioned by Chalmers, in which researchers will need to hike strange lands and bushwhack unfamiliar foliage, the physicalist expedition will likely offer less exotic surprises. The challenge will not be to survey an alien world, but to map our own—the brain—with unprecedented detail. It is the familiarity of the terrain that will make a successful journey so wondrous. Requiring no supra-scientific spark, invoking no novel qualities of matter, consciousness would simply emerge. Ordinary stuff, governed by ordinary laws, carrying out ordinary processes, would have the extraordinary capacity to think and feel.

I have encountered many people who resist this perspective. People who feel that any attempt to subsume consciousness within the physical description of the world belittles our most precious quality. People who suggest that the physicalist program is the ham-fisted approach of scientists blinded by materialism and unaware of the true wonders of conscious experience. Of course, no one knows how all this will play out. Perhaps a hundred or a thousand years from now the physicalist program will look naïve. I doubt it. But in acknowledging this possibility, it is also important to counter the presumption that by delineating a physical basis for consciousness we devalue it. That the mind can do all it does is extraordinary. That the mind may accomplish all it does with nothing more than the kinds of ingredients and types of forces holding together my coffee cup makes it more extraordinary still. Consciousness would be demystified without being diminished.

Consciousness and Quantum Physics

Over the decades, a frequent suggestion has been that quantum physics is essential for understanding consciousness. In one sense this is surely true. Material structures, the brain included, are made of particles whose behavior is governed by the laws of quantum mechanics. Quantum mechanics thus underpins the physical basis of everything, including the mind. But when consciousness meets the quantum, it is not uncommon for commentators to suggest deeper connections. Many of these are motivated by a gap in our understanding of quantum mechanics that has resisted a century of thought by some of the world's most accomplished scientific and philosophical minds. Let me explain.

Quantum mechanics is the most accurate theoretical framework for describing physical processes ever developed. There has never been a prediction of quantum mechanics that has been contradicted by replicable experiments, and the results of some of the most detailed quantum mechanical calculations agree with experimental data to better than one part in a billion. If you are not into quantitative figures, most of the time it's fine to just let them wash over you. But not now. Take in the number I just quoted: **quantum mechanical calculations, based on Schrödinger's equation, agree with experimental measurements to better than nine digits after the decimal point.**[35] Trumpets should blare and the species should take a bow because that represents a triumph of human understanding.

Nevertheless, there is a puzzle at the core of quantum theory.

The primary new feature of quantum mechanics is that its predictions are probabilistic. The theory might assert that there is a 20 percent chance that an electron will be found here, a 35 percent chance that it will be found over there, and a 45 percent chance way over there. If you then measure the electron's position in a great many identically prepared versions of the same experiment, you will find to impressive accuracy that in 20 percent of your measurements the electron **is** here, in 35 percent of them it **is** over there, and 45 percent of the time it **is** way over there. That is why we have confidence in quantum theory.

Now, quantum theory's reliance on probabilities may not sound particularly exotic. After all, when you flip a coin we also use probabilities to describe the possible outcome—there's a 50 percent chance that the coin lands heads and a 50 percent chance that it lands tails. But here is the difference, familiar to many yet still deeply shocking: in the ordinary classical description, after you flip the coin but before you look, the coin is either heads or tails, you simply don't know which. By contrast, in the quantum description, prior to examining the whereabouts of a particle like an electron that has a 50 percent chance of being here and a 50 percent chance of being there, the particle is not **either** here or there. Instead, quantum mechanics says the particle is hovering in a fuzzy mixture of being **both** here and there. And if the probabilities give the electron a nonzero chance to be at a variety of different locations, then according to quantum mechanics it would be hovering in a fuzzy mixture of being simultaneously situated at all of them. This is so fantastically strange, and so counter to experience, that you might be tempted to dismiss the theory out of hand. And if it weren't for quantum mechanics' unmatched capacity to explain experimental data, that reaction would be both widespread and justified. However, the data force us to treat quantum mechanics with utmost respect, and so we scientists have worked tirelessly to make sense of this counterintuitive feature.[36]

The problem is, the more we've worked, the weirder things have become. There is nothing in the quantum equations that shows how reality transitions from the fuzzy mixture of many possibilities to the single definite outcome you witness upon undertaking a measurement. In fact, if we assume—as seems utterly sensible—that the same successful quantum equations apply not just to the electrons (and other particles) you may be studying but also to the electrons (and other particles) that make up your equipment, and those making up you, and those making up your brain, then according to the mathematics the transition shouldn't happen at all. If an electron is hovering both here and there, then your equipment should find that it is both here and there, and upon reading the equipment's display, your brain should think the electron is both here and there. That is, after you perform a measurement, the quantum fuzziness of the particles you are studying should infect your equipment, your brain, and presumably your conscious awareness, causing your thoughts to hover in a fuzzy mixture of multiple outcomes. And yet, after each and every measurement, you report nothing of the sort. You report that you witnessed a single, definite result. The challenge, known as the **quantum measurement problem,** is to resolve the puzzling disparity between the fuzzy quantum reality described by the equations and the sharp familiar reality you consistently experience.[37]

As far back as the 1930s, physicists Fritz London and Edmond Bauer,[38] and a few decades later Nobel laureate Eugene Wigner,[39] suggested that consciousness might be the key. After all, the puzzle becomes puzzling only when you report on your conscious experience of a definite reality, yielding a mismatch between what you say and what the mathematics of quantum mechanics predicts. Imagine, then, that the rules of quantum mechanics apply all along the chain, from the electron that's being measured, to the particles in the equipment performing the measurement, to the particles constituting the readout on the equipment's display. But when you look at the readout and the sensory data flows into your brain something changes: the standard quantum laws cease to apply. Instead, when conscious awareness is brought to bear, some other process takes over—a process that ensures you become cognizant of a single definite result. Consciousness would thus be an intimate participant in quantum physics, dictating that as the world evolves all but one of the many possible futures are eliminated, either from reality itself or at least from our cognitive awareness.

You can see the appeal. Quantum mechanics is mysterious. Consciousness is mysterious. How fun to imagine that their mysteries are related, or are the same mystery, or that each mystery resolves that of the other. But in my decades of immersion in quantum physics, I have not encountered a mathematical argument or experimental data that have shifted my long-held assessment of the purported link: extraordinarily unlikely. Our experiments and observations support the view that when a quantum system is prodded—whether the prodder is a conscious being or a mindless probe—the system snaps out of the probabilistic quantum haze and assumes a definite reality. Interactions—not consciousness—coax the emergence of a definite reality. Of course, to verify this, or anything else for that matter, I need to bring my consciousness to bear; I can't be aware of a result without my conscious mind participating in the process. So there is no foolproof argument that consciousness does not play a special quantum role. Still, even in the most refined approaches, which have gone well beyond a superficial identification of two apparently distinct mysteries, the proposed quantum-consciousness connections are tenuous.

As our understanding of quantum mechanics deepens, so too will our accounting of the microphysical processes underlying the functions of everything, including body and brain. From a physicalist stance, consciousness is among such functions and so will one day be included within a quantum accounting. However, barring a stunning surprise, quantum mechanics textbooks of the near future or far will not include special directives on how to use the equations in the presence of consciousness. Magnificent though it is, consciousness will be understood as another physical quality that arises in a quantum universe.

Free Will

Few of us take pride in how our pancreas produces chymotrypsin or the trigeminal nerve network facilitates a sneeze. We don't feel a vested interest in our autonomic processes. If I'm asked who I am, I turn to the thoughts, sensations, and memories that I can access with my mind's eye or interrogate with my inner voice. Everyone's pancreas synthesizes chymotrypsin and everyone sneezes but, I like to imagine, there's something deeply, fully, and intrinsically me in what I think, in what I feel, in what I do. Bound up in this intuition is a belief so common that many of us never give it a second thought, let alone a first: We have a will that's free. We are autonomous. We call our own shots. We are the ultimate source of our actions. But are we?

This question has inspired more pages in the philosophical literature than just about any other conundrum. Two thousand years ago, Democritus's lean worldview consisting of atoms and the void was a prescient nod to nature's unity, jettisoning the capricious whim of gods in favor of immutable laws. But whether comings and goings are fully controlled by divine power or by physical law, we are left asking where, if anywhere, is there room for freely willed actions?[40] A century later, Epicurus, who had rejected divine intervention, bemoaned the fact that scientific determinism was smothering free will. If we grant that gods hold authority, at least there is a chance that our steadfast reverence may be rewarded with an allocation of freedom. But natural law, immune to all flattery, is incapable of loosening the reins. To solve the dilemma, Epicurus imagined that every so often atoms spontaneously execute a random swerve, defying their lawful fate and allowing for a future not determined by the past. While surely a creative move, far from everyone found the arbitrary insertion of chance into the laws of nature a convincing source for human freedom. And so across the ensuing centuries the problem of free will continued to furrow the brows of a pantheon of revered thinkers—Saint Augustine, Thomas Aquinas, Thomas Hobbes, Gottfried Leibniz, David Hume, Immanuel Kant, John Locke—and on through a lineage too long to list, including many who currently think about such things in philosophy departments around the world.

Here is a modern version of the argument that knocks free will back on its heels. Your experiences and mine appear to confirm that we influence the unfolding of reality through actions that reflect our freely willed thoughts, desires, and decisions. Yet, maintaining our physicalist stance, you and I are nothing but constellations of particles[41] whose behavior is fully governed by physical law. Our choices are the result of our particles coursing one way or another through our brains. Our actions are the result of our particles moving this way or that through our bodies. And all particle motion—whether in a brain, a body, or a baseball—is controlled by physics and so is fully dictated by mathematical decree. The equations determine the state of our particles today based on their state yesterday, with no opportunity for any of us to end-run the mathematics and freely shape, or mold, or change the lawful unfolding. Indeed, following this chain ever further back, the big bang is the ultimate source of all particles, and their behavior over cosmic history has been dictated by the nonnegotiable and insensate laws of physics, which determine the structure and function of everything that exists. Our sense of individuality, value, and esteem rest on our autonomy. But faced with the intransigence of physical law, autonomy withdraws. We are no more than playthings knocked to and fro by the dispassionate rules of the cosmos.

The central question, then, is whether there is any way to avoid this apparent dissolution of free will into the motion of servile particles. Many thinkers have tried. Some have forsworn reductionism.

Although voluminous data confirm that we have a deep understanding of the laws governing individual particles (electrons, quarks, neutrinos, and so on), perhaps when a hundred billion billion billion particles are arranged into a human body and brain, they are no longer governed, or at least not fully governed, by the fundamental laws of the microworld. And perhaps, this line of thinking imagines, this allows for phenomena on macroscopic scales—notably, free will—that the microscopic laws would prohibit.

Admittedly, no one has ever carried out the mathematical analysis required to make predictions for the lawful progression of particles constituting a person. The complexity of the math would be fantastically beyond our most refined computational capacities. Even predicting the motion of a far simpler object like a pool ball can elude us because small inaccuracies in determining the ball's initial speed and direction can be exponentially amplified as the ball ricochets off the banks of the table. So my focus here is not on predicting your next move. My focus is on the existence of laws that govern your next move. And even though the calculations exceed our current abilities, there has never been the slightest mathematical, experimental, or observational indication that these laws exert anything but total control. Unexpected and impressive phenomena can surely emerge from the coordinated motion of a great many microscopic ingredients—typhoons to tigers—but all evidence suggests that were we able to work out the math for such large groups of interacting particles, we would be able to predict their collective behaviors. And so, while it is logically conceivable that we will one day learn that collections of particles constituting bodies and brains are released from the rules governing inanimate collections, this possibility contravenes all that science has so far revealed about the workings of the world.

Other researchers have placed their bets on quantum mechanics. After all, classical physics is deterministic: provide the mathematics of classical physics—Newton's equations—with the precise locations and speeds of all particles at any one moment and the equations will tell you their locations and speeds at any future moment. With such rigidity, with the future fully determined by the past, how can there be any room for free will? The state of your particles right now, reading these words and contemplating these ideas, was determined by their configuration long before you were even born and so, surely, could not have been selected by your will. But in quantum physics, as we have seen, the equations predict only the **likelihood** of how things will be at any future moment. By inserting an element of probability—chance—quantum mechanics seems to provide a modern and experimentally motivated version of the Epicurean swerve, slackening the deterministic reins. However, loose language can be deceptive. The mathematics of quantum mechanics, Schrödinger's equation, is just as deterministic as the mathematics of classical Newtonian physics. The difference is that whereas Newton takes as input the state of the world now and produces a unique state for the world tomorrow, quantum mechanics takes as input the state of the world now and produces a unique table of probabilities for the state of the world tomorrow. The quantum equations lay out many possible futures, but they deterministically chisel the likelihood of each in mathematical stone. Much like Newton, Schrödinger leaves no room for free will.

Yet other researchers have turned to the unresolved quantum measurement problem. Understandably. A gap in scientific knowledge is an alluring place to hide something deeply valued, at least until the gap is closed. That gap, you'll recall, is that there is still no consensus on how the world transitions from the probabilistic account provided by quantum mechanics to the definite reality of common experience. How is one unique future selected from quantum mechanics' list of possibilities? And, of particular interest here, might free will be lurking in the answer? Unfortunately, no. Consider an electron that according to quantum mechanics has a 50 percent chance of being **here** and a 50 percent chance of being **there.** Can you freely pick the outcome—**here** or **there**—that an observation of its position will reveal? You can't. The data attest to the outcome being random, and random outcomes are not freely willed choices. The data also confirm that results accumulated over many such experiments have a statistical regularity: in this example, half of the results will find the electron **here** and half of the results will find it **there.** A freely willed choice is not constrained, even in a statistical sense, by mathematical rules. But as the evidence demonstrates in this instance and all others too, the math **does** rule. So although the passage from quantum probabilities to experiential certainties remains puzzling, it is clear that free will is not part of the process.

To be free requires that we are not marionettes whose strings are pulled by physical law. Whether the laws are deterministic (as in classical physics) or probabilistic (as in quantum physics) is of deep significance to how reality evolves and to the kinds of predictions science can make. But for assessing free will, the distinction is irrelevant. If the fundamental laws can continually churn, never grinding to a halt for lack of human input and applying all the same even if particles happen to inhabit bodies and brains, then there is no place for free will. Indeed, as is affirmed by every scientific experiment and observation ever conducted, long before we humans came on the scene the laws ruled without interruption; after we arrived, they continued to rule without interruption.

To sum up: We are physical beings made of large

collections of particles governed by nature's laws. Everything we do and everything we think amounts to motions of those particles. Shake my hand and particles constituting your hand push up and down against those constituting mine. Say hello, and particles constituting your vocal cords jostle particles of air in your throat, setting off a chain reaction of colliding particles that ripples through the air, knocking into the particles constituting my eardrums, setting off a surge of yet other particles in my head, which is how I manage to hear what you're saying. Particles in my brain respond to the stimuli, yielding the thought **that's a strong grip,** and sending signals carried by other particles to those in my arm, which drive my hand to move in tandem with yours. And since all observations, experiments, and valid theories confirm that particle motion is fully controlled by mathematical rules, we can no more intercede in this lawful progression of particles than we can change the value of pi.

Our choices **seem** free because we do not witness nature's laws acting in their most fundamental guise; our senses do not reveal the operation of nature's laws in the world of particles. Our senses and our reasoning focus on everyday human scales and actions: we think about the future, compare courses of action, and weigh possibilities. As a result, when our particles do act, it seems to us that their collective behaviors emerge from our autonomous choices. However, if we had the superhuman vision invoked earlier and were able to analyze everyday reality at the level of its fundamental constituents, we would recognize that our thoughts and behaviors amount to complex processes of shifting particles that yield a powerful sense of free will but are fully governed by physical law.

And yet, to conclude our discussion here would be to overlook a variation on the theme of freedom that not only squares with our understanding of physical law but captures a quality so essential that you can take it as a defining characteristic of what it means to be human.

Rocks, Humans, and Freedom

Imagine that you and a rock, each minding your own business, are idly sitting next to each other on a park bench. As I walk by, you suddenly see that a hefty tree limb has snapped and is hurtling toward me. You leap from the bench and tackle me with great force, thrusting us both out of harm's way. What is the explanation for your heroic, lifesaving act? All the particles making up you and all of those making up the rock are subject to the very same laws, and so neither you nor the rock has free will. Yet it is you who jumped from the bench while the rock just sat there. How do we explain this?

You saved me but the rock didn't because your particles are so spectacularly ordered, so breathtakingly configured, that they can undertake exquisitely choreographed motions that are not possible for the particles constituting the rock.[42] As I walk by, you can wave, or say hello, or tell me you've solved the equations of string theory, or do jumping jacks, or save me from a falling branch, or a gazillion other possibilities. Photons that bounce off my face and enter your eyes, sound waves vibrating from a cracking branch that enter your ears, tactile influences from a strong breeze that blows against your skin, as well as a vast array of other stimuli external and internal, set off particle cascades throughout your body carrying signals that generate a wealth of sensations, thoughts, and behaviors, which are themselves yet other particle cascades. Thankfully for me, the specific particle cascade in response to the stimuli of the snapping branch thrust your particles into immediate action. By comparison, the rock's responses to stimuli are more muted. Impinging photons, sound waves, and tactile pressures generate the simplest of reactions. The rock's particles may jitter slightly, their temperature may increase slightly, or for an especially strong wind the positions of the entire lot may shift slightly. That's it. Within the rock there's just not a whole lot going on. What makes you special is that your sophisticated internal organization allows for a rich spectrum of behavioral responses.

The point, then, is that when evaluating free will there is much to be gained by shifting attention from a narrow focus on ultimate cause to a broader perusal of human response. Our freedom is not from physical laws that are beyond our ability to affect. Our freedom is to exhibit behaviors—leaping, thinking, imagining, observing, deliberating, explaining, and so on—that are not available to most other collections of particles. Human freedom is not about willed choice. Everything science has so far revealed has only strengthened the case that such volitional intercession in the unfolding of reality does not exist. Instead, human freedom is about being released from the bondage of an impoverished range of response that has long constrained the behavior of the inanimate world.

This notion of freedom does not require free will. Your lifesaving act, while duly appreciated, arose from the action of physical law and hence was not freely willed. But the fact that your particles were able to jump from the bench, and later, to reflect on their action and to be moved by their reflection, is utterly astonishing. The particles clustered into a rock cannot do anything remotely like this. And it is these capabilities manifesting as the wondrous sweep of thought, feeling, and behavior that captures the essence of being human—the essence of human freedom.

My use of the term "free" to describe behaviors that according to the laws of physics are not freely willed may seem like a linguistic bait and switch. But the point, as the compatibilist school of philosophy has long suggested, is that when it comes to freedom and physics, all is not lost; there is great benefit in considering alternative kinds of freedom that comport with physical law. There are various proposals for how to accomplish this, but it's as if such theories gloomily deliver the bad news, "When it comes to the traditional sort of free will, you are no different from a rock," but then, just as you turn away to sulk, they exclaim, "But cheer up! There's this **other** variety of freedom, gratifying in its own right, that you have in abundance."[43] In the approach I am advocating, such freedom is found in liberation from a restricted range of behaviors.

Personally, I take great comfort in this variety of freedom. As I sit here, typing out my thoughts, I am unfazed by the realization that at the level of fundamental particles everything I'm thinking and everything I'm doing constitutes the unfolding of physical laws that are beyond my control. What matters to me is that unlike my desk and unlike my chair and unlike my mug, my collection of particles is able to execute an enormously diverse set of behaviors. Indeed, my particles just composed this very sentence and I'm pleased they did. Sure, that reaction, too, is nothing but my particle army carrying out their quantum mechanical marching orders, but that doesn't diminish the reality of the feeling. I am free not because I can supersede physical law, but because my prodigious internal organization has emancipated my behavioral responses.

Relevance, Learning, and Individuality

Giving up the traditional concept of free will may still seem to require relinquishing much of what we value. If the unfolding of reality, including that of sentient beings, is set by physical law, do our behaviors matter? Can we simply sit back, do nothing, and let physics run its course? Is there any place for individuality? How can capacities we greatly value, like learning and creativity, play any role?

Let's take this last question first. And in doing so, it's useful to think about a Roomba. Does a Roomba possess the traditional quality of free will? Don't strain. This is not a trick question. Most of us would agree that it doesn't. Yet, as the Roomba glides along your living-room floor, encountering walls and columns and furniture, its internal particulate configuration rearranges—its navigation maps and internal instructions are updated—and these changes modify the Roomba's subsequent behavior. The Roomba **learns.** Indeed, as the Roomba faces the challenge of navigating around objects it has encountered, the solutions it employs—avoid those stairs, circle around that table leg, and so on—display rudimentary creativity.[44] Learning and creativity do not require free will.

Your internal organization, your "software," is more refined than the Roomba's, facilitating your more sophisticated capacity for learning and creativity. At any given moment, your particles are in a specific arrangement. Your experiences, whether external encounters or internal deliberations, reconfigure that arrangement. And such reconfigurations impact how your particles will subsequently behave. That is, such reconfigurations update your software, adjusting the instructions that guide your ensuing thoughts and actions. An imaginative spark, a blundering error, a clever line, an empathic hug, a dismissive remark, a heroic act all result from your personal particle constellation progressing from one arrangement to another. As you observe how everyone and everything responds to your actions, your particle constellation shifts again, reconfiguring its pattern to further adjust your behavior. At the level of your particulate ingredients, this **is** learning. And when the resulting behaviors are novel, the reconfiguration has generated creativity.

This discussion highlights one of our central themes: the need for nested stories that explain distinct but interconnected layers of reality. Were you content with a story that describes the unfolding of reality solely at the level of particles, you would not be motivated to introduce concepts like learning and creativity (or, for that matter, entropy and evolution). All you would need to know is how collections of particles continually rearrange their configuration, and that information is delivered by the fundamental laws (and a specification of the state of the particles at some moment in the past). But most of us are not content with that sort of story. Most of us find it enlightening to tell additional stories, compatible with the reductionist account, but focused on larger and more familiar scales. It is in these stories, whose main characters are aggregates of particles like you and me and the Roomba, that concepts including learning and creativity (and entropy and evolution) provide an indispensable language. While the reductionist story describing the Roomba would catalog the motion of billions of billions of particles, the higher-level story might explain that the Roomba's sensors recognized that it was on the edge of a flight of stairs, stored that dangerous location in memory, and reversed course to avoid a potentially catastrophic drop. The two stories are fully compatible even though one uses the language of particles and laws while the other uses the language of stimuli and responses. And because the Roomba's responses include the ability to modify future behavior by updating its internal instructions, the concepts of learning and creativity are essential to the higher-level story.

Such nested stories are yet more relevant when it comes to you and me. The reductionist account, which describes us both as collections of particles, provides important but limited insights. We recognize, for example, that we are made of the same stuff and governed by the same laws as all material structures. But the higher-level story, the human story, is the one by which we live our lives. We think and deliberate, we struggle and strive, we succeed and fail. Stories told in this familiar language must, again, be fully compatible with the reductionist accounts told in terms of particles. But in the service of everyday life, these higher-level stories are incomparably more illuminating. When I have dinner with my wife, I am just not that interested in listening to an account of the motion carried out by her hundred billion billion billion particles. However, when she tells me about the ideas she is developing, places she is going, and people she is meeting, I am all in.

Within such higher-level accounts, we speak as though our actions have relevance, our choices have impact, our decisions have significance. In a world progressing via resolute physical law, do they? Yes. Of course they do. When my ten-year-old self struck a match within a gas-filled oven, that action had consequences. That action set off an explosion. The higher-level account that lays out a series of connected events—feeling hungry, putting pizza in the oven, turning on the gas, waiting, striking the match, being engulfed by flames—is accurate and insightful. Physics does not negate this story. Physics does not drain this story of relevance. Physics augments this story. Physics tells us that there is another account, underlying the human-level story, told in the language of laws and particles.

What's remarkable, and to some disturbing, is that these underlying accounts reveal that a common belief pervading our higher-level stories is faulty. We feel that we are the ultimate authors of our choices, decisions, and actions, but the reductionist story makes clear that we are not. Neither our thoughts nor our behaviors can break free from the grip of physical law. Nonetheless, the causally connected sequences at the heart of our higher-level stories—my sensation of hunger causing me to insert a pizza in the oven, leading me to check on its temperature, resulting in my striking a match—are manifest and are real. Thoughts, responses, and actions matter. They yield consequences. They are the links in the chain of physical unfolding. What's unexpected based on our experiences and intuitions is that such thoughts, responses, and actions emerge from antecedent causes funneled through the laws of physics.

Responsibility has a role too. Even though my particles, and hence my behaviors, are under the full jurisdiction of physical law, "I" am in a very literal if unfamiliar way responsible for my actions. At any given moment, I **am** my collection of particles; "I" is nothing but a shorthand that signifies my specific particulate configuration (which, although dynamic, maintains sufficiently stable patterns to provide a consistent sense of personal identity[45]). Accordingly, the behavior of my particles is **my** behavior. That physics underlies this behavior through its control of my particles is surely interesting. That such behavior is not freely willed is worthy of acknowledgment. But these observations do not diminish the higher-level description which recognizes that my specific particle configuration—the way my particles are arranged into an intricate chemical and biological network including genes, proteins, cells, neurons, synaptic connections, and so on—responds in a manner that is unique to me. You and I speak differently, act differently, respond differently, and think differently because our particles are arranged differently. As my particle arrangement learns and thinks and synthesizes and interacts and responds, it imprints my individuality and stamps my responsibility on every action I take.[46]

The human capacity to respond with great variety is testament to the core principles that have guided our exploration thus far: the entropic two-step and evolution by natural selection. The entropic two-step explains how orderly clumps can form in a world that is becoming ever more disordered, and how certain of these clumps, stars, can remain stable over billions of years as they produce a steady output of heat and light. Evolution explains how, in a favorable environment such as a planet bathed by a star's steady warmth, collections of particles can coalesce in patterns that facilitate complex behaviors, from replication and repair, to energy extraction and metabolic processing, to locomotion and growth. Collections that acquire the further capacities to think and learn, to communicate and cooperate, to imagine and predict, are better equipped to survive and hence to produce similar collections with similar capacities. Evolution thus selects for these abilities and, generation upon generation, refines them. In time, some collections conclude that their cognitive powers are so remarkable that they transcend physical law. Some of the most thoughtful of these collections are then bewildered by the conflict between the freedom of will they experience and the unyielding control of physical law they recognize. But the fact is there is no conflict because there is no transcending of physical law. There can't be. Instead, the collections of particles need to reassess their powers, focusing not on the laws that govern particles themselves but on the high-level, thoroughly complex, and extraordinarily rich behaviors each collection of particles—each individual—can exhibit and experience. And with that reorientation, the particle collections can tell an illuminating story of wondrous behaviors and experiences, suffused with wills that feel free and speak as though they have autonomous control, and yet are fully governed by the laws of physics.

Some will balk at this conclusion. I surely have. Although I am convinced intellectually by the argument I have presented, that doesn't undo my deep and strong impression that I freely control what's happening inside my head. But the strength of that impression rests in large part on its familiarity. And as many who have experimented with mind-altering substances can attest, when the identity of particles coursing through the brain is even modestly modified, the familiar can shift. The balance of power in the brain can change. The mind can seemingly have a mind of its own. Decades ago, in the beautiful city of Amsterdam, such an experience resulted in one of the most terrifying nights of my life. My mind created an internal world in which there were endless copies of me, each hell-bent on undermining the reality experienced by the others. As one of me was lulled into thinking he was experiencing the "true" reality, the next me would reveal the artifice of that world, wiping out everything and everyone the initial me cared about, and in the process revealing another "true" reality, which the next me would confidently inhabit—only to have the nightmarish sequence repeat. And repeat.

From the standpoint of physics, I had merely introduced into my brain a small collection of foreign particles. But that change was enough to eliminate the familiar impression that I freely control the activities playing out in my mind. While the reductionist-level template remained in full force (particles governed by physical laws), the human-level template (a reliable mind endowed with free will navigating through a stable reality) was upended.

Of course, I am not presenting a mind-altering moment as an argument for or against free will. But the experience made visceral an understanding that would otherwise have remained abstract. Our sense of who we are, the capacities we have, and the freedom of will we seemingly exert all emerge from the particles moving through our heads. Fiddle with the particles, and those familiar qualities can fall away. It's an experience that helped align my rational grasp of the physics with my intuitive sense of the mind.

Everyday experience and everyday language are filled with references, implicit and explicit, to free will. We speak of making choices and coming to decisions. We speak of actions that depend on those decisions. We speak of the implications that these actions have on our lives and the lives of those we touch. Again, our discussion of free will does not imply that these descriptions are meaningless or need to be eliminated. These descriptions are told in the language appropriate to the human-level story. We **do** make choices. We **do** come to decisions. We **do** undertake actions. And those actions **do** have implications. All of this is real. But because the human-level story must be compatible with the reductionist account, we need to refine our language and assumptions. We need to set aside the notion that our choices and decisions and actions have their ultimate origin within each of us, that they are brought into being by our independent agencies, that they emerge from deliberations that stand beyond the reach of physical law. We need to recognize that although the **sensation** of free will is real, the capacity to exert free will—the capacity for the human mind to transcend the laws that control physical progression—is not. If we reinterpret "free will" to mean this sensation, then our human-level stories become compatible with the reductionist account. And together with the shift in emphasis from ultimate origin to liberated behavior, we can embrace an unassailable and far-reaching variety of human freedom.

As with life's origin, there is no sharply defined moment when consciousness emerges or self-reflection arises or the sensation of free will sets in. But the archaeological record suggests that by one hundred thousand years ago, perhaps earlier, our ancestors had begun to have these experiences. Early humans had long since stood up. Now we could look around and wonder.

What, then, did we do with such powers?

6

LANGUAGE
AND STORY

From Mind to Imagination

P attern is central to human experience. We
survive because we can sense and respond to
the rhythms of the world. Tomorrow will be
different from today, but beneath the myriad comings and goings we rely on enduring qualities. The
sun will rise, rocks will fall, water will flow. These
and an uncountable collection of allied patterns we
encounter from one moment to the next profoundly
influence our behavior. Instincts are essential and
memory matters because patterns persist.

Mathematics is the articulation of pattern. Using
a handful of symbols we can encapsulate pattern
with economy and precision. Galileo summed it
up by declaring that the book of nature, which he
believed revealed God just as surely as the Bible,

is written in the language of mathematics. During the centuries that followed, thinkers have debated a secular version of the sentiment. Is mathematics a language humankind developed to describe patterns we encounter? Or is mathematics the source of reality, rendering the world's patterns the expression of mathematical truth? My romantic sensibilities lean toward the latter. How wonderful to imagine that our mathematical manipulations touch the very foundation of reality. But my less sentimental assessment allows for mathematics to be a language of our own making, developed in part by overindulging our predilection for pattern. After all, much mathematical analysis plays little role in promoting survival. Rare was the meal, and rarer still was the opportunity to reproduce, that our ancestors secured by contemplating prime numbers or squaring the circle.

In the modern era, Einstein's capacities set an unmatched standard for tapping into nature's rhythms. And yet, although his legacy can be summarized by a handful of mathematical sentences—terse, precise, and sweeping—Einstein's forays into the far recesses of reality did not always begin with equations. Or even with language. "I often think in music,"[1] is how he described it. "I very rarely think in words at all."[2] Perhaps your process mirrors Einstein's. Mine doesn't. On occasion, when struggling with a difficult problem, I have had a sudden flash of insight reflecting some or other brain process beneath conscious awareness. But when I'm cognizant, even when using mental imagery to see my way toward a solution, it would be a stretch to say words are absent or to draw an association with music. More often than not, I make progress in physics by fiddling with equations and collecting conclusions in ordinary sentences I write out longhand in notebooks that fill one shelf after another. When I concentrate, I often talk to myself, usually silently, occasionally audibly. Words are essential to the process. Although I find Wittgenstein's summary, "The limits of my language mean the limits of my world"[3] too broad in its scope—I have no doubt that there are vital qualities of thought and experience that stand outside language, a point we will return to later—without language my capacity for certain kinds of mental maneuvers would diminish. Words not only express reasoning, they vitalize it. Or, as said with incomparable grace by Toni Morrison, "We die. That may be the meaning of life. But we do language. That may be the measure of our lives."[4]

Save for the singular genius, and perhaps even there too, language is essential for unleashing imagination. With language we can articulate a vision in which the real world provides an impoverished glimpse of far richer possibility. We can conjure imagery, authentic and fanciful, in minds remote and proximal. We can pass on hard-earned knowledge, substituting the ease of instruction for the difficulty of discovery. We can share plans and align intentions, facilitating coordinated action. We can combine our individual creative capacities into an immensely consequential communal force. We can look into ourselves and recognize that though shaped by evolution we are able to soar beyond the needs of survival. And we can marvel at how a carefully arranged collection of grunts and glides and fricatives and stops can convey insight into the nature of space and time or provide an affecting portrait of love and death: "Wilbur never forgot Charlotte. Although he loved her children and grandchildren dearly, none of the new spiders ever quite took her place in his heart."

With language, we embark on writing a collective narrative, an overlay in story, to make sense of experience.

First Words

Notwithstanding the apocryphal palindrome "Madam, I'm Adam," no one knows when we began to speak or why. Darwin speculated that language emerged from song and imagined that those endowed with Elvis-like talents would more readily attract mates and thus more abundantly seed subsequent generations of gifted crooners. Given enough time, their melodious sounds would gradually transform into words.[5] Alfred Russel Wallace,

Darwin's lesser-feted codiscoverer of evolution by natural selection, saw things differently. He was convinced that natural selection could not shed light on the human capacities for music, art, and, in particular, language. In the competitive arena of survival, our singing, painting, and chattering ancestors were, in Wallace's view, no better off than their less flamboyant cousins. Wallace could see only one way forward: "We must therefore admit the possibility," he wrote in the widely read **Quarterly Review,** "that in the development of the human race, a Higher Intelligence has guided the same laws for nobler ends."[6] The otherwise blind laws of evolution must have been harnessed by a divine power and directed toward the development of communication and culture. When Darwin read Wallace's article, he was aghast, responding with a heavily emphasized "no"[7] in the margin and noting to Wallace: "I hope you have not murdered too completely your own & my child."[8]

In the intervening century and a half, researchers have developed a variety of theories for the origin and early development of language, but like tag-team wrestling, each seemingly convincing proposal has been met with a fresh opponent. There is far greater consensus on the early development of the universe. Odd as it may sound, this makes sense. The birth of the universe left a treasure trove of fossils. The birth of language didn't. The pervasive microwave background radiation, the particular abundances of simple atoms like hydrogen and helium, and the motion of distant galaxies provide direct imprints of processes that took place during the universe's earliest epoch. Sound waves, the earliest manifestation of language, rapidly disperse to oblivion. A moment or two after they're produced, they vanish. Absent tangible relics, researchers have latitude in reconstructing the early history of language, with the result, no surprise, being a profusion of different, often conflicting, theories.

Even so, there is wide agreement that human language differs profoundly from any other variety of communication in the animal kingdom. Were you an average vervet monkey, you'd be able to sound the alert, warning others in your tribe that an approaching predator was a leopard (short high-pitched whine), an eagle (repeated low-pitched snort), or a python (onomatopoeically labeled "chutters").[9] But you'd be at a total loss to discuss the terror you felt when a python slithered by yesterday or articulate your plan for raiding a nearby bird nest tomorrow. Your language skills would draw on a small, closed collection of specific, fixed-meaning utterances, all centered on what's happening right here and right now. Much the same holds for communication evident within other species. As Bertrand Russell summarized it, "A dog cannot relate his autobiography; however eloquently he may bark, he cannot tell you that his parents were honest but poor."[10] Human language is completely different. Human language is open. Rather than using fixed and limited phrases, we combine and recombine a finite collection of phonemes to yield intricate, hierarchical, and virtually unlimited sequences of sounds conveying a virtually unlimited spectrum of ideas. We can just as easily talk about yesterday's snake or tomorrow's nest as we can describe a delightful dream of flying unicorns or our deepening disquiet as night spills across the horizon.

Drill down farther, and we strike controversy. How is it that within a few short years after birth, without formal instruction, we become fluent in one or even multiple languages? Are our brains specifically configured to acquire language, or does cultural immersion together with our general propensity to learn new things offer an adequate explanation? Did human language begin as collections of vocalizations with fixed meanings, like the vervet monkey's alarm calls, which then splintered into words, or did language begin as elementary sounds that grew into words and phrases? Why do we have language? Did evolution directly select for language because it provides a survival advantage, or is language a by-product of other evolutionary developments like larger brain size? And across all these thousands of years, what in the world have we all been talking about? And why?

Noam Chomsky, among the most influential of all modern linguists, has argued that the human capacity to acquire language relies on us each possessing a hardwired universal grammar—a concept with a rich historical lineage wending its way back to thirteenth-century philosopher Roger Bacon, who concluded that many of the world's languages share a common structural foundation. In modern usage the term has been subject to various interpretations, and over the years Chomsky too has refined its meaning. In its least contentious form, universal grammar proposes that there is something in our innate neurobiological makeup that provides a language primer, a species-wide brain boost that propels us all to listen, to understand, and to speak. How else, the reasoning goes, could children, subject to the haphazard, fragmented, and freewheeling linguistic assault of daily life, possibly internalize a wealth of precise grammatical constructs and rules other than by possessing a formidable mental arsenal standing at the ready to process the verbal onslaught? And because any child can learn any language, the mental arsenal cannot be language-specific; the mind must be able to latch on to a universal core common to all languages. Chomsky has proposed that a singular neurobiological event, a "slight rewiring of the brain" perhaps eighty thousand years ago, may have resulted in our ancestors acquiring this capacity, sparking a cognitive big bang that blasted language clear across the species.[11]

Cognitive psychologists Steven Pinker and Paul

Bloom, pioneers of a Darwinian approach to language, suggest a less bespoke history, one in which language emerged and developed through the familiar pattern of a gradual buildup of incremental changes that each conferred a degree of survival advantage.[12] As our hunter-gatherer forebears roamed the plains and forests, the capacity to communicate—"Group of wild boar grazing at eleven o'clock," or "Watch out for Barney, he's got his eye on Wilma," or "Here's a better way of attaching that sharpened stone to the handle"—was vital for effective group functioning and essential for sharing accumulated knowledge. Brains capable of communicating with other brains thus had an edge in the competitive arena of survival and reproduction, impelling linguistic capacities to refine and spread widely. Still other researchers identify a suite of adaptations including breath control, memorization, symbolic thinking, awareness of other minds, formation of groups, and so on that may have worked in tandem to yield language even though language itself may have had little to do with the survival value of the adaptations themselves.[13]

Uncertain too is how long we've been speaking. Linguistic evidence from the remote past is virtually nonexistent, but by examining plausible archaeological proxies, researchers have suggested timeframes for when language may have first emerged. Artifacts like hafted tools (chiseled stones or bones securely attached to a handle), cave art, geometric engravings, and beadwork provide evidence that our ancestors at least as far back as one hundred thousand years engaged in planning, symbolic thinking, and advanced social interactions. As we are inclined to link such sophisticated cognitive capacities to language, we can imagine that as our ancestors sharpened their spears and axes or crawled through dark caves to paint bird and bison, they were prattling on about tomorrow's hunt or last night's campfire.

More direct evidence for the capacity to speak is gleaned from a different collection of archaeological insights. Scientists tracing the growth of cranial cavities and structural changes in the mouth and throat conclude that if our ancestors were so inclined they may have had the physiological capability to converse well over a million years ago. Molecular biology provides clues too. Human speech requires a high degree of vocal and oral dexterity, and in 2001 researchers identified what may be an essential genetic basis for such abilities. Studying a British family with a speech disorder spanning three generations—difficulty with grammar and with coordinating the complex movements of mouth, face, and throat necessary for normal speech—researchers homed in on a genetic mishap, a change to a single letter in a gene called **FOXP2** sitting on human chromosome 7.[14] The instructional misprint is shared by the afflicted family members and has thus been strongly implicated in the disruption of both language and speech. Early press coverage of the discovery dubbed **FOXP2** the "grammar gene" or the "language gene," headline-grabbing descriptions that irked informed researchers, but oversimplified hyperbole aside, the **FOXP2** gene does appear to be one essential component for normal speech and language.

Intriguingly, close variations of the **FOXP2** gene have been identified in many species, from chimps to birds to fish, allowing researchers to trace how the gene has changed over evolutionary history. For chimps, the protein encoded by their **FOXP2** gene differs from ours by only two amino acids (out of more than seven hundred), while that of Neanderthals is identical to ours.[15] Did our Neanderthal cousins speak? No one knows. But this line of sleuthing suggests that a genetic basis for speech and language may have been set sometime after we split from chimps, a handful of millions of years ago, but before we separated from Neanderthals, about six hundred thousand years ago.[16]

The proposed links between language and each of the historical markers—ancient artifacts, physiological structures, genetic profiles—are clever but tentative. Consequently, studies based on these markers yield a broad span for when the world's first words may have debuted, from tens of thousands to a few million years ago. As skeptical researchers have also noted, it's one thing to have the physical capability and mental agility to engage in conversation and quite another to actually do so.

What, then, may have motivated us to speak?

Why We Spoke

There is no shortage of ideas for why our early ancestors broke the silence. Linguist Guy Deutscher notes that researchers have fingered the first words emerging "from shouts and calls; from hand gestures and sign language; from the ability to imitate; from the ability to deceive; from grooming; from singing, dancing and rhythm; from chewing, sucking and licking; and from almost any other activity under the sun,"[17] a delightful list that likely reflects creative theorizing more than it does language's historical antecedents. Still, one or perhaps a combination of these may tell a relevant story, so let's look at a few of the suggestions for where our first words came from and why they stuck.

In times of old, prior to the innovation of coiling material into a baby sling, a mom tending to a two-handed task would set her baby down. Those that cried and babbled pulled mom's attention back and, plausibly, mom's response may have been vocal too—cooing, humming, grunting—supported by soothing facial expressions, hand gestures, and gentle touching. Baby's babbling and mom's TLC would have resulted in higher infant survival rates, selecting for vocalization and, according to this proposal, setting our forebears on the trajectory for words and language.[18]

Or, if motherese doesn't do it for you, note that gestures provide a direct means for communicating basic yet vital information—nodding toward this object or pointing at that location. Some of our nonhuman primate cousins, although lacking spoken language, can be adept at communicating rudimentary ideas through hand and body gestures. And in controlled research settings, chimps have learned hundreds of hand signs standing for various actions, objects, and ideas. Perhaps, then, our spoken language emerged from an earlier phase of gesture-based communication. As our hands became increasingly occupied with the construction and use of tools, and as more complex gatherings made gesturing inefficient or clumsy—difficult to see at night; difficult to see everyone's hands and bodies in groups that are hunting or foraging—vocalization might have offered a more effective means for sharing information. As I'm among those whose hands jump into action every time they speak, and sometimes before, this explanation strikes me as particularly plausible.

Yet, should gesturese leave you skeptical, consider evolutionary psychologist Robin Dunbar's proposal that language emerged as an efficient substitute for the widely practiced activity of social grooming.[19]

If you were a chimpanzee, you'd make friends and establish alliances by carefully picking nits, flaking skin, and other detritus off the fur of others in your community. Some members of your in-group would return the favor, while those ranking higher in status would note your service but would leave your nits intact. The grooming ritual is an organizational activity, fostering and maintaining the group's hierarchy, cliques, and coalitions. Early humans may have engaged in similar social grooming, but as group sizes grew, servicing relationships individually would have required a burdensome investment of time. Friendships, couplings, and alliances are vital, but so is ensuring that there's enough food to eat. What to do? Well, says Dunbar, this dilemma may have sparked the emergence of language. At some point our ancestors may have substituted verbal exchange for manual grooming, allowing them to quickly share information—who's doing what to whom, who's being deceitful, who's engaged in subversive plotting, and so on—off-loading hours of picking nits in favor of minutes of dishing dirt. Recent studies have shown that as much as 60 percent of our conversation today is devoted to gossip, a staggering number (especially to those of us who've hardly mastered small talk) that some researchers argue reflects the primary purpose of language at its inception.[20]

Linguist Daniel Dor develops the social role of language yet further. In a compelling and wide-ranging analysis, he proposes that language is a communally constructed tool with a specific and profoundly important function: to give individuals the power to guide each other's imaginations.[21] Before the emergence of language our social commerce was dominated by our shared experiences. If we both saw something or heard something or tasted something, we could reference it with gestures, sounds, or pictures. But it would have been difficult to communicate about experiences we hadn't shared, not to mention the daunting challenge of airing abstract thoughts and inner sensations. With language, we surmounted these challenges. With language the market for our social exchange swelled enormously: you could use language to describe experiences I might have never had; through words you could conjure them in my mind. I could do the same for you. Over the millennia, as the welfare of our prelinguistic forebears became ever more dependent on coordinated communal action—cooperative hunting of large prey, building controlled fires, cooking for large groups, shared caring and instructing of the young[22]—they breached the limits of nonverbal exchange, brought language into the world, and established a vastly augmented social arena encompassing not just our shared experiences but our shared thoughts.

These and almost every other proposal for the origin of language emphasize the spoken word, the external manifestation of language. In his characteristically iconic way, Chomsky does an about-face, proposing that in its earliest incarnation language may have facilitated internal thought.[23] Processing, planning, predicting, evaluating, reasoning, and understanding are but some of the essential tasks the inner voice between our forebears' ears could accomplish with cool confidence once thought was able to leverage language. Spoken language, in this view, was a subsequent development, like the addition of audio speakers to early model personal computers. It's as though long before they spoke, our ancestors were the deep and silent type, deliberating hard on their daily tasks but keeping the cogitation to themselves. Chomsky's position is contentious. Researchers have pointed to intrinsic features of language that appear designed for mapping internal concepts to the spoken word (notably, phonology and much of grammatical structure), suggesting that from the get-go language has been about external communication.

Although language's origin remains enigmatic, what's unquestionable, and of most relevance as we head onward, is that language and thought provide a potent mix. Whether or not an internal version of language preceded its external vocalization, and whether or not that vocalization was prompted by song or infant care or gesticulation or gossip or communal discourse or possessing a big brain or something else entirely, once the human mind had language, our species' engagement with reality was poised for radical change.

That change would ride on one of the most pervasive and influential of human behaviors: telling stories.

Storytelling and Intuition

George Smith was in a hurry. The fingers on his right hand gently but persistently tapped the inlaid ebony border of the long mahogany table. He'd just learned that Robert Ready, the museum's master stone restorer, would not return for several days. **Several days.** How could he wait? For three years, he had thrown on his coat, grabbed his carefully made sandwich of marmalade and Stilton, and dodged crowds and carriages as he raced to the British Museum, where he would spend the remaining minutes of his lunch break poring over fragments of hardened clay tablets recovered from an archaeological dig in Nineveh. His family was poor. He'd left school at fourteen to apprentice as a bank engraver. His prospects seemed limited. But George was a genius. He had taught himself ancient Assyrian and become expert at reading cuneiform writing. The museum's curators, who had taken a liking to the strange kid who hung around at noon, soon realized he was more adept at deciphering the cuneiform carvings than any of them and so brought George into their enclave as a full-time employee. Now, but a handful of years later, George had culled the thousands of clay pieces to assemble the first complete tablet and had already deciphered much of it. He'd found, or believed he'd found, a magnificent secret told by the series of triangular cuts and wedges—reference to a deluge myth preceding the account of Noah in the Old Testament—but he needed Robert Ready to delicately scrub the layer of crust obscuring an essential section of the text. George could taste victory. He shivered as he imagined the discovery elevating him to a new life. He couldn't contain himself. George decided to risk scrubbing the tablet himself.

OK, I'm getting carried away. The real George Smith waited. Days later, Robert Ready returned and plied his skills, and so was revealed the most ancient of our species' recorded stories, the Mesopotamian **Epic of Gilgamesh,** composed as far back as the third millennia BC. My free-form recounting does what storytellers—we humans—have long done: rework reality (what's known about the historical George Smith[24]), sometimes moderately (as here), sometimes aggressively, sometimes for heightened drama, sometimes for posterity, sometimes for the pure joy of spinning a good yarn. The artistic motivation of those who wrote **Gilgamesh,** a tale likely shaped by many voices over many generations, is unknown. But in this story full of battles and dreams, arrogance and jealousy, corruption and innocence, the characters and their concerns speak to us clearly across the millennia.

And that, really, is what's so striking. In the perhaps five thousand years since **Gilgamesh** was set down, history has witnessed transformation upon transformation of how we eat and shelter, how we live and communicate, how we medicate and procreate, and yet we immediately recognize ourselves in the unfolding narrative. Gilgamesh and his brother-in-arms Enkidu set out on a quest that would test their courage, their morality, and ultimately their sense of who they were—a Neolithic **Thelma and Louise.** Late in the journey, as Gilgamesh hovers over the lifeless Enkidu, he laments in wrenching but all too familiar terms: "He covered, like a bride, the face of his friend, like an eagle he circled around him. Like a lioness deprived of her cubs, he paced to and fro, this way and that. His curly [hair] he tore out in clumps, he ripped off his finery, [like] something taboo he cast it away."[25] Like many, I have known this place. Decades ago, charging from room to room in my tiny walk-up apartment, not knowing where to turn, I frantically sought to escape the news that my father had suddenly died. Even at a remove of hundreds if not thousands of generations, there is much we share with our forebears.

And it is not only that we humans consistently

grieve and mourn and thrill and delight and explore and wonder. We also share the urge to express all this and to process all this through story. **Gilgamesh** may be the oldest extant written story, but if our species was writing stories five thousand years ago, then long before that we surely were telling stories. It's what we do. And what we've long done. The question is why? Why would we eschew hunting additional bison and boar or gathering extra roots and fruit to spend time imagining escapades with petulant gods or journeys to fanciful worlds?

You might answer, because we like story. Yes. Of course. Why else would we steal off to the movies even though that report is due tomorrow? Why else would it feel like a guilty pleasure to set aside "real work" and carry on with that novel we've been reading or series we've been watching? Yet that's the beginning of an explanation, not the end. Why do we eat ice cream? Because we like ice cream? Yes, sure. But, as evolutionary psychologists have convincingly argued, the analysis can go deeper.[26]

Those of our forebears who enjoyed loading up on rich sources of energy like fleshy fruit and ripened nuts could better cope when days turned lean, thus producing more progeny and propagating a genetic predilection for sweets and fats. Today's craving for pistachio Häagen-Dazs, no longer praised as a health promoter, is a modern relic of yesteryear's vital scavenging for calories. It's Darwinian selection manifested at the level of behavioral inclination. Not that genes determine behavior. Our actions result from a complex amalgam of biological, historical, social, cultural, and all manner of chance influences that are imprinted on our particle arrangement. But our tastes and instincts are an essential part of that mix, and in the service of enhanced survival evolution had a strong hand in shaping them. We can learn new tricks but, genetically and hence instinctually speaking, we are old dogs.

The question, then, is whether Darwinian evolution can illuminate not only our culinary but also our literary tastes. Why were our ancestors drawn to expend precious resources of time, energy, and attention telling stories that, at first blush, don't seem to enhance our survival prospects? Fictional stories are particularly puzzling. What evolutionary utility could arise from following the exploits of imaginary characters facing make-believe challenges in nonexistent worlds? With its relentless random walk through the adaptive landscape, evolution is effective at sidestepping extravagant behavioral predispositions. A genetic mutation that led us away from the storytelling instinct, freeing up time for sharpening a few extra spears or scavenging a couple of extra buffalo carcasses, would seem to offer a survival advantage that, over time, would win. But it didn't. Or, for some reason, it's an opportunity evolution missed.

Researchers have tried to understand why, but

the clues are scarce. There is precious little evidence for establishing either the prevalence or the utility of storytelling among forebears stretching back thousands of generations. This highlights a general challenge that pervades research seeking an evolutionary basis for behavior, one that we will encounter in various forms in the chapters that follow. From the standpoint of natural selection, what matters is the impact this or that behavior would have had on the survival and reproductive prospects of our forebears during the bulk of their history. A trustworthy account thus requires a refined understanding of the ancient mind-set as it negotiated the ancestral environment. But recorded history provides information for only the final quarter of 1 percent of the roughly two million years stretching back to the earliest human migrations out of Africa. Researchers have developed indirect probes of the past, including detailed examination of ancient artifacts, extrapolations of ethnographic analyses of today's remaining hunter-gatherer groups, and studies of the brain's architecture in search of cognitive echoes of ancient adaptive challenges. The patchwork of evidence constrains theorizing but still allows for a variety of perspectives.

One such perspective holds that to seek an adaptive role for storytelling is to look for enhanced fitness in the wrong place. A given behavioral predisposition may be a mere by-product of other evolutionary developments—developments that did enhance survival and thus did evolve in the usual way by natural selection. The general directive, emphasized colorfully in a famous paper by Stephen Jay Gould and Richard Lewontin, is that you can't cherry-pick evolution.[27] Evolution sometimes offers only package deals. Big brains of the grey-white human variety, chock-full of densely connected neurons, are really good for survival, but perhaps something intrinsic to their design ensures that they revel in story. Consider, for instance, that our success as social beings relies in part on having good intel—who's up, who's down, who's strong, who's vulnerable, who's trustworthy. Because of the adaptive utility of such information, we are inclined to pay attention when it is available. And when in possession of such information, it is not uncommon to share it in exchange for burnishing our social status. As fiction is rife with information of this sort, our adaptively molded minds may be primed to perk up, listen, and repeat, even though the narrative is fanciful. Natural selection would thus smile on brains as they grew more adept at social living while rolling its eyes when listening to their obsessive storytelling.

Convinced? Many—and I count myself in this group—don't find it plausible that for all its capacity to innovate, the brain got locked into a thoroughly pervasive, utterly central, and yet adaptively irrelevant behavior. Aspects of the storytelling experience may be part of an evolutionary package deal, but if telling stories and listening to stories and telling those stories again amounted to sideline chatter, evolution would, one anticipates, have found a way to shed this wasteful tic. How then might storytelling earn its adaptive keep?

In seeking an answer, we must be mindful of the rules of the game. For many behaviors it is all too easy to concoct after-the-fact adaptive roles. And because we can't test such suggestions by rerunning the evolutionary unfolding, there is a danger that we're left with a collection of "just so" stories. The most convincing proposals are ones that start with a given adaptive challenge—one that if surmounted would result in greater reproductive success—and argue that a particular behavior (or suite of behaviors) is intrinsically well designed for meeting that challenge. The Darwinian explanation of our sweet tooth is exemplary. Humans require a minimum number of calories to survive and reproduce. Faced with the potential of a devastating shortfall in caloric intake, a preference for foods densely packed with sugars has manifest adaptive value. If you were designing the human mind, aware of the human body's physiological needs and the nature of the ancestral environment, it is easy to imagine that you would program the human brain to encourage its body to eat fruit whenever available. That natural selection arrived at this very strategy is thus not at all surprising. At issue is whether there are analogous adaptive considerations that might lead you to program the human mind to create, tell, and listen to stories.

There are. Storytelling may be the mind's way of rehearsing for the real world, a cerebral version of the playful activities documented across numerous species which provide a safe means for practicing and refining critical skills. Leading psychologist and all-around man of the mind Steven Pinker describes a particularly lean version of the idea: "Life is like chess, and plots are like those books of famous chess games that serious players study so they will be prepared if they ever find themselves in similar straits."[28] Pinker imagines that through story we each build a "mental catalogue" of strategic responses to life's potential curveballs, which we can then consult in moments of need. From fending off devious tribesmen to wooing potential mates, to organizing collective hunts, to avoiding poisonous plants, to instructing the young, to apportioning meager food supplies, and so on, our forebears faced one obstacle after another as their genes sought a presence in subsequent generations. Immersion in fictional tales grappling with a wide assortment of similar challenges would have had the capacity to refine our forebears' strategies and responses. Coding the brain to engage with fiction would thus be a clever way to cheaply, safely, and efficiently give the mind a broader base of experience from which to operate.

Some literary scholars have pushed back, noting that strategies pursued by fictional characters facing pretend challenges are not, generally speaking, portable to real life, or at least, not advisedly so.[29] "You might end up running around like the comically insane Don Quixote or the tragically deluded Emma Bovary—both of whom go astray because they confuse literary fantasy with reality," is how Jonathan Gottschall sportively summarizes the critique.[30] Pinker, of course, was not suggesting that we copy actions we encounter in stories but rather that we learn from them—an approach, as Gottschall notes, that is perhaps conveyed more fully by a modest shift in metaphor to one introduced by psychologist and novelist Keith Oatley:[31] Instead of mental file think flight simulator. Stories provide fabricated realms in which we shadow characters whose experiences far outstrip our own. Through borrowed eyes protected by the tempered glass of story, we intimately observe an abundance of exotic worlds. And it is through these simulated episodes that our intuition expands and refines, rendering it sharper and more flexible. When faced with the unfamiliar, we don't initiate cognitive look-ups that search a Dear Abby of the mind. Instead, through story we internalize a more nuanced sense of how to respond and why, and that intrinsic knowledge guides our future behavior. Cultivating an innate sense of heroic passion is a far cry from tilting at windmills—and that was my take, and that of many others too, on turning the last page in the adventures of Alonso Quijano.

With the flight simulator as our metaphor for the adaptive utility of story, how would we program the simulator itself? What kinds of stories would we have it run? We can take the answer from the first page of the Creative Writing 101 curriculum. An axiom of storytelling is the need for conflict. The need for difficulty. The need for trouble. We are drawn in by characters pursuing outcomes that require clearing treacherous hurdles, external and internal. Their journeys, literal and symbolic, keep us on the edge of our seats or furiously turning pages. To be sure, the most captivating of stories invoke surprising, entertaining, even awe-inspiring approaches to characters, plot, and the storytelling technique itself, but for many, remove the conflict and the story fizzles. It is no coincidence that the same goes for the Darwinian utility of the content running on the narrative flight simulator. Without conflict, without difficulty, without trouble, the adaptive value of story would fizzle too. A Josef K. who is happy to confess to an unnamed crime and dutifully serve an unjustified punishment would be a quick read. And with no other narrative adjustments, a less than impactful one. As would following a Dorothy who cheerfully hands over the ruby slippers, steps off the yellow brick road, and assimilates into Munchkinland. Clear skies, textbook-perfect engines, and model passengers are not the simulations that improve pilot readiness. The usefulness of rehearsing for the real world is encountering situations that would be challenging to respond to without preparation.

It's a perspective on story that may also shed light on why you and I and everyone else spend a couple of hours each day concocting tales that we rarely remember and more rarely share. By day I mean night, and the tales are those we produce during REM sleep. Well over a century since Freud's **The Interpretation of Dreams,** there is still no consensus on why we dream. I read Freud's book for a junior-year high school class called Hygiene (yes, that's really what it was called), a somewhat bizarre requirement taught by the school's gym teachers and sports coaches that focused on first aid and common standards of cleanliness. Lacking material to fill an entire semester, the class was padded by mandatory student presentations on topics deemed loosely relevant. I chose sleep and dreams and probably took it all too seriously, reading Freud and spending after-school hours combing through research literature. The wow moment for me, and for the class too, was the work of Michel Jouvet, who in the late 1950s explored the dream world of cats.[32] By impairing part of the cat brain (the locus coeruleus, if you like that sort of thing), Jouvet removed a neural block that ordinarily prevents dream thoughts from stimulating bodily action, resulting in sleeping cats who crouched and arched and hissed and pawed, presumably reacting to imaginary predators and prey. If you didn't know the animals were asleep, you might think they were practicing a feline kata. More recently, studies on rats using more refined neurological probes have shown that their brain patterns when dreaming so closely match those recorded when awake and learning a new maze that researchers can track the progress of the dreaming rat mind as it retraces its earlier steps.[33] When cats and rats dream it surely seems they're rehearsing behaviors relevant to survival.

Our common ancestor with cats and rodents lived some seventy or eighty million years ago, so extrapolating a speculative conclusion across species separated by tens of thousands of millennia comes with ample warning labels. But one can imagine that our language-infused minds may produce dreams for a similar purpose: to provide cognitive and emotional workouts that enhance knowledge and exercise intuition—nocturnal sessions on the flight simulator of story. Perhaps that is why in a typical life span we each spend a solid seven years with eyes closed, body mostly paralyzed, consuming our self-authored tales.[34]

Intrinsically, though, storytelling is not a solitary medium. Storytelling is our most powerful means for inhabiting other minds. And as a deeply social species, the ability to momentarily move into the mind of another may have been essential to our survival and our dominance. This offers a related design rationale for coding story into the human behavioral repertoire—for identifying, that is, the adaptive utility of our storytelling instinct.

Storytelling and Other Minds

Professional discourse among physicists generally involves specialized jargon articulated in a confetti of equations. Not the kind of material that would draw those huddled around the campfire to lean in. Yet if you know how to read the equations and interpret the jargon, the stories they tell can be stirring. In November 1915, when an exhausted Albert Einstein, on the verge of completing his general theory of relativity, plied the equations to explain the long-standing enigma of Mercury's orbit deviating slightly from Newtonian predictions, he was so moved that he experienced palpitations of the heart. He had been navigating through the treacherous waters of complex mathematics for nearly a decade, and the result of that calculation was akin to the first sighting of land. To paraphrase Alfred North Whitehead's later assessment, it meant that Einstein's bold quest had arrived safely on the shores of understanding.[35]

I have never had a discovery that monumental. Few have. But even more prosaic discoveries can provide a similar heart-pounding thrill. At those moments, there is a sense of deep connection with the cosmos. That, truly, is what the stories embedded in the abstract mathematics and the specialized language are all about. The stories give an intimate accounting of the universe, or something within the universe, as it is born, as it ages, as it transforms. The stories provide a means for experiencing the universe from a perspective that is otherwise unattainable. They provide a gateway to realms of reality that, in the most gratifying of examples, are wholly unexpected. Through mathematics, confirmed by experiment and observation, we are given leave to commune with a strange and wondrous cosmos.

The stories we have been telling in natural languages for thousands of years play an analogous role. Through story we break free from our usual singular perspective and for a brief moment inhabit the world in a different way. We experience it through the eyes and imagination of the storyteller. The flight simulator of story is our portal to the idiosyncratic worlds playing out in nearby minds. In the words of Joyce Carol Oates, reading "is the sole means by which we slip, involuntarily, often helplessly, into another's skin; another's voice; another's soul . . . to enter a consciousness not known to us."[36] Without story, the nuances of other minds would be as opaque as the microworld without knowledge of quantum mechanics.

Is there an evolutionary consequence to this distinctive quality of story? Researchers have imagined so. We prevailed, in large part, because we are an intensely social species. We are able to live and work in groups. Not in perfect harmony, but with sufficient cooperation to thoroughly upend the calculus of survival. It is not just safety in numbers. It is innovate, participate, delegate, and collaborate in numbers. And essential to such successful group living are the very insights into the variety of human experience we've absorbed through story. As psychologist Jerome Bruner noted, "We organize our experience and our memory of human happenings mainly in the form of narrative,"[37] leading him to doubt that "such collective life would be possible were it not for our human capacity to organize and communicate experience in narrative form."[38] Through narrative we explore the range of human behavior, from societal expectation to heinous transgression. We witness the breadth of human motivation, from lofty ambition to reprehensible brutality. We encounter the scope of human disposition from triumphant victory to heartrending loss. As literary scholar Brian Boyd has emphasized, narratives thus make "the social landscape more navigable, more expansive, more open with possibilities," instilling in us a "craving for understanding our world not only in terms of our own direct experience, but through the experiences of others—and not only real others."[39]

Whether told through myths, stories, fables, or even embellished accounts of daily events, narratives are the key to our social nature. With math we commune with other realities; with story we commune with other minds.

When I was a kid, I'd often watch the original **Star Trek** series with my dad, a tradition I've repeated with my own son. Morality tales and space opera have a strong pull on those who enjoy heroic exploration served up with doses of philosophical pondering. One of the most riveting episodes, "Darmok," from the **Next Generation** spinoff, depicts an extraordinary role for story in the fashioning of civilization. The Tamarians, an alien race of humanoids, communicate solely through allegory, and so Captain Picard's direct use of language is as baffling to them as their constant reference to an oeuvre of unfamiliar stories is to him. Picard finally grasps their allegory-based worldview and establishes a cross-species meeting of the minds by recounting **The Epic of Gilgamesh.**

To the Tamarians, the patterns of life and community are imprinted in a collection of shared stories. Our mental template is less single-minded, but even so, narrative provides one of our primary conceptual schemas. Anthropologist John Tooby and psychologist Leda Cosmides, pioneers of evolutionary psychology, suggest why: "We evolved not so long ago from organisms whose sole source of (non-innate) information was the individual's own experience."[40] And experience, whether contending with the crowds of today's Times Square or coordinating a group hunt on the plains of Cenozoic Africa, delivers information in storylike packets. If we possessed the fanciful, superhuman, particle-revealing vision I invoked in the previous chapter, the packets of experience might have a different character: perhaps we would organize our thoughts and memories in terms of particle trajectories or quantum wave functions. But with ordinary human perceptions, the palette of experience is colored in narrative, and so our minds are adapted for painting the universe in story.

Note, though, that form is one thing and content another. While experience has instilled an enchantment with the structure of story, we use narrative to organize our understanding well beyond the bounds of human encounters. Scientific advances provide a prime example. Tales of a lone species setting out to conquer the grand mysteries of reality and returning with some of the most startling insights can indeed be the stuff of drama and heroics. But the standard of success for the scientific content of these stories is poles apart from the measures we bring to bear on our human odysseys. The raison d'être of science is to pull back the veil obscuring an objective reality, and so scientific accounts must conform to standards of logic and be tested through replicable experimental scrutiny. That is the power but also the limitation of science. By rigorously adhering to a standard that minimizes subjectivity, science fixes on results that transcend any given member of the species. Schrödinger's profoundly important quantum equation tells us a lot about electrons—and how thrilling to have an equation that delineates the comings and goings of these wispy particles with greater precision than any account of any other happening on the planet—but the math doesn't tell us much about Schrödinger or the rest of us. It's a price science pays proudly for a quantum chronicle that may prove relevant well beyond our little corner of reality, perhaps holding sway across all space and throughout all time.

The stories we tell of the comings and goings of characters, whether real or fictional, have a different concern. They illuminate the richness of our ineluctably circumscribed and thoroughly subjective existence. Ambrose Bierce's breathtaking tale, occupying a brief moment in a military execution at Owl Creek Bridge, distills what Ernest Becker described as the "excruciating inner yearning for life."[41] Through story, we witness an amplified version of that yearning. And as we envision the exhausted but elated Peyton Farquhar reaching out to grasp his wife, and the noose sharply jolts him, and us, from his imagined escape, our sense of what it means to be human ramifies. Through language, story explodes the limits that would otherwise be imposed by our own narrow experiences. As the masterfully chosen words direct our imagination, we gain a deeper sense of our common humanity and a more nuanced understanding of how to survive as a social species.

Whether dealing with fact or fiction, the symbolic or the literal, the storytelling impulse is a human universal. We take in the world through our senses, and in pursuing coherence and envisioning possibility we seek patterns, we invent patterns, and we imagine patterns. With story we articulate what we find. It is an ongoing process that is central to how we arrange our lives and make sense of existence. Stories of characters, real and fanciful, responding to situations familiar and extraordinary, provide a virtual universe of human engagement that infuses our responses and refines our actions. Sometime in the far future, if we finally play host to visitors from a distant world, our scientific narratives will contain truths they will have likely discovered too, and so will have little to offer. Our human narratives, as with Picard and the Tamarians, will tell them who we are.

Mythic Tales

Within the community of scientists, research findings gain currency by explaining puzzling data, or by offering resolutions to thorny theoretical problems, or by allowing us to accomplish feats previously beyond reach. The vast majority of scientific developments remain the province of experts, but some do manage to rise above the others and achieve a broad cultural impact. For the most part, these are developments relevant to grand concerns that transcend nitty-gritty scientific detail: How did the universe begin? What is the nature of time? Is space what it appears to be? If you absorb science's most refined answers to such big questions, your perspective on reality will almost certainly shift. That we are a minor planet orbiting an average star formed in the aftermath of a stupendous swelling of primordial space is a realization that constantly informs my thoughts regarding how we fit into the grand picture. That time elapses at a different rate for me than it does for anyone else who is not moving precisely with me is a stunning fact that I reflect on endlessly. That our apparently three-dimensional reality may be a thin slice through a grander spatial expanse is a thrilling possibility that I delight in imagining.

Across millennia, cultures too have produced particular stories that have also managed to rise above the others and achieve a broad impact on their community's view of reality. These are a culture's myths—stories held in sufficient regard to garner a sense of the sacred. It is notoriously difficult to define myth, but we will take it to denote stories that invoke supernatural agents to explore culture's grand concerns: its origin, its long-practiced rituals, its particular ways of imposing order on the world. Through their longevity, wide appeal, and portfolio of fundamental explanations, myths become the basis of a shared heritage, a corpus of tragedy and triumph, of chronicle and fantasy, of adventure and reflection that defines a people and shapes a society.

There is a long history of scholars developing insightful ways to read and interpret myth. Early in the twentieth century, anthropologist Sir James Frazer proposed that myths emerge from attempts to explain the otherwise puzzling phenomena of life and nature encountered by our ancient brethren. Psychoanalyst Carl Jung believed that through archetypes—universal patterns that, he surmised, are inherent to the unconscious mind—myths express shared qualities of the human experience. Joseph Campbell argued for a "monomyth," a master template for mythological stories in which a reluctant character receives a call to action, undertakes an adventure rife with danger and death-defying rites of passage, and finally returns home, a born-again hero whose journey gives our sense of reality a hearty shake.[42] More recently, philologist Michael Witzel has suggested that a universal template emerges most clearly not at the level of individual myths but only when we consider the collective myths of entire traditions—a concatenated story line, he suggests, that extends all the way from the world's beginning to its final demise. Invoking linguistics, population genetics, and archaeology, Witzel argues that common qualities in these narratives can be traced to an earlier form of mythology that originated in Africa, perhaps as far back as one hundred thousand years ago.[43]

These proposals, and others too numerous to mention, incite controversy and impassioned critiques. They have their proponents and detractors; they rise and fall. Some scholars suggest that while the allure of a single overarching explanation for myth is strong—it would help identify the pervasive qualities that shaped our ancient heritage—the complexity of human life as it played out through a dimly lit and uncertain history may not lend itself to a singular explanation. For our purpose here, the explanatory sweep can be more limited. Religious scholar and author Karen Armstrong has offered the sparest of summaries, noting that myths are "nearly always rooted in the experience of death and the fear of extinction,"[44] and even if we're a touch more conservative and soften "nearly always" to "often" or "in many instances," we still have a strong guiding light to lead us onward.

A few examples: When Gilgamesh hears of a man to whom the gods apparently granted immortality, he'll stop at nothing—journeying across a vast wilderness, staring down scorpion monsters, negotiating the Waters of Death—to learn the secret for escaping the otherwise inevitable end. Death is central to the Hindu tale of the goddess Kali, whose perfection so enrages her divine compatriots that they sever her head with a bolt of lightning;[45] death is at the core of the Kono creation myth in which Sa, the deity of death, believes his daughter has been abducted by the god Alatangana and for revenge decrees mortality for all humankind; it's a significant theme in the Oceania story of Ma-ui, who passes through the ferocious jaws of the sleeping Goblin-Goddess, Great Hina-of-the-Night, intent on securing immortality by ripping out her heart—but Hina awakes and with her razor-sharp teeth tears him to shreds.[46] Randomly crack open your favorite anthology of world myths, and you won't have to tread far before you're toeing death's door. These tales of characters fighting for their lives and bringing death into the world are echoed through the many stories that tell of the annihilation of the entire world. As Witzel notes, such destruction "may take place as a final world-wide conflagration—the **Götterdämmerung** or Ragnarök in the Edda, molten metal in Zoroastrian myth, Śiva's destructive dance and fire in India, fire in Munda myth, fire/water and so on in Maya and other Mesoamerican myths, and Atum's final destruction of the earth in Egypt."[47] And if that leaves you hankering for more, there are numerous stories telling of other destructions that make generous use of ice, endless winters, and, popular the world over, floods.

What's going on here? Why so much danger, death, and destruction? Narrative invites conflict and trouble; unless we are committed to upending narrative norms, without these elements we would be hard-pressed to find a story to tell. Blend that with the larger-than-life concerns at the core of myth—origins of place or people and rationales for ways of being—and the dilemmas inherent to story are pushed to an extreme. The progression could hardly be otherwise. By the time we have language and by the time we tell stories, we have acquired the capacity to live beyond the moment. We are able to navigate past and future with ease. We are able to plan and design, to coordinate and communicate, to anticipate and prepare. The utility of these capacities is manifest, but with such mental agility we also live with the memory of those who were but are no more. We infer the pattern, never breached, that each life ends. We recognize that life and death are locked in an uncleavable embrace. They are dual qualities of existence. To reflect on origins is to rouse questions of endings. To reflect on how to live a life is to reflect on the absence of life. The inevitability of death is a commanding realization for us in the here and now, and, one can imagine, only more so during epochs when the end could come yet more capriciously. It is little wonder that death and destruction garnered thematic prominence.

But why populate these ancient tales with manic giants, fire-breathing serpents, bull-headed men, and the like? Why terrifying fantastic tales instead of terrifying realism? Why go **Poltergeist** and **The Exorcist** instead of **Saving Private Ryan** and **Reservoir Dogs**? Cognitive anthropologist Pascal Boyer, building on early work of cognitive scientist Dan Sperber,[48] suggests an answer. For a concept to grab hold of our attention with enough force that we remember it and transmit it to others, the concept must be sufficiently novel to offer surprise but not so outrageous that we immediately deem it ridiculous. Boyer argues that a given musing lands in the cognitive sweet spot when it is "minimally counterintuitive"—which means that it violates one or perhaps as many as two of our deeply ingrained expectations.[49] Invisible people? Sure, so long as invisibility is the only counterintuitive feature. A river that answers calculus problems by singing them to the theme of **M*A*S*H**? Silly, and so is dismissed by most everyone and quickly forgotten. Aligning with the larger-than-life themes of mythic tales, the protagonists we encounter are larger than life but minimally counterintuitive constructs of the human imagination. No surprise that these protagonists have physical forms, thought processes, and even personality profiles that, at the very least, are thoroughly familiar, even if their powers exceed expectations based on anything we have ever encountered.

Language provides another cylinder powering myth's creative engine. Once we have the capacity to describe the structure of ordinary things—raging storms, burning trees, slithering snakes, and so on—language provides a ready-made narrative Mr. Potato Head, allowing us to mix and match freely. Giant rocks and talking people are but one swap from the more captivating linguistic mash-up of talking rocks and giant people. Language unleashes the cognitive capacity to imagine all manner of unrehearsed combinations that guide us toward novelty.[50] Minds that acquired this power were minds capable of seeing old problems in a new way. They are minds that would innovate. They are minds that, in time, would control and reshape the world.

Seeding the creative swirl, too, is our theory of mind—our innate tendency to ascribe a mind to anything we encounter that even hints of having agency. As in our earlier discussion of conscious-ness, when we encounter other people, even at a distance and without direct engagement, we im-mediately endow them with minds more or less like our own. Evolutionarily speaking, that's a good thing. Other minds can generate behaviors that we're better off anticipating. Same goes for animals, and so we instinctively ascribe intents and desires to them too. But sometimes, as emphasized by the psychologist Justin Barrett and the anthropologist Stewart Guthrie, we overdo it.[51] Evolutionarily speaking, that can be a good thing too. To mistake a distant moonlit shrub for a resting lion, no big deal. To think that the noise we just heard was a

windblown branch when it's an approaching leopard, deadly. When assigning agency in the wild,
better to overendow than underendow (up to a
point, of course), a lesson that successful molecules
of DNA and the storytelling vehicles they inhabit
have taken to heart.

Decades ago, during what for me was a fairly rare
camping expedition, I was challenged to undertake
a brief solo period in the woods. Equipped with a
tarp, sleeping bag, three matches, a small can, a pen,
and a journal, I found myself more deeply alone
than I had ever been. By any measure practical
or psychic, I was not prepared. I managed a low
makeshift roof by impaling the tarp on judiciously
chosen branches, but I used up all the matches on
my first unsuccessful attempt to build a fire. As the
sun began to set and terror started to rise, I rolled
out the sleeping bag, scurried in, and stared at the
tarp hovering close above my face. I was just this
side of panic. To my city-habituated ears and over-
worked imagination, every gust and every creak
was a bear or mountain lion. I had no illusions of
heroism, but each seemingly interminable second
felt like my own death-defying rite of passage. I
took out my pen and scratched two circular eyes,
a splotchy nose, and a crooked mouth, slightly
turned up at the corners; pen on tarp is not ideal,
but the broken blue lines and indented plastic were
enough. I was still alone but didn't feel it as fully.
If each of the night's forest noises was endowed

with a mind, then so too for my etching. I would be cast away for only three days, but I'd created my own Wilson.

Evolution instilled a tendency for us to imagine our surroundings chock-full of things that think and feel, sometimes envisioning them offering help and counsel, but more often conceiving of them as plotting and planning, crossing and double-crossing, attacking and avenging. Overendowing the world's sounds and stirrings with minds bent on danger and destruction can save your life. Having the cognitive flexibility to mash up elements of reality into concoctions of the fantastic can seed innovation. Empowering otherwise ordinary protagonists with surprising supernatural qualities grabs attention and facilitates cultural transmission. In combination, these elements illuminate the kinds of stories that captivated our ancestors' imagination and provided narrative guidance for navigating the ancient world.

Over time the most enduring of these mythic tales would seed one of the world's most transformative forces: religion.

7

BRAINS
AND BELIEF

From Imagination to the Sacred

I imagine that when we finally make contact with intelligent beings beyond earth, they too will recount a history replete with attempts to find meaning. Life capable of building telescopes, of fashioning spacecraft, of reaching out to the cosmos and listening in on its chatter is life that has the capacity for self-reflection. As intelligence matures, the very same impulse to explore and to understand manifests as an urge to infuse experience with significance. Answer enough how questions and why questions quickly follow. Here on earth, survival forced our early brethren to be technicians. They needed to learn to fashion stone, bronze, and iron. They needed to master the techniques of hunting, gathering, and farming. But while servicing essential survival needs, our ancestors struggled with the very same questions we do—questions of origin, meaning, and purpose. To survive is to kindle the search for why survival matters. Technicians inevitably become philosophers. Or scientists. Or theologians. Or writers. Or composers. Or musicians. Or artists. Or poets. Or devotees of thousands of variations and combinations of systems of thought and creative expression that promise insight into the very questions that gnaw at our insides long after our stomachs are full.

As our enduring stories and myths make clear, the most persistent of such questions are existential. How did the world begin? How will it end? How can we be here one moment and yet be gone the next? Where do we go? What other worlds might be out there?

Imagining Other Worlds

About one hundred thousand years ago, somewhere in the Lower Galilee region of present-day Israel, a child who was four, maybe five years old, maybe playing quietly, maybe making mischief, suffered a traumatic blow to the head. The child's gender is unknown, but let's imagine she's a little girl. The cause of the injury is obscure too. Stumbling down a steep rocky hill, falling from a tree, receiving excessive punishment? What we do know is that the impact gashed the front right side of her skull, causing brain damage, which she endured until the age of twelve or thirteen, when she died. These facts have been gleaned from skeletal remains found at Qafzeh, one of the most ancient of all burial sites, whose excavation began in the 1930s. Although the remains of twenty-six others were also found at the site, the burial of the young girl is distinctive. Antlers from two deer were laid across the girl's chest with one end resting on her palm, an arrangement according to the researchers that provides evidence of a ceremonial burial. Could the antlers be an unintentional ornament? Possibly. But it is easy to follow the research team's judgment and envision Qafzeh 11, as the child is known, being laid to rest in a ritual enacted a hundred millennia ago by early humans who were reflecting on death, struggling to grasp what it means, and, perhaps, thinking about what might follow.[1]

Tentative though conclusions about events so distant surely are, excavations of burials from later eras make the interpretation yet more plausible. In 1955, in the village of Dobrogo, about two hundred kilometers northeast of Moscow, Alexander Nacharov was operating an excavator for the Vladimir Ceramic Works when he noticed that intermingled with the yellowish brown loam he'd scooped up were bones. They turned out to be the first of many that would be unearthed over the next few decades at Sunghir, one of the most celebrated burial sites of the Paleolithic era. One grave is particularly stunning: a boy and a girl, ages approximately ten and twelve at death, were buried head-to-head in what looks like an eternal melding of two young minds. Interred more than thirty thousand years ago, their remains are adorned by one of the most elaborate collections of grave goods ever discovered. Headgear made from decorated arctic fox teeth, ivory armbands, more than a dozen ivory spears, perforated ivory disks, and—bringing a smile to fans of Liberace— more than ten thousand carved ivory beads that were likely sewn into the children's burial garb. Researchers have estimated that at the furious pace of one hundred hours per week, it could easily have taken an artisan more than a year to make these ornaments.[2] The investment provides at the very least a strong hint that ritual burials were part of a strategy to transcend the finality of death. The body might cease, but some vital quality, which might be enhanced or appeased or honored or gratified by elaborate burial accessories, would carry on.

Nineteenth-century anthropologist Edward Burnett Tylor argued that dreams were a persuasive influence guiding early humans to this very conclusion.[3] We can well imagine that nightly escapades, from the curious to the outré, would have provided a persistent suggestion of a world beyond what's available to open eyes. Whether feeling comforted or frightened, to awake from a visit with a departed friend or relative is to be left with the sense that they still exist. Not in the way they once did. Not here, clearly. But in some ethereal way they are close by. Written accounts, although not available until much later, support the speculation through abundant instances of dreams providing windows onto unseen realities. Ancient Sumerians and Egyptians interpreted dreams as directives from the divine; throughout the Old and New Testaments, divine will is frequently revealed through dreams. And in the modern era, studies of isolated hunting societies like Australian aborigines reveal the essential role of Dreamtime, an eternal realm from which all life originates and to which all life will return. Dreamlike trance states are also common to a number of traditions that engage in rituals driven by percussive music and strenuous dance, which can proceed for hours and induce hypnotic reveries that participants have described as being transported to distinct planes of reality.[4]

During waking hours, too, there would have been no shortage of episodes suggesting a reality beyond the visible: powerful forces at work on earth and in the skies; capricious happenings of daily existence; frequent life-threatening and life-ending dangers. Evolutionary success in a social setting primed our brains to attribute common experiences to the actions of fellow beings. When lightning struck or the floods came or the earth shook, we continued to imagine that a thinking being was responsible.

Faced with it all, we can imagine our forebears implicitly acknowledging the limits of their influence in an uncertain world and in response conjuring characters inhabiting an invisible realm that would wield the very powers they lacked.

Unwitting or not, it was a spectacularly clever response. It allowed us to write otherwise random events into coherent stories: To imagine unseen realms populated with characters familiar and fabricated. To provide names and faces, real and fantastic, who keep tabs on what we do and exert ultimate control over our fate. To recast mortality as a gateway traversed by Qafzeh 11, her two dozen cave companions, as well as generations of ancestors, en route to these invisible but ascendant worlds. To tell and retell their stories, and with these narratives invoke the personalities, foibles, grudges, jealousies, and all manner of human demeanor playing out in nearby worlds to explain the otherwise unexplained happenings in our own.

Our ancient artistic forays provide further hints of an otherworldly preoccupation. On rock walls the world over explorers have found tens of thousands of painted images, some dating farther back than forty thousand years. They reveal a menagerie from lion to rhino to creative hybrids including deer meets woman and bird meets man. The human form takes a secondary role, often executed as a rudimentary sketch if it appears at all. Collections of human handprints are plentiful, depicted as chaotic overlapping stenciled outlines whose meaning we can only guess—straining to touch another realm, longing to acquire the rock's seemingly infinite durability, imprinting exuberant ornamentation, leaving early versions of "Kilroy was here"? Intentions fade and so we are left to wonder. As we do, we recognize in the dancing sorcerer and dying bison the earliest efforts of a creative force that seems like our own. Looking just beneath the rock surface, we catch a glimpse of ourselves staring back.

Therein lies both thrill and pitfall. The allure of encountering our ancient cultural kin may beguile us into attributing undue meaning to their creative works. Perhaps cave art is nothing more than mindless doodles of the early conscious mind. Or, in a more elevated description, perhaps cave art demonstrates an ancient aesthetic drive, what some have called "art for art's sake."[5] Inferring the inspiration of those who lived hundreds of centuries ago is a risky business, and so we're well advised not to overreach. But when you consider the ordeal required to reach at least some of these sites—archaeologist David Lewis-Williams describes how explorers now and, presumably, cave artists then "crouched and crawled underground along a narrow, absolutely dark passage for more than a kilometre, slid along mud banks and waded through dark lakes and hidden rivers"[6]—an art-for-art's-sake explanation seems less plausible. Even those of our ancient brethren with an especially strong bohemian commitment would likely have chosen easier ways to satisfy a purely artistic impulse.

Perhaps, then, our artistic forebears were undertaking magical ceremonies to assure the success of the hunt, an idea promoted in the early 1900s by archaeologist Salomon Reinach.[7] What's a little spelunking and painting if it can assure a delightful and necessary dinner?[8] Or, as suggested by Lewis-Williams, developing earlier ideas discussed by historian of religion Mircea Eliade, perhaps cave art derives from shamanic head trips. As mythic narratives acquired increased followings, shamans—spiritual leaders who gained prominence by convincing others, and perhaps themselves too, of their ability to travel to the unseen realms of nearby realities—became intermediaries between this world and the next. The inspiration for Paleolithic paintings may then have been trancelike visions experienced by shamans negotiating with mythological characters or channeling imaginary animals.

Striking similarities between compositions separated by continents and millennia seem to gesture toward a single sweeping explanation for cave art. But even if that is too ambitious a vision, there is one characteristic of which archaeologist Benjamin Smith is thoroughly convinced: "Caves were far from just 'canvases.' They were places in which rituals were conducted, where people communicated with spirits and ancestors dwelling in another realm, they were places loaded with meaning and resonance."[9] According to Smith and many like-minded researchers, our forebears believed deeply that through art and ritual they could influence spiritual forces. That confident conclusion not-withstanding, as we look back twenty-five, fifty, perhaps even a hundred thousand years, details are hazy, and so it is unlikely that we will ever defini-tively know what motivated our ancient brethren. Even so, a consistent if tentative picture comes into focus. We see our forebears engaging in ceremonial burials, ritualized send-offs to other worlds; creat-ing art that imagines realities beyond experience; telling mythic narratives that invoke powerful spirits, immortality, and the afterlife—in short, the strands of what later generations would label religion are coming together, and we don't have to strain to see recognition of life's impermanence entwined in the braiding.

Evolutionary Roots of Religion

Can we parlay ancient burgeoning religiosity into an explanation of the wide adoption of religious practice the world over? Proponents of the cogni-tive science of religion such as Pascal Boyer argue that we can. Even across the broadest spectrum of religious engagement, he suggests, there is a uni-formly applicable evolutionary basis:

The explanation for religious beliefs and behaviors is to be found in the way all human minds work. I really mean all human minds, not just the minds of religious people . . . because what matters here are properties of minds that are found in all members of our species with normal brains.[10]

The thesis is that features inherent to human brains, shaped over eons by the relentless battle for evolutionary supremacy, prime us for religious conviction. Not that there are god genes or devotional dendrites. Instead, Boyer draws on an understanding of the brain developed in recent decades by cognitive scientists and evolutionary psychologists that refines the familiar metaphor of mind as computer. Rather than likening the brain to a general-purpose computer awaiting whatever programming it acquires through experience, the brain is likened to a special-purpose computer, hardwired with programming designed by natural selection to bolster the survival and reproductive prospects of our forebears.[11] These programs support what Boyer refers to as "inference systems," dedicated neural processes that are adept at responding to the kinds of challenges—from throwing spears to courting mates to establishing alliances—which would have determined whose genes successfully migrated to the next round and whose did not. Boyer's central point is that these inference systems are readily coopted by the very qualities intrinsic to religion.

We have already encountered one such inference system: our theory of mind, by which we impute the kind of agency we each experience internally to entities we encounter in the external world. The adaptively beneficial tendency to overendow such agency clarifies why we so readily imagine our surroundings—whether beneath the earth or up in the sky—being inhabited by attentive minds. Other inference systems include our intuitive grasp of psychology and of physics: without formal instruction, we all have a basic understanding of the capacities of minds and bodies. Join these inference systems with our attraction to minimally counterintuitive concepts (recall that these are concepts that violate a small number of our intuitive expectations) and there is little mystery in why we latch on to notions like spirits and gods (agents endowed with humanlike minds, but differing from expectations in their corporeality and their powers, both psychological and physical). Normal brains also have social inference systems that, for example, keep track of relationships, ensuring that the bearer receives a fair shake. If I do something for you, you're going to have to do something for me, and make no mistake, I'm keeping a running tab. This reciprocal variety of altruism may be the source of the transactional nature of the relationship adherents typically have with the supernatural beings that populate religious traditions: I'll sacrifice, I'll pray, I'll do good, but come tomorrow's combat, you've got my back. On the flip side, when bad things happen, we're all too ready to chalk it up to our individual or collective failure to meet divine expectations.

In his book **Religion Explained,** Boyer develops these ideas fully; other researchers have developed variations on similar themes.[12] But my sketch conveys the gist of the approach: the brain's evolution was shaped by the battle for survival, and the victorious brain that emerged has qualities that embrace religion with open arms. It's an example of what I earlier referred to as an evolutionary package deal. A predilection for religious belief may have no adaptive value of its own but comes bundled with a suite of other brain qualities that **were** selected because of their adaptive functions. This doesn't mean we will all be religious any more than our naturally selected sweet tooth means we will all indulge in glazed donuts. It does mean that the brain's inference systems are particularly responsive to the kinds of features that show up in the world's religions. Indeed, such resonance is the very reason such features have persisted in the world's religions. Be they ghosts or gods, demons or devils, saints or souls, religious conceits are virtuoso conductors of the evolving human mind. We are attentive to them, we act on them, we promulgate them, and thus they spread widely.[13]

So is that it? Survival of the fittest outfitted our minds, and fit minds are readily inculcated with a religious sensibility? What about the role we imagine that religion must have played (and for many, continues to play) in explaining the seemingly unexplainable from the origin of life and the universe to the meaning of death? Boyer and many others advancing similar perspectives do not deny religion's role in addressing these issues, but they argue that such considerations are insufficient for explaining why religion arose and why it has the features it does. The elephant in the religious room is the human mind, and without a primary focus on the mind's evolved nature we leave out the dominant force.

The case developed by Boyer and fellow researchers is compelling and insightful. But as with all theorizing in the spectacularly complex arena of brain, mind, and culture, definitive conclusions that convince all modern minds, or at least those minds thinking carefully about the issues at hand, are hard to come by. Moreover, even if the cognitive science of religion succeeds in revealing that we have an inherent susceptibility to religious thought, there remains ample room for religion to be more than an evolutionary appendage, more than a mere by-product of earlier cognitive adaptations. As other researchers have argued, religion may be ubiquitous because it has provided its own contribution to our adaptive fitness.

Take One for the Team

As the sizes of their clans grew, hunter-gatherer tribes faced a critical problem. How do you ensure cooperation and loyalty among increasingly large collections of individuals? For groups of kin, an idea going back to Darwin and developed over subsequent decades by a number of renowned scientists, including Ronald Fisher, J. B. S. Haldane, and W. D. Hamilton, suggests that evolution by natural selection solves the problem without breaking a sweat.[14] I'm loyal to my siblings, my children, and other close relatives because we share a meaningful portion of our genes. By saving my sister from a charging elephant, I'm increasing the likelihood that genetic segments identical to mine will persist and will be passed on to subsequent generations. Not that I need to know this. And during my gallant feat, I'm surely not calculating relative abundances in the future gene pool. But by the standard Darwinian logic, my instinctive inclination to protect my kin, and even sacrifice myself for groups of my kin, will be naturally selected, fostering the continuance of such behavior in progeny who share a significant percentage of my genetic profile. The reasoning is straightforward but raises the question, When groups outgrow a collection of kin, is there a genetic carrot that wields the cooperative stick?

If you could find a way to make me think or at least act as though members of the larger group are part of my extended family, the problem might be solved. But how do you accomplish that? Earlier, we discussed how story, by enhancing our understanding of other minds, may have facilitated communal living. Some researchers, like evolutionary biologist David Sloan Wilson, developing ideas championed near the turn of the twentieth century by sociologist Émile Durkheim, take this adaptive role much farther.[15] Religion **is** story, enhanced by doctrines, rituals, customs, symbols, art, and behavioral standards. By conferring an aura of the sacred upon collections of such activities and by establishing an emotional allegiance among those who practice them, religion extends the club of kinship. Religion provides membership to unrelated individuals who thus feel part of a strongly bound group. Even though our genetic overlap is minimal, we are primed to work together and protect one another because of our religious attachment.

Such cooperation matters. Deeply. As we have seen, humans prevailed in no small part because our species has the capacity to pool brain and brawn, to live and work in groups, to divvy up responsibilities and effectively meet the needs of the collective. The greater social cohesion of those in a religiously bound group would have made them a more formidable force in the ancestral world, and according to this line of argument, securing an adaptive role for religious affiliation.

It is a perspective that has generated decades of debate. Some researchers throw up their hands whenever group cohesion is trotted out as an evolutionary explanation, viewing it as a hackneyed fallback to explain putatively prosocial behaviors whose adaptive value has otherwise proven elusive.[16] Moreover, the adaptive value of cooperation is itself a complex business: In any group of cooperative individuals, selfish members can game the system. By taking advantage of affable comrades, selfish individuals can acquire an undue allotment of resources and thus unfairly increase the likelihood of surviving and reproducing. Passing on their selfish tendencies, their progeny will tend to do the same, over time driving their trusting companions—together with their religious sensibilities—into extinction. So much for religion's adaptive coup.

Proponents of the religious basis for social cohesion acknowledge the issue but stress that it is only half the story. Within the confines of an isolated group of cooperative members, selfish infiltrators will surely win. But the groups of interest—hunter-gatherers in the Pleistocene—were not isolated. They interacted. They fought. And according to one reading of the archaeological record, their battles were deadly. A collection of cooperative members, each devoted to the well-being of the group, would tend to fare better. As Darwin himself put it, "When two tribes of primeval man, living in the same country, came into competition, if (other circumstances being equal) the one tribe included a great number of courageous, sympathetic and faithful members, who were always ready to warn each other of danger, to aid and defend each other, this tribe would succeed better and conquer the other."[17] Moreover, those whose service was inspired by devotion to departed ancestors or watchful deities would have been even more reliable and fervent in their commitment to the cause.[18] And so to determine which genetic traits would have swum broadly through the gene pool, we must not only take account of within-group dynamics, favoring the selfish, but also between-group dynamics, favoring the cooperative. If we assume that across many thousands of generations between-group success dominated the calculus of survival, allegiance to the group would hold sway, and so religion's social cohesion would triumph.

The victory thus imagined remains tentative because it depends on that very assumption—the dominance of between-group over within-group forces—and far from everyone is convinced that it provides an accurate portrayal of life and death throughout our hunter-gatherer past. Emboldening the skeptics further, an explanation for cooperative behavior can emerge from more down-to-earth considerations: the mathematics of game theory. Between the extremes of selfish and selfless behavior, there are innumerable strategies an individual member of a group may pursue. Perhaps I lean toward being selfless, but if you cross me one too many times my selfish side will emerge with a vengeance. Perhaps once you've lost my trust, I'll never give you another chance—or perhaps, do me a few good turns and I'll offer you a shot to earn it back. And so on. In a large group populated by individuals committed to a range of different strategies, what happens? Well, different cooperative strategies confer different survival value and so across the generations will themselves be subject to Darwinian selection. Using mathematical analyses and computer simulations, researchers have pitted various strategies against one another and found that one in particular—"I'll do something good for you so long as you do something good for me in return, but you do something underhanded and I'll quickly retaliate"—reliably trumps other variants, including those far more selfish. The theoretical analysis thus suggests that qualified cooperation of this sort aids survival.[19] To the detractors, this demonstrates that cooperation can arise organically and spread via natural selection, with no need for participants to hold a common religious belief.

After decades of wrangling, some researchers now claim that these disputes have finally been settled. But since such evaluations have been issued by proponents on both sides, the assessment of religion's role as the Pleistocene's survival-promoting social glue continues to elude consensus. It is a complex problem. Binding among other seductive qualities the enchantment of story, the inclination to endow agency, the comfort of ritual, the appetite for explanation, the security of community, and the cognitive appeal of countering expectations, religion is a rich and intricate human development whose genesis is from a time so remote that hard data, from ancient practice to intragroup conflict, is scarce. The debate will no doubt continue.

Another possibility entirely is that in evaluating religion's potential adaptive function, the argument over group cohesion is missing an essential part of the story. Various researchers have suggested that religion's adaptive impact is most directly evident at the level of the individual.

Individual Adaptation and Religion

During our inquiry into the origin of language, one proposal featured the role of gossip in maintaining hierarchies and fostering alliances. Frivolous as such conversation may be viewed in the modern age, psychologist Jesse Bering places gossip at the nexus of religion's adaptive role in the ancient world. Before we acquired the capacity to speak, a rogue in our midst might misbehave—stealing food, borrowing sexual partners, hanging back during the hunt—but if the witnesses to the transgression were small in number and weak in status the culprit could get away scot-free. Once language took hold, that changed. With even a single but widely discussed infraction, the culprit's reputation would suffer and reproductive opportunities would plummet. Bering's suggestion is that if a would-be transgressor imagines that there is always a powerful witness—hovering in the wind, or in trees, or in the sky—he would be less likely to transgress, less likely to be fodder for unfavorable gossip, and less likely to become a social outcast. Consequently, he would be more likely to have offspring and pass on his god-fearing instincts. A predisposition for religion protects his genetic lineage and so becomes self-perpetuating.[20]

Supporting evidence comes from experiments Bering has run in which children are presented with a challenging task and then left alone to accomplish it. In the absence of oversight, the researchers found what you would expect. Many kids will cheat. However, those children who are told that there is an invisible witness in the room, a friendly but fully attentive presence, are far more likely to adhere to the rules. This holds even for those kids who claim that they don't actually believe there is an invisible being at all. Bering's conclusion is that the young mind, which he plausibly argues provides a more direct window on our inherent human nature compared with older minds that have been subject to greater cultural influence, is predisposed to act in accordance with an invisible presence constantly monitoring behavior. In ancient times, it was this very priming that encouraged the prosocial behavior that protected reputations, increased reproductive opportunities, and thus further spread the priming itself—a priming, that is, for a religious sensibility.

A different adaptive role for religion has been developed by experimental social psychologists who have spent decades furthering the vision of Ernest Becker, whose **Denial of Death** set us on our way in chapter 1. The terror of knowing we are going to die, these researchers argue, "would have rendered our ancestors quivering piles of biological protoplasm on the fast track to oblivion."[21] What may have saved us, they suggest, was the promise of life beyond physical death, either literal or symbolic. Becker himself made a persuasive case that addressing mortality awareness by invoking the supernatural was a wondrous human innovation. To alleviate the distress of transience requires a palliator with unqualified and unlimited durability, something impossible to achieve in the real world of material things.

Granted, you may find the image of our physically robust forebears huddled on the savanna in an anxiety-induced paralysis hard to fathom. Yet through shrewd psychosocial experiments, researchers have argued that even here in the modern age we are demonstrably if unwittingly affected by mortality awareness. In one such experiment, court judges in Arizona were tasked with recommending a fine for defendants accused of a misdemeanor. In the written instructions the judges were provided, which included a standard personality profile questionnaire, half were asked a couple of additional questions that required reflection on their own mortality (for example, What emotions does thinking about your own death arouse?). The researchers anticipated that because the legal code is part of society's concerted effort to assert control over an otherwise anarchic reality—providing a bulwark against the dangers lurking just beyond the bounds of civilization—those judges who had been reminded of the ultimate danger, their own demise, would more vehemently enforce legal statutes. The predictions were right on target. But even the researchers found the size of the disparity in the fines recommended by the two groups of judges remarkable. On average, the fines issued by the mortality-primed judges were **nine** times that of the control group.[22]

As the researchers emphasize, if the diligently trained judicial mind steeped in the standard of dispassionate fairness can be so affected by shining a little additional conscious light on mortality, we should pause before dismissing a similar but equally stealthy influence at work within each of us. Indeed, hundreds of subsequent studies (varying the subjects, their country of origin, their

purported tasks, the manner in which mortality
awareness is stimulated, and so on) have demonstrated that such influences can be measured
and manifest widely, from the voting booth, to
xenophobic prejudice, to creative expression, and
religious affiliation.[23] Becker maintained, and these
studies support, that culture has evolved in part
to mitigate the potentially debilitating effects that
would otherwise accompany mortality awareness.
Accordingly, from this perspective, if you scoff
at such a possibility it is because culture is doing
its job.

Pascal Boyer, with whom we began our discussion on the evolutionary roots of religion, rejects
this role for religion, noting that "a religious world
is often every bit as terrifying as a world without
supernatural presence, and many religions create
not so much reassurance as a thick pall of gloom."[24]
But rather than bracing a rattling bag of bones, in
the spirit of Becker's adherents, and far from cast-
ing dark shadows across its devoted followers, as
envisaged by Boyer, a religious sensibility may have
provided a more modest benefit to a less dispirited
patient. Perhaps ancient religious activities illuminated death in a softer light and set everyday
experience within a more enduring narrative—a
beneficial consequence of religious experience that
William James described as providing "an assurance
of safety and a temper of peace" while instilling a
"new zest which adds itself like a gift to life, and

takes the form of either lyrical enchantment or of appeal to earnestness and heroism."[25]

Clearly, there is as yet no consensus on why religion arose nor on why it has so tenaciously remained. And not for lack of ideas: coopting the naturally selected brain, driving group cohesion, calming existential anxiety, protecting reputations and reproductive opportunities. The historical record may be too spotty for us ever to build a definitive case; religion may play roles too varied to submit to all-embracing explanations. I remain partial to religion's relevance to our singular recognition of our finite lives; as Stephen Jay Gould summarized it, "A large brain allowed us to learn . . . the inevitability of our personal mortality"[26] and "all religion began with an awareness of death."[27] But whether religion then took hold because it transformed that awareness into an adaptive advantage is a wholly different question.

The brain's exquisite order allows it to generate copious thoughts and actions, some directly linked to survival, others not. Indeed, it is this very capacity, our extensive behavioral repertoire, that provides the foundation for the variety of human freedom we discussed in chapter 5. What is unassailable is that through these actions, we have steadfastly kept religion with us, developing it over the millennia into institutions whose influence pervades the planet.

A Sketch of Religious Roots

During the first millennium BC, across India, China, and Judea, tenacious and inventive thinkers reexamined ancient myths and ways of being, entailing among other developments what philosopher Karl Jaspers described as the "beginnings of the world religions, by which human beings still live."[28] Scholars debate the degree of relatedness of these far-flung developments, but there is accord on the outcome. Religious systems became increasingly organized as adherents set down stories, culled insights, and synthesized directives that, having been channeled through anointed prophets and passed orally from one generation to the next, had garnered a stamp of the sacred. There is great variation in the content of the resulting texts, of course, but they hold in common a fascination with the very questions guiding our exploration in these pages: Where did we come from? And where are we going?

Among the earliest surviving written records are the Vedas, composed in Sanskrit on the Indian subcontinent, with portions that date from as far back as 1500 BC. Together with the Upanishads, a rich body of commentary likely written sometime after the eighth century BC, the Vedas are a voluminous collection of verse, mantra, and prose that constitutes the sacred texts of what would become the Hindu religion—now practiced by one in seven inhabitants of the earth, about 1.1 billion people. Before I was yet ten years old, I had a personal entrée to these works.

It was the late 1960s. Peace, love, and Vietnam were in the air as my father, sister, and I strolled on a bright sunny day through Central Park. We paused at the Naumburg Bandshell just off of Poet's Walk, where a large gathering of Hare Krishna devotees were energetically drumming, chanting, and dancing. One adherent, eyes bulging and tears streaming, was expressing an impassioned astral communion by pulsing to the beat while staring intently into the sun. Shockingly, at least for me, I suddenly realized that one of the drummers, outfitted in flowing robes and sporting a shaved head, save for a single tuft atop, was my brother. I thought he was away at college. The outing, apparently, was my father's way of introducing us to the new direction that my brother's life had taken.

In the decades that followed, communication with my brother was episodic, but in each encounter the Vedas were either central or circling nearby. It's hard to say whether my own interests were sparked by these encounters or whether the conversations naturally emerged from siblings approaching similar questions from widely different perspectives. It was surely enriching to learn of ancient and for me unfamiliar ruminations into cosmic origins: "There was neither non-existence nor existence then; there was neither the realm of space nor the sky which is beyond. What stirred? Where? In whose protection? Was there water bottomlessly deep? There was neither death nor immortality then. There was no distinguishing sign of night nor of day. That one breathed, windless, by its own impulse. Other than that there was nothing beyond."[29] I was moved by the universality of the human need to feel the rhythms of reality. But to my brother, the Vedas were more than that. They provided a grander vision of the cosmology I was studying mathematically. As poetry, the words artfully capture the enigma of a beginning to the beginning. As metaphor, they speak to the perplexing nature of a time before time. As a meditation, perhaps a communal immersion around a crackling fire enveloped by an awe-inspiring but utterly mysterious inky-black star-filled canopy, the lines convey the seeming paradox of how there can be a universe at all. But ancient hymns and verse, imaginative stories of the thousand-headed Purusha dismembered to create the sun, earth, and moon, as well as the many other evocative and lofty offerings, do not account for the origin of the universe. The words reflect our pattern-seeking, explanation-craving, survival-attuned minds developing a vivid story to provide a symbolic framework for living—how we came to be, how we should behave, the consequences of our actions, and the nature of life and death. What became apparent to me through these sporadic fraternal brushes is that the Vedas seek something stable, some kind of constant quality underlying the shifting sands of familiar reality. It is a description that I, and many of my colleagues, would happily use in characterizing the charge of fundamental physics. The disciplines share a common urge to see beyond appearances available to everyday experience. Yet the nature of the explanations each discipline deems capable of advancing this charge are thoroughly distinct.

In the middle of the sixth century BC, Siddhārtha Gautama, a prince born in present-day Nepal who had been brought up studying the Vedas, became distraught as the life of luxury he'd been handed confronted the anguish endured by those leading a more common existence. As the famous story goes, Gautama decided to forgo privilege and wander the world in search of a way to alleviate the misery of human suffering. The resulting insights, developed and promulgated by his followers largely after his death, constitute Buddhism, now practiced by one in every dozen inhabitants of earth, about half a billion people. As Buddhist thought spread, numerous sects developed, but all share in the belief that perception is an illusory guide to reality. There are qualities of the world that may seem stable but, in truth, all things always change. Deviating from its Vedic origins, Buddhism denies that there is an immutable substrate underlying existence and attributes the root of human suffering to the failure of recognizing the impermanence of everything. The Buddha's teachings outline a way of life that promises an unvarnished, more clearly perceived view of truth, and as with the Vedas, the path to such enlightenment involves a series of rebirths, with the endgame seeking to conclude the cycles of reincarnation by reaching an eternal state of bliss that stands beyond desire, beyond suffering, and beyond self. If humanity's earlier imagining of realms where life continued beyond this life was a remarkable mental maneuver for addressing the enigma of mortality, the Hindu and Buddhist stances are more remarkable still. Death is reimagined as a new beginning in a cyclical process whose very goal is an ultimate and permanent release from life. The conclusion of the cycles, once attained, leads to a dominion where the concept of distinct existence disappears. Our impermanence becomes a sacred rite of passage en route to the timeless.

Because Hinduism and Buddhism seek a reality beyond the illusions of everyday perception, a characterization that also describes many of the most surprising scientific advances of the last hundred years, a small industry has produced articles, books, and films that purport to establish links with modern physics. While one can find similarities in perspective and language, I have never encountered more than a metaphorical resonance between distinct ideas vaguely construed. Descriptions of modern physics provided in popularizations, mine as well as those of others, usually suppress mathematics in favor of more accessible accounts, but, unequivocally, mathematics is the anchor of the science. Words, however carefully chosen and crafted, are only a translation of the equations. Invoking such translations as the basis for contact with other disciplines will almost never rise above the level of a poetic alliance.

This judgment is consonant with at least some of the spiritual disciplines' leading voices. Some years ago I was invited to participate in a public forum with the Dalai Lama. During the discussion, I noted the preponderance of books explaining how modern physics is recapitulating discoveries made in the Far East thousands of years ago, and I asked the Dalai Lama whether he considered these claims valid. His forthright answer left a significant impression on me: "When it comes to consciousness, Buddhism has something important to say. But when it comes to material reality, we need to look to you and your colleagues. You are the ones penetrating deeply."[30] I remember thinking, How wonderful to imagine religious and spiritual leaders worldwide following his simple, fearless, and honest example.

During roughly the same era that the Buddha was wandering in India, the Jewish people in the Kingdom of Judah were being trounced by the Babylonians and forced into exile. In an effort to codify their identity, Jewish leaders gathered disparate written accounts and oversaw the transcription of oral histories, yielding early versions of the Hebrew Bible—a document that would continue to evolve and become a sacred text of the Abrahamic religions, now practiced by more than one of every two inhabitants on earth, about four billion people.[31] The God of Judaism, Christianity, and Islam is the all-powerful, all-knowing, everywhere-present, singular creator of everything—a conception that, for many worldwide, is the dominant image they conjure when there's talk, secular or sacred, of religion.

The Old Testament tells its own widely known origin story. Well, it tells two such stories. The first takes six days, begins with the formation of the heavens and the earth and concludes with the creation of man and woman; the second fills only a single day, with man created early on; during his first nap, woman enters the scene. Generation upon generation quickly follow, but the Old Testament is less than forthcoming regarding where protagonists go when they die. Save a couple of brief references to resurrection, there is no commitment to an afterlife. Jewish mystics and interpreters subsequently developed numerous ideas involving immortal souls awaiting another world, but there is no single interpretation that reconciles the myriad sources and commentaries. Half a millennium later, that uncertainty would be wiped away as Christianity developed a theological doctrine infused with eternal souls that maintain their identities well beyond their time on earth. Half a millennium beyond that, Islam would introduce its own extensive body of belief addressing similar themes, aligning with Christianity in its reverence for an approaching day of judgment when the dead would be raised and those deemed worthy would receive eternal heavenly reward while all others would endure eternal damnation.

The handful of religions we have briefly surveyed are collectively followed by more than three out of every four inhabitants of planet earth. With billions of adherents, the nature and style of religious engagement varies considerably and, if we include the more than four thousand smaller religions currently practiced around the world, the range of commitments and the specifics of doctrinal content broaden yet more widely. Even so, there are common qualities, such as exalted figures who have seen further or been granted access to stories that purport to explain how it all began, how it will all end, where we will all go, and how best to get there. Deeper still is a prevalent expectation that adherents will assume a sacred mind-set. The world is full of stories that can inform how we live. The world is full of pronouncements that can guide how we behave. Those stories and pronouncements that are bound into a religious doctrine are elevated above all others because in the mind of the faithful they elicit some variety of **belief.**

The Urge to Believe

Some years ago, while I was in the final chaotic days of an all-consuming project, an invitation arrived to deliver a keynote speech at a gathering in Washington State. Distracted, I accepted the invitation without ensuring that the organization had been properly vetted. A few months later, when the talk rolled around, I realized I was slated to speak at Ramtha's School of Enlightenment, an organization led by Judy Zebra Knight, who claims to channel a thirty-five-thousand-year-old warrior, Ramtha, hailing from the lost land of Lemuria (which, apparently, was frequently at war with the lost land of Atlantis). A quick search turned up revealing video clips, including one from an old episode of **The Merv Griffin Show** in which Knight throws her head back, snaps it forward, goes into a trance, drops her voice, takes on a manner of speech partway between Yoda and the Queen, and, she would have us believe, embodies the Lemurian sage. My little daughter, watching over my shoulder, tried not to giggle. She failed. I would have giggled too if I wasn't mortified that I'd accepted the invitation. But it was the day before the presentation so too late to back out gracefully.

On arrival, my first encounter was with hundreds of blindfolded people, arms extended, all milling about a large grassy enclosure. My guide explained that pinned to each was a card on which they'd written their life's dream, and the exercise was to "feel" one's way to an identical card that had been planted somewhere on the field. He noted that success is a key step toward ensuring that the dream would be realized. "How's that going?" I asked. "Oh, wonderfully. Already in this session one participant found her matching card." Next up were the blindfolded archers. I kept a healthy distance and demurred at the entreaties to participate, all the more when I noted that a photographer had quietly joined the tour. The blindfolded archers were about as successful as the blindfolded seekers. Finally, I was joined by a young woman, probably in her twenties or thirties, whose telepathic talent allowed her to name successive cards in a shuffled deck. "Seven of diamonds," she predicted. "Darn, six of clubs. But I was only off by one. Nine of spades. Oh, it's a three of diamonds. Aha, **there's** that diamond." And so it went. She told me she practices many hours each day and knows she needs to train harder.

To those who'd gathered around, and later at the keynote, I couldn't help offering a few basic observations, many that we have touched on in these pages. I explained that we are a species that looks upon the world and sees patterns. And for the most part that's a good thing. Over many generations, natural selection equipped us to identify patterns in how people and objects appear and move, allowing us to identify them rapidly with just a few visual cues. We detect patterns in animal behavior, allowing us to anticipate when it's safe to approach and when it's best to head in another direction. We grasp patterns in how objects from rocks to spears fly when thrown, an ability that was particularly useful to our ancestors seeking to subdue the next meal. Through pattern we develop the means to communicate and thus join together in groups—tribes to nations—that exert the world's most powerful influences. In short, the capacity for recognizing pattern is how we survive. But, I continued, sometimes we go overboard. Sometimes, our naturally selected pattern detectors are so primed, so ready to announce that a signal has been found, that they see patterns and envision correlations that are not there. Sometimes we assign meaning to the meaningless. From basic math we know that on average, one out of every four times you'll correctly guess a card's suit; one out of every thirteen times, you'll guess its rank. But that pattern reveals nothing about telepathic ability. Once in a blue moon—well, less often than that—you'll randomly walk the field and find your matching card, but that says nothing about the fulfillment of dreams. How often, I asked, do you notice that a remarkable coincidence did **not** happen?

The attendees, by now all packed into a cavernous barn, cheered their approval. Many rose to a standing ovation, which, as I said to all assembled, was appreciated but confusing. I'm telling you that your approach to finding a deeper reality and the methods you're practicing lead nowhere. Another ovation.

Later, at the book signing, a number of participants, speaking sotto voce, offered clarification. "Many of us don't buy into a lot of the stuff that goes on here and it's important for someone to call that out. But there is **something** else out there, we can feel it, and we come to the school because we need to be around others who have the same urge to seek a deeper truth." I can relate. I understand the urge. The history of physics is a collection of episodes in which time and time again heroic mathematical and experimental explorations have revealed that there **is** something else out there—often something strange and wondrous that requires us to rework our picture of reality. There is every reason to believe that our current understanding, even with its capacity to explain copious data with uncanny precision, is provisional, and so we physicists anticipate that going forward, this rhythm of revision will repeat many times over. However, it is through centuries of effort that we have refined our investigative tools, and these are the mathematical and experimental methods that constitute the rigorous body of scientific practice. These are the methods that we pass on to our students and research fellows. These are the methods that have proven their capacity for reliably accessing hidden qualities of reality.

I am open to unconventional claims. If data collected in carefully designed and replicable experiments investigating, say, the ability to sense hidden cards in a deck, revealed better than random success, or if robust data established that a member of our species was able to channel an ancient sage hailing from a long-lost land, I'd be interested. Extraordinarily interested. But in the absence of such data, and in the absence of any reason whatsoever to anticipate that such data might be forthcoming, and in the absence of any argument as to why such claims are not in flat-out contradiction with all we demonstrably know about the workings of reality, there quickly comes a point when we should conclude that there is no basis for holding a belief in any such claims.

Which raises the question: Is there any basis for believing in an invisible, all-powerful being who created the universe, listens and responds to our prayers, keeps track of what we say and do, and doles out rewards and punishments? In developing an answer, it is worthwhile to flesh out the concept of belief more fully.

Belief, Confidence, and Value

Almost to a person, those who inquire about my belief in God invoke "belief" in the very same way they would if asking about my belief in quantum mechanics. In fact, I'm often asked the two questions in tandem. I tend to phrase my response in terms of confidence—a measure of certainty—noting that my confidence in quantum mechanics is high, because the theory accurately predicts features of the world, such as the electron's magnetic dipole moment, with a precision beyond the ninth decimal place, while my confidence in the existence of God is low because of the paucity of rigorous supporting data. Confidence, as these examples illustrate, emerges from dispassionate, essentially algorithmic judgment of evidence.

Indeed, when physicists analyze data and announce a result, they quantify their confidence using well-established mathematical procedures. The word "discovery" is generally used only when the confidence crosses a mathematical threshold: the probability of being misled by a statistical fluke in the data must be less than about one in 3.5 million (an arbitrary-looking number but one that naturally emerges in statistical analyses). Of course, even such high levels of confidence do not ensure that a "discovery" is true. Data from subsequent experiments may require us to adjust our confidence; in this case, too, mathematics provides an algorithm for calculating the update.

While few of us live by such mathematical methods, we arrive at many of our beliefs through similar if less overtly analytical reasoning. We see Jack with Jill, and wonder if they might be a couple; we see them together again and again, and our confidence in that conclusion grows. Later, we learn that Jack and Jill are siblings, and so we discount our previous assessment. And on it goes. It is an iterative process that you might anticipate converging on beliefs that reflect the true nature of the world. But that need not be the case. Evolution did not configure our brain processes to form beliefs that align with reality. It configured them to favor beliefs that generate survival-promoting behaviors. And the two considerations need not coincide. If our forebears had carefully investigated every swish and rustle that caught their attention, they would have found that most could be explained without invoking a volitional agent. But from the standpoint of adaptive fitness, their burdensome investment in seeking the truth would have had little going for it. Across tens of thousands of generations, our brains eschewed greater accuracy for a rough-and-ready understanding. Nimble responses often beat considered assessments. Verity is an important character in the drama of belief but is easily upstaged by survival and reproduction.

Thickening the plot further, evolution added

another cast member: emotions. In 1872, more than a dozen years after announcing evolution by natural selection, Darwin published **The Expression of the Emotions in Man and Animals,** exploring his conviction that the biologically adapted brain, not culture, is the primary driver of emotional expression. Drawing on close observations of his children, widely disseminated questionnaires, and cross-cultural data he gathered during his long expeditions, Darwin made the case that, for example, the tendency to smile when pleased or blush when embarrassed was universal. You could count on those very responses clear across the world's cultures. In the century and a half since, researchers have taken Darwin's lead and sought the adaptive roles that might explain various human emotions as well as investigated the neural systems that might be responsible for generating them. Fear, the research has shown, is indeed primal—from the get-go, there was significant adaptive value in rapid behavioral and physiological responses to danger. Parental love, which drives essential care for helpless progeny, is likely an ancient adaptation too. Embarrassment, guilt, and shame, which are particularly relevant for conducive behavior within larger groups, are adaptations that likely came later as group sizes grew.[32] The relevance for us here is that much as adaptive pressure shaped the language-possessing, storytelling, myth-making, ritual-practicing, art-creating, and science-pursuing human mind, adaptive pressure also shaped our rich emotional capacities. Emotion has been enmeshed throughout our evolutionary development. Beliefs have thus emerged from a complex calculus synthesizing reasoned analyses and emotional responses within a mind acquiring a talent for survival.[33]

Our belief calculus is also dependent on a range of factors including social influences, political forces, and brute expediencies. Early in one's life, belief is strongly biased by parental authority. Mom or Dad says it's true? Then it's true. As Richard Dawkins has noted, natural selection favors parents who pass their children information that enhances survival, and so to believe what Mom or Dad says makes evolutionary sense. Later on, many initiate their own belief calculus—investigating, discussing, reading, and challenging—one that itself is frequently biased by preexisting expectations and exposure to the beliefs of others. Most of us also extend the list of authorities deemed trustworthy—teachers, leaders, friends, officials, and other anointed experts. We have to. No one can rediscover, or even verify, thousands of years of accumulated knowledge. I once had a dream, a nightmare really, that I was back in my PhD dissertation defense, and the examiner, chuckling under his breath, told me that all the experiments and all the observations supporting the quantum mechanical "laws" of physics had been concocted. I was the brunt of an elaborate practical joke, having been misled by a pantheon of authorities I respected and a community of peers I trusted. Unlikely as the dream's scenario may be, the fact is I have personally verified the results from but a tiny fraction of the discipline's essential experiments. You could say I've taken most results on faith.

My confidence derives from decades of firsthand experience, witnessing how physicists minimize human subjectivity by focusing on carefully accumulated data, relentlessly interrogating hypotheses, and discarding all but those that meet a rigorous set of universal standards. But even with such diligent attention, historical contingency and emotionally driven human biases find ways of seeping in. One of the dominant approaches to quantum mechanics (called the **Copenhagen interpretation**) can be traced in part to powerful personalities that held sway during the theory's inception. I'll refer you to one of my other books, **The Hidden Reality,** for a discussion, but I suspect that had quantum mechanics been developed by a different cast of characters, the formal science would exist all the same but this particular interpretive perspective would not have enjoyed the same dominant position across so many decades. The beauty of science is that through continued research, the doctrines of one age are carefully rethought by the next and so are nudged ever closer to the goal of objective truth. But even in a discipline designed for objectivity, it takes a process. And it takes time.

Little wonder that in the messy, haphazard, emotionally laden realm of everyday human ventures the spectrum of belief is wide and imaginative, if at times confusing and frustrating. In forming beliefs, some look to science, both in content and for strategy. Some rely on authority, others on community. Some are coerced, sometimes subtly, sometimes overtly. Some place their utmost trust in tradition. Others give full jurisdiction to intuition. And in the mind's subterranean, generally unmonitored processing centers, we each employ an idiosyncratic and highly variable combination of all these tactics. What's more, there is nothing that prevents us from holding incompatible beliefs or from undertaking actions that suggest we do. I am comfortable admitting that every now and then I knock on wood or speak to the departed or seek heavenly reinforcement. None of this fits within my rational beliefs about the world, and yet I am perfectly content with my occasional apotropaic leanings. In fact, there is a certain delight in momentarily stepping beyond rational strictures.

Note too that while professional philosophers are paid to scrutinize belief—to reveal hidden assumptions and bring attention to faulty inferences—that's not how most of us now, or our ancestors then, go about it. Many beliefs in most lives go unexamined.

Perhaps this is its own variety of adaptation. Navel gazers tend to overlook that food stocks are low or that a tarantula is making a stealthy approach. Which means that in evaluating how it could be that so and so believes such and such, envisioning belief as having emerged from intense consideration and thorough cross-examination is often wide of the mark. As Boyer points out, "We assume that notions of supernatural agents . . . are presented to the mind and that some decisionmaking process accepts these notions as valid or rejects them." But because these ideas tickle a great many of the brain's inference centers—from agency detection, to theory of mind, to relationship tracking, and so on—and because natural selection has equipped these centers to perform their own diagnostics well below the threshold of awareness, the rational judge and jury model "may be a rather distorted view of how such concepts are acquired and represented."[34]

Even the very things to which the concept of belief can be sensibly applied change from epoch to epoch. As Karen Armstrong notes, those carrying out the rites of the ancient Eleusinian mysteries "would have been puzzled if they had been asked whether they believed that Persephone really had descended into the earth, in the way that the myth described."[35] It would be tantamount to asking whether you believe in winter. "Believe in winter?" you'd rightly reply. "The seasons, well, they just are." Similarly, Armstrong imagines, our forebears would embrace Persephone's travels "because wherever you looked you saw that life and death were inseparable, and that the earth died and came to life again. Death was fearful, frightening and inevitable, but it was not the end. If you cut a plant, and threw away the dead branch, it gained a new sprout."[36] Myth did not supplicate for belief. It did not elicit a crisis of faith that through painstaking deliberation was resolved by its beholders. Myth provided a poetic schema, a metaphorical mind-set, which became inseparable from the reality it illuminated.

Perhaps, too, there is an analogy with what happens in the long-term development of natural language.[37] In striving for emphasis and creative expression, speakers sprinkle their sentences with one metaphor after another. I just did, but more than likely you hardly noticed. We sprinkle salt on stews; we sprinkle sugar on pastries. And yet the sprinkling I invoked is a so thoroughly banal metaphor that rare is the reader for whom the phrase evokes a hand gently dispersing words upon a feast of freshly cooked sentences. Over time, metaphors become so overused that any poetic quality they may have initially possessed gradually evaporates (water evaporates, not poetry) and they become everyday workhorse words (horses work, not words). In a word, they become literal. Perhaps an analogous process plays out with mythic-religious notions. Perhaps such notions begin as evocative, poetic, metaphorical ways of looking out on the world that, over an expanse of time, gradually lose their poetry, shed their metaphorical meaning, and transition into literalism.

The closest I come to such literalism is acknowledging that some or other god may exist. I recognize that no one can ever rule out this possibility. As long as a purported god's influence does not in any way modify the progression of reality that is well described by our mathematical laws, then that God is compatible with all we observe. But there is an enormous gulf between mere compatibility and explanatory necessity. We invoke the equations of Einstein and Schrödinger, the evolutionary framework of Darwin and Wallace, the double helix of Watson and Crick, and a long list of other scientific achievements not because they are compatible with our observations, which of course they are, but because they provide a powerful, detailed, and predictive explanatory structure for understanding our observations. On this measure, religious doctrines do not register; of course, many among the faithful deem this measure irrelevant. The snag is that a literal perspective precludes that assessment. A religious assertion interpreted as a literal claim about the world that contravenes established scientific law is false. Full stop. In such cases, espousing a literal interpretation is on par with accepting the existence of Ramtha.

Nevertheless, religious doctrine (or even that of

Ramtha) can remain fully part of rational discourse if we are willing to move away from literalism, cherry-picking scripture, disregarding elements we find offensive or outmoded, interpreting stories and statements poetically or symbolically or, yet more simply, as elements of a fictional account. There are many reasons we might be drawn to do so. We might find joy or comfort in seeing our lives play out within a larger and, to some, more fulfilling narrative, giving scant regard to religion's supernatural qualities or metaphysical claims. We might derive value from reading religious stories as a deeply moving archive that symbolically captures essential qualities of the human condition. We might savor the challenge of developing an interpretive system that squares particular religious doctrines with scientific understanding. We might find it rewarding to overlay a sacred sensibility on our engagement with the world, adding a veneer that enhances experience but does not negate rationality. We might benefit from the support and solidarity of religious affiliation. We might find it emotionally enriching to participate in religious rituals, consecrating life passages and marking sacred days that connect us with a venerable tradition. Such varieties of religious engagement can provide activity, motivation, community, and guidance that, for some, lay out a path toward a richer life endowed with greater meaning. Such varieties of religious engagement do not require a belief in the factual nature of religious content; they reflect a belief in the value of such content, regardless of whether the content is veridical.

Over a century ago, William James offered a perceptive and heartfelt analysis of religious experience, one that resonates with the Dalai Lama's observation regarding physics and consciousness. James emphasized that while science cultivates an objective, impersonal approach it is only by considering our inner worlds—"the terror and beauty of phenomena, the 'promise' of the dawn and of the rainbow, the 'voice' of the thunder, the 'gentleness' of the summer rain, the 'sublimity' of the stars, and not the physical laws which these things follow"[38]—that we can ever hope to develop a full account of reality. Much like Descartes, James was underscoring that our inner experience is, in fact, our **only** experience. Science may seek an objective reality, but our only access to that reality is through the mind's subjective processing. The human mind thus relentlessly interprets an objective reality by producing a subjective one.

And so, if religious practice—or perhaps a better label here would be spiritual practice—is undertaken as an exploration of the mind's inner world, an inward-directed journey through the inescapably subjective experience of reality, then questions of whether this or that doctrine reflects an objective truth become secondary.[39] The religious or spiritual quest need not seek demonstrable aspects of the outer world; there is a whole inner landscape to explore, from the terror and beauty, promise and voice, gentleness and sublimity that James referenced to the vast list of other human constructs including good and evil, awe and dread, wonder and gratitude that we have invoked throughout the ages to ordain value and find meaning. However hard we may stare at nature's individual particles, however diligently we may pursue nature's fundamental mathematical rules, we won't catch sight of these concepts. They emerge only when particular complex arrangements of particles evolve the capacities to think, feel, and reflect. And how spectacular and how gratifying that there can be such collections of churning particles, operating under the inflexible control of physical law, yet capable of bringing these qualities into the world.

For me, the analogy with language's sharp metaphors worn smooth by age brings out an essential point, obvious yet telling: many of the world's religions are old. That is vital. It tells us that for centuries, if not millennia, a religious practice has held a people's attention and in various combinations provided the structure of ritual, informed their sense of place in the world, guided their moral sensibility, inspired the creation of artistic works, offered participation in a larger-than-life narrative, promised that death is not the end, and, of course, also intimidated with harsh penalties, emboldened some to violent battle, justified the enslaving and killing of transgressors, and so on. Some good, some bad, some utterly awful. But through it all, religious traditions have hung on. While decidedly not providing insight into a verifiable basis of material reality—the purview of science—religion has provided some of its adherents with a sense of coherence that has given life context, placing the familiar and exotic, the joys and the travails within a grander story. And because of that, the world's venerable religions provide lineages that connect followers clear across the ages.

I was raised Jewish. As a family, we attended services on major holidays, and I was enrolled in a local Hebrew school. The annual influx of new students meant the class restarted each year with the Hebrew alphabet, so I would quietly sit on the side and thumb through the Old Testament. I complained bitterly to my parents, but truth be told I enjoyed reading of Samuel and Absalom and Ishmael and Job, and all the rest. As the years passed, I grew more distant from the religion, feeling little need for formal participation. Then, during a break from my graduate studies at Oxford, I took a trip to Israel. An overzealous rabbi somehow caught wind that a young American physicist was wandering the streets of Jerusalem. He tracked him down, surrounded him with Talmudic scholars who were "also studying the origin of the universe," and convinced—well, pressured—the unduly deferential mid-twenty-year-old student to visit his temple and wrap his arms and forehead with the traditional leather accoutrement of the tefillin ritual. To the rabbi, this was God's will in action. The student was destined to be brought back into the fold. To the student, it was heavy-handed coercion to engage in a sacred practice in the absence of an inner conviction. When the student finally unwrapped the leather bands and left the temple, he knew he was done.

Yet, when my father died, the daily arrival in our living room of a minyan of observant Jews to recite the Kaddish prayer was of great comfort. My dad, not a religious man himself, was being embraced by a tradition reaching back thousands of years, experiencing a ritual administered to countless before him. The religious words the men chanted hardly mattered. They were in Aramaic, a collection of ancient sounds, a tribal poetry imprinted in cadence and rhythm, and I had no interest in a translation. What mattered to me for those brief moments—the nature, if you will, of my belief—was history and connection. That, to me, is the grandeur of heritage. That, to me, is the majesty of religion.

8

INSTINCT AND CREATIVITY

From the Sacred to the Sublime

On May 7, 1824, Ludwig van Beethoven appeared onstage at the Theater am Kärntnertor in Vienna for the premiere of his ninth and final symphony. It was Beethoven's first public performance in nearly a dozen years. The program announced that Beethoven would only assist in the direction, but as the theater filled and anticipation swept through the audience, he could not contain himself. According to first violinist Joseph Böhm, "Beethoven himself conducted, that is, he stood in front of a conductor's stand and threw himself back and forth like a madman. At one moment he stretched to his full height, at the next he crouched down to the floor, he flailed around with his hands and feet as if he wanted to play all the instruments himself and sing for the whole chorus."[1] Beethoven suffered from severe tinnitus—what he described as a roar in his ears—and by this time in his life was almost entirely deaf. Consequently, as the orchestra rang out their final triumphant note, he had unwittingly fallen a few measures behind and was still fiercely conducting. The contralto gently took hold of Beethoven's sleeve and turned him around to face the audience, handkerchiefs waving and loudly cheering. Beethoven wept. How could he have known that sounds he had heard only in his mind would strike a universal chord in the heart of humanity?

Our myths and religions reveal how our forebears collectively tried to make sense of the world. Embracing story, ritual, and belief, our traditions have sought—sometimes with compassion, sometimes with untold brutality—a narrative to explain the journey so far and to urge us onward from here. As individuals, we've been trekking the same path, relying on instinct and ingenuity to safeguard survival while seeking rhyme and reason for why we should care. Some on this journey would capture the coherence of reality in new and startling ways, offering reflections through works in literature, art, music, and science that would redefine our sense of self and enrich our relation to the world. The creative spirit, which had long since been chiseling figurines, coloring cave walls, and telling stories, was poised for flight.

Magnificent minds—rare but arising in every age, all shaped by nature and some by imagined inspiration from the divine—would discover new ways for articulating the transcendent. Their creative odysseys would express a variety of truth standing beyond derivation or validation, giving voice to defining qualities of human nature that remain silent until experienced.

To Create

Sensitivity to pattern ranks among our most potent survival skills. As we have seen repeatedly, we observe patterns, we experience patterns, and, most importantly, we learn from patterns. Fool me once, shame on you. Fool me twice, and while it may be premature to declare shame on me, by the third or fourth time, such a shift of responsibility is justified. To learn from pattern is an essential survival talent imprinted by evolution on our DNA. Alien visitors dropping by earth may subsist on different biochemistry, but they will likely have no difficulty grasping the concept; almost certainly, pattern analysis is central to how they have prevailed too.

Nevertheless, such intergalactic interchange may not be a perfect meeting of minds. Certain of our cherished patterns might leave our alien visitors baffled. Arrange particular pigments on a white canvas or chip away particular chunks from a marble mass or generate particular vibrations across jostling air molecules—yielding particular patterns of light and texture and sound—and upon encountering such patterns we humans can feel reality open in ways we never imagined possible. For a brief but seemingly boundless moment we can sense our place in the world shift as if we have been transported to another realm. If the aliens have had these types of experiences, they'll get what we're talking about. But when we recount our inner response to creative works, there is a chance they will stare at us blankly. And as language can go just so far in describing these experiences, the aliens may sport a bemused expression as they glance from continent to continent and see vast numbers of our species, some by themselves, others in groups, intently concentrating and absorbing and tapping and gyrating as they envelop themselves in worlds of art and music.

Baffled by our response to artistic expression, the alien visitors are likely to be just as baffled, perhaps more so, by the creation of such works. The blank page. The pristine canvas. The unformed mass of marble. The lump of clay. The unwritten score that awaits the composer's inspiration, or, once composed, waits to be played. Or sung. Or danced. Some of our species spend their days and nights imagining shapes to extract from the formless and sounds to pour into the silence. Some will expend the core of their life's energy realizing these imaginative visions, producing patterns in space and time that may be revered, or abhorred, or ignored, or deemed the very essence of existence. "Without music," said Friedrich Nietzsche, "life would be a mistake."[2] And, in the words of George Bernard Shaw's Ecrasia, "Without art, the crudeness of reality would make the world unbearable."[3] But what sparks the imaginative impulse? Is it catalyzed by behavioral instincts shaped by natural selection? Or have we long been expending precious resources of time and energy on artistic pursuits that have little connection to survival and reproduction?

We are thrust into the world without consultation. Once here, we are granted leave to embrace life for merely a moment. How elevating to grab the reins of creation and fashion something we control, something intrinsically ours, something that is a reflection of who we are, something that captures our peculiar take on human existence. While many among us would decline an opportunity to switch places with Shakespeare or Bach, Mozart or van Gogh, Dickinson or O'Keeffe, plenty would jump at the chance to be infused with their creative mastery. To illuminate reality with beacons of our own making, to move the world with works that flow through our particular molecular makeup, to craft experiences that can stand the test of time—well, it all sounds thoroughly romantic. For some, there is magic in the creative process, an irrepressible drive for self-expression. Others see an opportunity to elevate their status and esteem. For others still, there is a nod toward eternity; our artistic creations, as Keith Haring once said, are a "quest for immortality."[4]

If creating and consuming works of the imagination were a recent addition to human behavior, or if these activities were only rarely practiced across human history, it is unlikely that they would reveal universal qualities of our evolved human nature. After all, some things—like bell-bottoms and fried bananas—arise from contingent peculiarities, and so teasing out the details of their historical lineage offers only limited enlightenment. But the fact is, far into the past and clear across lands inhabited, we have been singing and dancing and composing and painting and sculpting and carving and writing. Cave paintings and elaborate burial goods, as encountered in the previous chapter, date from as far back as thirty to forty thousand years ago. Etchings and artifacts that show evidence of artistic expression have been discovered from a few hundred thousand years earlier.[5] We are faced with a behavior that is pervasive and yet, unlike eating and drinking and procreating, doesn't wear its survival value on its sleeve.

With a modern sensibility, this may not strike you as puzzling. To experience a work that enlivens the soul or moves us to tears is to go beyond the humdrum of the everyday, and who wouldn't thrill to an experience like that? But as with the superficial observation that we eat ice cream because we like sweet things, this explanation is focused solely on our proximate responses and hence is limited to the most immediate impetus for creative inclinations. Can we go deeper? Can we gain insight into why our forebears were so willing to turn from the all-too-real challenges of survival and expend precious time, energy, and effort engaging the imaginative?

Sex and Cheesecake

When we encountered our early brethren telling stories, we considered a similar question, and the most convincing answer invoked the flight simulator metaphor: through the creative use of language we have experienced perspectives familiar and foreign, allowing us to broaden and refine our responses to encounters in the real world. By telling stories and hearing stories and embellishing stories and repeating stories, we played with possibility without suffering consequences. We followed trail upon trail that began with "What if?" and, through reason and fantasy, explored a wealth of possible outcomes. Our minds freely roamed the landscape of imagined experience, giving us a newfound nimbleness of thought that, plausibly, proved valuable for survival.

As we consider more abstract forms of art, this explanation needs to be revisited. It's one thing to envision the mind burnishing the ideals of courage and heroism through riveting tales of hard-won battles or spellbinding accounts of treacherous journeys. It's seemingly quite another to argue that the mind exercised an adaptive muscle by listening to the Pleistocene's Édith Piaf or Igor Stravinsky. There is a seemingly yawning chasm between experiencing music—or, for that matter, painting or dancing or sculpting—and surmounting challenges encountered in the ancestral world.

Darwin himself considered the potential adaptive function of an innate artistic sense motivated by the famous evolutionary puzzle of the peacock's tail. A large and brightly colored tail makes it a challenge for a peacock to hide and, when chased by a fast-approaching predator, makes it a challenge to escape. Why would such a grand, beautiful, but apparently maladaptive structure evolve? The answer, Darwin concluded after much consternation, is that while the peacock's tail can be a ball and chain in the struggle for survival, the tail is nevertheless an essential part of the peacock's reproductive strategy. It is not only we humans who find the peacock's tail appealing. Peahens do too. They are attracted to sprightly plumes, and so the more impressive his tail, the more likely the peacock will mate. The resulting progeny, in turn, stand a good chance of inheriting dad's traits and mom's tastes, propagating a genetic war in which battles are won not by acquiring more food or ensuring greater safety but by growing more resplendent tails.

It is an example of **sexual selection,** a Darwinian evolutionary mechanism whose cogs are driven by reproductive access. A peacock who dies young will fail to reproduce, the very reason natural selection favors those who survive. But the same reproductive failure will befall a peacock who lives long and prospers yet is shunned by all potential mates. To influence the biological makeup of subsequent generations, survival is necessary but not sufficient. Producing offspring is what matters, and so characteristics that promote mating will enjoy a selective advantage, sometimes even at the expense of safety.[6] Such costs cannot be astronomical—there is a limit to how unwieldy tails can be before survival would be utterly imperiled—but need not be free. And though the peacock's tail is the go-to example, similar considerations are applicable across a great many species. White-bearded manakins strut their moves in raucous mosh pit dances to entice potential mates; fireflies flash hypnotic courtship displays with success turning on the finesse of their flitting light shows; male bowerbirds construct elaborate bachelor dens, entwining twigs, leaves, shells and even colorful candy wrappers in an ostentatious display that apparently serves no other purpose than to seduce a future Mrs. Bowerbird.[7]

When Darwin first described sexual selection in his 1871 two-volume book **The Descent of Man,**

and Selection in Relation to Sex, the proposal was not an instant hit. To many of his contemporaries, it seemed inconceivable that behavior in the brutish realm of nonhuman animals might hinge on aesthetic responses.[8] Not that Darwin was imagining birds or frogs lost in poetic reverie, gazing at the sun's reddish rays as it dips below the horizon. The aesthetic sense he proposed was focused solely on mate selection. Even so, Darwin's ascription of a "taste for the beautiful"[9] to a broad swath of the animal kingdom seemed cavalier. Heck, to Alfred Russel Wallace, who viewed human aesthetic sensibilities to be a gift from God, it was unseemly.[10]

But if we don't invoke an innate sensitivity to beauty, how do we explain the lavish bodily adornments, creative displays, and physical constructions that are integral to myriad mating games playing out in the animal kingdom? Well, there is a less lofty approach. Consider again the peacock's tail. While we humans may appreciate the aesthetics of a peacock's plumage, to a peahen it may arouse an instinctual response of considerable genetic importance. Peacocks adorned with dazzling plumage are strong and healthy, increasing the likelihood that they will sire hardy offspring. And since peahens, much as the females in most species, can produce far fewer progeny compared to their male counterparts, they have developed an especially strong preference for fit males; such unions enhance the success rate of each resource-consuming and hence precious fertilization.[11] With rich plumage being a visible demonstration of a potential mate's strength and vigor, peahens attracted to such tails are more likely to spawn robust peachicks. These peachicks, in turn, will on average be endowed with the very genes for desiring and acquiring resplendent plumage, facilitating the spread of such traits through future generations. Beauty, in this analysis of sexual selection, is a good deal more than skin-deep. Beauty amounts to publicly available credentials attesting to a potential mate's adaptive fitness.

In either case—whether mate choice is driven by aesthetic sensitivities or by health evaluations—the resulting preferences can provide a rationale for costly traits, bodily and behavioral, whose intrinsic survival benefits are questionable. As this description seems applicable to our species' long-standing and essentially universal artistic practices, perhaps sexual selection offers illumination. Darwin thought it might. He invoked sexual selection to explain the human penchant for bodily piercings and colorations and suggested as well that the powerful response music can elicit is an evolutionary outcome of sexual selection shaping human mating calls. Males who could best sing or dance, or had the most alluring tattoos or decorated garments, may have been the target of choosy females and so more readily sired artistically attuned progeny. In boy meets girl, artistic talent may have determined whether boy went home alone.

More recently, psychologist Geoffrey Miller, and also philosopher Denis Dutton, have developed this perspective further, suggesting that human artistic capacities provide a fitness indicator perused by discerning females.[12] Not only do expertly crafted artifacts, creative displays, and energetic performances demonstrate a mind and body that is firing on all cylinders, but such works also attest to the artist being generously endowed with the right stuff for survival. After all, the reasoning goes, only by virtue of possessing material resources and physical prowess could the artist afford the extravagance of expending time and energy on activities that lack survival value. (Artists of the Pleistocene, apparently, were anything but starving.) In this view, artistic undertakings amount to a self-promoting marketing strategy that results in unions between talented artists and discriminating mates, yielding progeny more likely than not to be endowed with similar traits.

Sexual selection as the evolutionary driver of human artistic activity is intriguing, but has generated more strife than accord. Researchers have raised many issues: Is artistic talent an accurate signal for physical health? Might artistic capacities be so entwined with raw intelligence and creativity, qualities with unassailable survival value, that artistic predilections spread via natural selection with no need to invoke sexual selection? With sexual selection's focus on male artists, how does the theory explain the artistic activities of females? And perhaps most challenging of all, public engagement with artistic activities during the Pleistocene as well as that era's courtship rituals and mating practices are largely a matter of conjecture. Sure, the conquests of Lucian Freud and Mick Jagger may be legendary, but what if anything does that tell us about the importance of artistic skill or stage presence for reproductive access among early hominins? In light of such concerns, Brian Boyd has offered a considered summary: "Sexual selection has been an extra gear for art, not the engine itself."[13]

Steven Pinker suggests a wholly different perspective on the adaptive utility of the arts. In a passage that has been quoted frequently by supporters as well as detractors, he argues that all but the language arts amount to nutritionally bankrupt desserts served up to pattern-obsessed human brains. Much as "cheesecake packs a sensual wallop unlike anything in the natural world because it is a brew of megadoses of agreeable stimuli which we concocted for the express purpose of pressing our pleasure buttons,"[14] the arts, according to Pinker, are adaptively useless creations designed to artificially excite human senses that evolved to promote the fitness of our ancestors. This is not a value judgment. Pinker's sharply crafted arguments, brimming with cultural allusions, make clear that he has a deep affection for the arts. Instead, this is a dispassionate assessment of whether the arts have played a role in one particular task: enhancing the prospect that in the ancestral world the genes of our forebears, and not those of our artless, tone-deaf, left-footed, philistine cousins, were passed on to the next generation. And it is toward this one end that Pinker argues that the arts are irrelevant.

Evolution has surely coaxed us toward a raft of behaviors aimed at increasing our biological fitness, from finding food, securing mates, and ensuring safety to establishing alliances, fending off adversaries, and instructing progeny. Heritable behaviors that, on average, resulted in greater reproductive success spread widely and became the go-to mechanisms for surmounting particular adaptive challenges. In shaping some of these behaviors, one carrot evolution wielded was pleasure: if you find particular survival-promoting behaviors pleasurable, you will be more likely to undertake them. And by virtue of their survival-promoting qualities, these behaviors will increase the likelihood that you'll stick around long enough to reproduce, endowing future generations with similar behavioral tendencies. Evolution thus generates a collection of self-reinforcing feedback loops that renders pleasurable those behaviors that enhance fitness. In Pinker's view, the arts cut the feedback loops, sever adaptive benefits, and directly stimulate our pleasure centers, yielding gratifying experiences that from an evolutionary perspective are unearned. We like how the arts can make us feel, but neither creating nor experiencing them makes us more fit or appealing. From the standpoint of survival, the arts are junk food.

Music is Pinker's poster child, the genre of the arts whose adaptive irrelevance he lays out most fully. He suggests that music is an auditory parasite, free riding on emotionally evocative aural sensitivities that long ago provided survival value to our forebears. For example, sounds whose frequencies are harmonically related (frequencies that are multiples of a common frequency) indicate a single and potentially identifiable source (basic physics reveals that when a linear object vibrates, whether a predator's vocal cords or a weapon made from hollowed bone, the vibrational frequencies tend to fill out a harmonic series). Those of our forebears who responded more pleasurably to such organized sounds would have paid them more attention and thus garnered greater awareness of their environment. The heightened cognizance would have tilted the survival scales in their favor, enhancing their well-being and promoting the further development of auditory sensitivity. Increased receptivity to other information-rich sounds, from thunder to footfalls to cracking branches, would have further sharpened attentiveness and thus filled out environmental awareness yet further. And so those of our ancestors who were more sonically attuned possessed a fitness advantage, promoting the spread of aural sensitivity throughout subsequent generations. According to Pinker, music hijacks such sonic sensitivity and takes it for a joyride of sensual pleasure that confers no adaptive value. Much as cheesecake artificially stimulates our ancient adaptive preference for foods with elevated caloric content, music artificially stimulates our ancient adaptive sensitivity to sounds with elevated information content.

Pinker's juxtaposition of guilty pleasure with rarefied experience is jarring. Intentionally so. The point is not to demean our experience of art but to broaden our assignment of significance. To be sure, there is something thoroughly satisfying in identifying an evolutionary basis for this or that human behavior, providing an indelible stamp of approval imprinted in our DNA. How gratifying to imagine that the arts, deemed by many to rank among humankind's most exalted achievements, have played an essential part in the very survival of the species? But however pleasing, such an explanation need not be true. Nor essential. Biological adaptation is not the sole standard for value. It is just as wonderful that we can lift ourselves above concerns for survival and use imagination to express something beautiful or disturbing or heartrending. Significance does not require adaptive utility. Years ago during a family dinner at a local restaurant, as a waiter delivered cheesecake to a nearby table, my mother, who was constantly dieting, felt compelled to stand and salute, a gesture of respect that can apply not only to the dessert itself but to pervasive human behaviors that, in Pinker's view, have garnered that dessert's adaptive classification.

Imagination and Survival

The recognition that the arts need feel no shame for lacking adaptive utility has not dissuaded researchers from continuing to seek straightforward Darwinian explanations for their endurance and ubiquity. Explanations, that is, that attempt to directly link artistic activities with the survival of our forebears. In this pursuit, anthropologist Ellen Dissanayake has stressed the need to consider the arts as they were practiced in ancestral contexts, arguing that across human history, art, and religion too, were not extracurricular diversions "to be indulged one morning each week or when there was nothing better to do, nor were they superfluous pastimes that could be rejected altogether."[15] Whether descending deep into the underworld to adorn a cave wall or wildly drumming, dancing, and singing into an otherworldly trance, art, like religion, was woven into the tapestry of ancient existence. And therein lies a potential adaptive role.

If aliens visited Paleolithic earth and wagered on who'd be top dog a million years later, genus **Homo** may not have inspired many bets. However, by pooling brawn and brain, we were able to prevail over forms of life larger, stronger, and faster, as well as those endowed with more refined senses of smell, sight, and sound. We triumphed because we are resourceful and creative, certainly, but above all because we are exceptionally social. In earlier chapters we discussed a number of mechanisms, from storytelling to religion to game theory, which may have facilitated our capacity to gather in productive groups. But because such behavior is as complex as it is influential, seeking a single explanation may be too narrow. Various amalgams of these mechanisms may have been important to our successful groupish tendencies, and as Dissanayake and other researchers have suggested, the list of prosocial influences should be extended to include art.

If you and I have confidence that we will each understand and anticipate the emotional responses of the other—even as we encounter unfamiliar challenges and pursue novel opportunities—there is a better chance we will cooperate successfully. The arts may have been essential to achieving this. Were you and I and others in our group frequent participants in the same ritualized artistic experiences, joining together through energetic rhythm, melody, and movement, the unity of such intense emotional journeys would have created a sense of communal solidarity. Anyone who has taken part in prolonged group drumming, singing, or movement knows the feeling; if you haven't, I highly recommend it. Intense and seemingly larger than life, these shared emotional episodes would have melded us into a more committed whole. As Noël Carroll, a philosopher who has also been at the forefront of these ideas, has emphasized, "Art has been about stirring up and shaping the emotions in a way that binds and inculcates those under its sway as participants in a culture."[16] And indeed, the very notion of culture—a broadly shared set of traditions, customs, and perspectives—relies on a common heritage of artistic practice and experience. Members of such emotionally attuned groups had a better chance of surviving and passing on a genetic tendency for such behaviors to subsequent generations.

Now, were you unmoved by group cohesion as an adaptive explanation for religion, you may be just as unmoved by group cohesion as an adaptive explanation for art. But much as in our discussion of religion, we don't need to solely focus on groups. Art may have had adaptive utility directly at the level of the individual, a perspective I find particularly compelling. The arts provide an arena unbounded by the strictures of flat-footed truth and everyday physical reality, allowing the mind to jump and twist and tumble as it explores all manner of imagined novelty. A mind that assiduously sticks to what's true is a mind that explores a wholly limited realm of possibility. But a mind that becomes accustomed to freely crossing the boundary between what's real and what's imagined—all the while keeping clear tabs on which is which—is a mind that becomes adept at breaking the bonds of conventional thinking. Such a mind is primed for innovation and ingenuity. History makes this manifest. We owe many of the greatest break-throughs of science and technology to a collection of individuals who were able to look at the very same problems that had confounded generations of previous thinkers and have the flexibility of thought to see those problems differently.

Einstein's essential step toward relativity was not driven by new experiments or data. He was work-ing with facts—to do with electricity, magnetism, and light—that were already well-known. Instead, Einstein's bold move was to break free from the widely held assumption that space and time were constant, which required the speed of light to vary, and in its place envision that the speed of light is constant, which required space and time to vary. This slogan-like summary is not meant to explain special relativity (for that, I refer you, for example, to chapter 2 of **The Elegant Universe**) but rather to note that the discovery relied on imagining a simple but fundamental rearrange-ment of the Lego pieces of reality, an inversion of symbolic patterns so familiar that most minds glided over the possibility entirely. It is a variety of creative maneuver that resonates with the highest levels of artistic composition. In the assessment of illustrious pianist Glenn Gould, the genius of Bach is demonstrated by his ability to devise melodic lines "which when transposed, inverted, made retrograde, or transformed rhythmically will yet exhibit . . . some entirely new but completely harmonious profile."[17] Einstein's genius rested on a similar, and similarly uncanny, ability to reconfigure the building blocks of understanding, looking anew upon concepts that had been scrutinized for decades, if not centuries, and combining them according to a novel blueprint. That Einstein described his intellectual process as thinking with music and that he frequently relied on visual explorations free of equations and words perhaps isn't all that surprising. Einstein's art was to hear rhythms and see patterns that revealed deep unity in the workings of reality.

Neither Einstein's relativity nor Bach's fugues are such stuff as survival is made on. Yet each is a consummate example of human capacities that were essential to our having prevailed. The link between scientific aptitude and solving real-world challenges may be more apparent, but minds that reason with analogy and metaphor, minds that represent with color and texture, minds that imagine with melody and rhythm are minds that cultivate a more flourishing cognitive landscape. Which is all just to say that the arts may well have been vital for developing the flexibility of thought and fluency of intuition that our relatives needed to fashion the spear, to invent cooking, to harness the wheel, and, later, to write the Mass in B Minor and, later still, to crack our rigid perspective on space and time. Across hundreds of thousands of years, artistic endeavors may have been the playground of human cognition, providing a safe arena for training our imaginative capacities and infusing them with a potent faculty for innovation.

Note too that the adaptive roles for art we've considered—sharpening innovation and strengthening social bonds—work in tandem. Innovation is the foot soldier of creativity. Group cohesion is the army of implementation. Success in the relentless battle for survival requires both: creative ideas that are successfully implemented. That the arts stand at the nexus of the two suggests an adaptive role beyond the mere pushing of pleasure buttons. Sure, it's possible that the arts are an adaptively inconsequential yet profoundly pleasing by-product of a large brain hosting a creative mind, but to many researchers that gives insufficient heed to art's capacity to sculpt our engagement with reality. Brian Boyd has made this point succinctly: "By refining and strengthening our sociality, by making us readier to use the resources of the imagination, and by raising our confidence in shaping life on our own terms, art fundamentally alters our relation to our world."[18]

I'm partial to the view that sharpening ingenuity, exercising creativity, broadening perspective, and building cohesion provides a template for how the arts mattered to natural selection. With this perspective the arts join language, story, myth, and religion as the means by which the human mind thinks symbolically, reasons counterfactually, imagines freely, and works collaboratively. Over the sweep of time, it is these capacities that have given rise to our culturally, scientifically, and technologically rich world. All the same, even if your view of art's evolutionary role veers toward creamy desserts, we can surely agree that myriad forms of art have been a steady and valued presence throughout human history. Which means that inner lives and social exchanges have embraced modes of engagement that do not place a premium on factual information conveyed through language.

What does this tell us about art and truth?

Art and Truth

About twenty years ago, on one of those gloriously sunny fall days with leaves turning red and burnt orange, I was driving alone on a highway from New York City to our family home upstate when, seemingly out of nowhere, a dog darted across the road. I slammed on the brakes, but a moment before the car finally stopped I felt a jarring thud closely followed by another, as the front and then rear wheels ran the dog over. Jumping from the car, I hoisted the dog, awake but hardly moving, into the passenger seat and sped off along country roads in search of a veterinarian. Minutes later, somehow, the dog sat up straight. I put my hand lightly on her head, which she pinned with her body against the seat as she slumped back. I pulled over. She looked up with an unblinking intensity. Pain. Terror. Resignation. A mixture of it all it seemed. Then, pressing her body harder against my hand, as if she couldn't bear leaving alone, she died.

I've had pets that have died. This was different. Sudden. Forceful. Violent. In time, the shock wore off, but the final moment stayed with me. My rational self knows that I'm reading undue meaning into an unfortunate but all-too-common occurrence. Still, the transition from life to death of an animal I encountered by chance and who had died by my own hand, albeit unintentionally, had an uncanny and unexpected pull on me. It carried with it a certain kind of truth. Not a propositional truth. Not a matter of fact. Nothing I could meaningfully measure. But in that moment, I felt something slightly shift in my sense of the world.

I can identify a small collection of other experiences that, each in their own distinctive way, have left me with a similar feeling. Holding my first child for the first time; crouching in a rocky crevice in the hills outside of San Francisco as a howling windstorm raged overhead; hearing my young daughter singing solo at a school gathering; suddenly solving an equation that had resisted months of previous attempts; watching from a bank along the Bagmati River as a Nepalese family performed the ritual burning of a deceased family member; skiing—no, flailing—down a double-diamond slope in Trondheim, and somehow surviving. You have your own list. We all do. Experiences that fully lock our attention and spark emotional responses we value even in the absence—or perhaps because of the absence—of a fully rational or linguistic description. What's curious, although likely common, is that while my own working process is thoroughly language based, I feel no urge to explore these experiences in words. When I think of them, I feel no lack of understanding calling out for linguistic clarification. They expand my world without need for interpretation. These are the times that my inner narrator knows it's time to take a break. An examined life need not be an articulated life.

The most arresting art can induce in us rarefied states of mind and body comparable to those produced by our most affecting real-world encounters, similarly molding and enhancing our engagement with truth. Discussion, analysis, and interpretation can further shape these experiences, but the most potent do not rely on a linguistic intermediary. Indeed, even for language-based arts, it is the imagery and sensations that, in the most moving experiences, leave the most lasting mark. As elegantly described by poet Jane Hirshfield, "When a writer brings into language a new image that is fully right, what is knowable of existence expands."[19] Nobel laureate Saul Bellow speaks too to art's singular capacity for expanding the knowable: "Only art penetrates what pride, passion, intelligence and habit erect on all sides—the seeming realities of this world. There is another reality, the genuine one, which we lose sight of. This other reality is always sending us hints, which without art, we can't receive." And without that other reality, Bellow notes, channeling thoughts set down by Proust, existence is reduced to a "terminology for practical ends which we falsely call life."[20]

Survival rests upon amassing information that accurately describes the world. And progress, in the conventional sense of increased control over our surroundings, requires a clear grasp of how these facts integrate into nature's workings. Such are the raw materials for fashioning practical ends. They are the basis for what we label objective truth and often associate with scientific understanding. But however comprehensive such knowledge may be, it will always fall short of providing an exhaustive account of the human experience. Artistic truth touches a distinct layer; it tells a higher-level story, one that in the words of Joseph Conrad "appeals to that part of our being which is not dependent on wisdom" and speaks instead to "our capacity for delight and wonder, to the sense of mystery surrounding our lives; to our sense of pity, and beauty, and pain; to the latent feeling of fellowship with all creation . . . in dreams, in joy, in sorrow, in aspirations, in illusions, in hope, in fear . . . which binds together all humanity—the dead to the living and the living to the unborn."[21]

Released from rigid verisimilitude and developing over the course of millennia, the creative instinct has amply explored the emotional range that marks Conrad's vision of the artistic journey and provides the vernacular in which Bellow's genuine reality whispers to us from just around the bend. Writers, in particular, have crafted world upon world of characters whose fictive lives provide heightened studies in human engagement. Odysseus and the fraught journey of vengeance and loyalty, Lady Macbeth and the claws of ambition and guilt, Holden Caulfield and the irrepressible rebellious instinct, Atticus Finch and the power of quiet but unshakable heroism, Emma Bovary and the tragedies of human connection, Dorothy and the winding road of self-discovery—the insights these works provide into the varieties of experience, the artistic truths they develop, add shadow and dimension to an otherwise rough sketch of human nature.

Visual and auditory works, in which language is not central, provide experiences that are more impressionistic. Yet, as with their literary counterparts, if not more so, they can spark the very same emotions that, as Conrad described, stand beyond wisdom; the voices inhabiting Bellow's genuine reality speak to us in varied ways. I can't listen to Franz Liszt's **Totentanz** without a visceral foreboding; Brahms's Third Symphony conjures a deep, unsated longing; the Bach **Chaconne** is an apotheosis of the sublime; the "Ode to Joy" finale of Beethoven's Ninth Symphony is for me and, of course, much of the world too, among the most optimistic statements the species has ever offered. Including music with lyrics, Leonard Cohen's "Hallelujah" praises the imperfect life with incomparable authenticity; Judy Garland's simple and exquisite rendition of "Over the Rainbow" captures the pure yearnings of youth; John Lennon's "Imagine" embodies the simple power of envisioning the possible.

As with life's punctuated moments, we each can bring to mind works, whether in literature or film, sculpture or choreography, painting or music, that in one way or another have moved us. Through these captivating experiences, we consume "megadoses" of essential qualities of human life on this planet. But far from empty calories, these heightened encounters provide insights that would be difficult if not impossible to otherwise acquire.

The lyricist Yip Harburg, author of many classics including "Over the Rainbow," said it simply: "Words make you think a thought. Music makes you feel a feeling. But a song makes you feel a thought."[22] **Feel a thought.** For me, that captures the essence of artistic truth. As Harburg emphasized, thinking is intellectual, feeling is emotional, but "to feel a thought is an artistic process."[23] It is an observation that rests on linking language and music but, really, it yokes the arts more generally. The emotional responses elicited by art ripple across the reservoir of churning thought that underlies conscious awareness. For works without words, these experiences are less directed and the feelings more open-ended. But all art has the capacity to make us feel thoughts, yielding a variety of truth we would be unlikely to anticipate from conscious deliberation or factual analysis. A variety of truth that does indeed stand beyond wisdom. Beyond pure reason. Beyond the reach of logic. Beyond the necessity for proof.

Make no mistake. We **are** all bags of particles— both mind and body—and the physical facts about the particles can fully address how they interact and behave. But such facts, the particulate narrative, shed only monochrome light on the richly colored stories of how we humans navigate the complex worlds of thought, perception, and emotion. And when our perceptions blend thought and emotion, when we feel thoughts as well as think them, our experience steps yet further beyond the bounds of mechanistic explanation. We gain access to worlds otherwise uncharted. As Proust emphasized, this is to be celebrated. Only through art, he noted, can we enter the secret universe of another, the only journey in which we truly "fly from star to star," a journey that cannot be navigated by "direct and conscious methods."[24]

Although focused on the arts, Proust's perspective resonates with my own long-held take on modern physics. "The only true voyage of discovery," he once said, "would be not to visit strange lands but to possess other eyes, to see the universe through the eyes of another, of a hundred others."[25] For centuries, we physicists have relied on mathematics and experiment to reshape our eyes, to reveal layers of reality untouched by generations of the past, to allow us to see familiar landscapes in shocking new ways. With these tools, we have found that the strangest of lands have emerged by intently examining the very realms we have long inhabited. All the same, to acquire such knowledge and to utilize the power of science more generally, we must follow the unshakable directive to look past the peculiarities of how each of our distinct collections of molecules and cells takes in the world, and home in on objective qualities of reality. For the rest, the all-too-human truths, our nested stories rely on art. As George Bernard Shaw put it, "You use a glass mirror to see your face, you use works of art to see your soul."[26]

Poetic Immortality

Not too infrequently, I'm asked for the single fact about the universe I find most mind-blowing. I don't have a stock answer. Sometimes I suggest relativity's malleability of time. On other occasions I suggest quantum entanglement, what Einstein called "spooky action at a distance." But sometimes I go simpler and suggest something most of us first encountered as schoolkids. When we look up at the night sky we see stars as they were many thousands of years ago. Using powerful telescopes, we see far more distant astronomical objects as they were millions or billions of years ago. Some of these astronomical sources may have long since died, and yet we continue to see them because light they long ago emitted is still in transit. Light provides an illusion of presence. And not just for stars. Undisturbed, reflected beams of radiation carry your imprint and mine across an arbitrary expanse of space and time, a poetic immortality racing across the cosmos at the speed of light.

Back here on earth, poetic immortality takes a different form. The yearning to hold on to life for as long as we choose has not been requited, at least not yet and perhaps never will be. But the creative mind, able to roam freely through imagined worlds, can explore the immortal, meander through eternity, and meditate on why we might seek or disdain or fear endless time. For millennia, artists have done just that. Some twenty-five hundred years ago, the Greek lyric poet Sappho lamented the inevitability of change, "You, children, pursue the violet-laden Muses' lovely gifts / and the clear-toned lyre so dear to song; / but for me—old age has now seized my once tender body," tempered by reference to the cautionary tale of Tithonus, a mortal granted immortality by the gods but still subject to the ravages of age, now endured for eternity. A final line that some scholars believe to be the true ending of the poem—"Eros has granted to me the beauty and the brightness of the sun"—suggests that through her passionate pursuit of life, expressed through her poetry, Sappho anticipated transcending decay and achieving ageless radiance; through her poetry, she imagined attaining a symbolic immortality.[27]

It is a version of a death-denying schema in which we mortals seek to live on through our heroic achievements, influential contributions, or creative works. The scale of such immortality requires an anthropocentric adjustment, from eternity to the duration of civilization—a significant cost, but one offset by the recognition that unlike its literal counterpart, the symbolic version of immortality is real. The only issue is one of strategy. Which lives will be remembered? Which works will last? And how to ensure that our lives and our works will be among them?

A couple of millennia after Sappho, Shakespeare contemplated the role of art and the artist in shaping what the world remembers. Addressing the subject of an epitaph he imagines composing, Shakespeare notes, "When all the breathers of this world are dead / You still shall live, such virtue hath my pen," a benefit, Shakespeare asserts, that he himself will not enjoy: "Your name from hence immortal life shall have / Though I, once gone, to all the world must die." Of course, we're in on Shakespeare's game: as it is the poet's words that will be read and recited, the subject of the epitaph is but a vehicle for the poet to achieve immortality, albeit symbolically. Indeed, centuries later, it is Shakespeare who lives on.

After leaving Freud's Vienna Circle, Otto Rank developed his thesis that the pursuit of symbolic immortality is a primary driver of human behavior. In Rank's view, the artistic impulse reflects the mind taking charge of its fate, having the courage to rework reality, and embarking on the lifelong project of shaping its own idiosyncratic self. The artist moves toward psychic health by accepting mortality—we're going to die, that's that, get over it—and shifting the urge for eternity onto a symbolic form carried by creative works. This perspective casts the clichéd image of the tortured artist in a different light. According to Rank, coping with mortality through creating art is a pathway to sanity. Or, as the writer and critic Joseph Wood

Krutch similarly described, "Man needs eternity, as the whole history of his aspirations bears witness; but the eternity of art is, in all probability, the only sort he will ever get."[28]

Could this dynamic have been at work tens of thousands of years ago, shedding light on why we diverted energy to activities that stand apart from the immediate needs of sustenance and shelter? Could it explain why, across millennia, artistic pursuits have remained central threads in the fabric of all human cultures? Yes and yes. Whether or not Rank's all-encompassing vision hits the bull's-eye, we can well imagine our ancient forebears sensing their own mortal nature, longing to clutch hold of their world and stamp it with something iconic, something self-authored, something lasting. We can well imagine that urge interrupting an otherwise diligent focus on survival and over time being reinforced and refined by the communal delight in joining the artist in imaginative worlds sprung from the human mind.

While the paucity of evidence reduces analysis of our distant past to informed guesswork, here in the modern age we encounter one work after another reflecting deeply on mortality and eternity.[29] Walt Whitman pondered the intolerability of granting finality to death: "Do you suspect death? If I were to suspect death I should die now. / Do you think I could walk pleasantly and well-suited toward annihilation? . . . / I swear I think there is nothing but immortality!" To William Butler Yeats, the ancient city of Byzantium was a destination where he might be released from his dying physical form, liberated from humanly concerns, and given leave to enter a timeless realm: "Consume my heart away; sick with desire / And fastened to a dying animal / It knows not what it is; and gather me / Into the artifice of eternity."[30] Herman Melville made plain that mortality sails along with us even when rough waters seem to have subsided: "All are born with halters round their necks; but it is only when caught in the swift, sudden turn of death, that mortals realize the silent, subtle, ever-present perils of life."[31] Edgar Allan Poe took death denial to a literary extreme giving voice to victims of premature burial fighting off death's most intimate embrace: "I shrieked with horror: I plunged my nails into my thighs and wounded them; the coffin was soaked in my blood; and by tearing the wooden sides of my prison with the same maniacal feeling I lacerated my fingers and wore the nails to the quick, soon becoming motionless from exhaustion."[32] Tennessee Williams, through the fictional patriarch Big Daddy Pollitt, noted that "ignorance—of mortality—is a comfort. A man don't have that comfort, he's the only living thing that conceives of death," and in consequence, "if he's got money he buys and buys and buys and I think the reason he buys everything he can buy is that in the back of his mind he has the crazy hope that one of his purchases will be life everlasting!"[33]

Dostoevsky, through his character Arkády Svidrigáylov, aired a different perspective, one weary of the reverence commanded by eternity: "Eternity is always presented to us as an idea that we can't grasp, as something enormous, enormous! Why does it have to be enormous? All of a sudden, instead of all that, imagine there'll be a little room, something like a country bathhouse, sooty, with spiders in all the corners, and that's the whole of eternity. You know, I sometimes imagine it like that."[34] It's a sentiment expressed too by Sylvia Plath, "O God, I am not like you / In your vacuous black / Stars stuck all over, bright stupid confetti / Eternity bores me, I never wanted it,"[35] and picked up lightheartedly by Douglas Adams through his accidental immortal, Wowbagger the Infinitely Prolonged, who plans to deal with his profound ennui by systematically insulting everyone in the universe, one by one, in alphabetical order.[36]

This range of dispositions, from longing to disdaining, demonstrates the larger point: our recognition of the limited time we are allotted has driven an artistically vibrant engagement with the concept of eternity. The examined life examines death. And for some, to examine death is to free the imagination to challenge its dominance, dispute its eminence, and conjure realms that lie beyond its reach. However intently researchers argue about their evolutionary utility, their role in building social cohesion, their necessity for innovative thinking, and their standing in the pantheon of primal urges, the arts provide our most evocative means for giving expression to the things we deem matter most—and among such things are life and death, the finite and the infinite.

For many, including me, the most concentrated of such expressions are provided by music. Music can offer an immersion so enveloping that within just a few brief moments it feels like we have stepped beyond time. Cellist and conductor Pablo Casals described the power of music to "inform ordinary activities with spiritual fervor, to give wings of eternity to that which is most ephemeral."[37] It is a fervor that makes us feel part of something larger, something that viscerally affirms Conrad's "invincible conviction of solidarity that knits together the loneliness of innumerable hearts."[38] Whether with the composer or fellow listeners or through a more abstract sort of communion altogether, music invites connection. And it is through such connection that the experience of music transcends time.

Back in the late 1960s, the third graders in Mrs. Gerber's class at P.S. 87 in Manhattan were asked to interview an adult of their choosing and write a short report explaining the interviewee's occupation. I took the easy way out and interviewed my dad—a composer and performer who was fond of citing his academic imprimatur, an "SPhD"

(Seward Park High School dropout). Partway through the tenth grade, my dad ditched the books and hit the road, singing, playing, and performing around the country. It has been more than a half century since that grade-school assignment, but one thing he mentioned has never left me. When I asked why he chose music, my dad answered, "To keep away the loneliness." He swiftly transitioned to a brighter tone, more suited to a third-grade report, but that uncensored moment was revealing. Music was his lifeline. It was his version of Conrad's solidarity.

Few composers move the world. My dad was not among them, a painful realization he slowly grew to accept. The melodies and rhythms handwritten across hundreds of yellowing manuscript pages, many from before I was born, are now of little interest to anyone but family. I am perhaps the sole remaining person who, from time to time, still listens to the ballads and songs and piano works he composed as far back as the 1940s and 1950s. For me, these compositions are a treasure, a connection that allows me to feel my dad's thoughts from a time when he was just beginning to make his way in the world.

Music has the remarkable power to create such profound connection even among those not bound by family, living in different times, inhabiting different realms. A moving description comes from Helen Keller, one of history's singular heroes. On

February 1, 1924, radio station WEAF in New York City broadcast the New York Symphony Orchestra's live performance of Beethoven's Ninth Symphony. At home, Helen Keller placed her hands on the diaphragm of an uncovered radio speaker and through the vibrations was able to sense the music, to experience what she called the "immortal symphony," even distinguish individual instruments. "When the human voice leaped up trilling from the surge of harmony, I recognized them instantly as voices. I felt the chorus grow more exultant, more ecstatic, upcurving swift and flame-like, until my heart almost stood still." And then, speaking to sounds that touch the spirit, music that reverberates to eternity, she concludes:

As I listened, with darkness and melody, shadow and sound filling all the room, I could not help remembering that the great composer who poured forth such a flood of sweetness into the world was deaf like myself. I marveled at the power of his quenchless spirit by which out of his pain he wrought such joy for others—and there I sat, feeling with my hand the magnificent symphony which broke like a sea upon the silent shores of his soul and mine.[39]

9

DURATION AND IMPERMANENCE

From the Sublime to the Final Thought

Every culture has a notion of the timeless, a revered representation of permanence. Immortal souls, sacred stories, illimitable gods, eternal laws, transcendent art, mathematical theorems. Yet, spanning categories from the other-worldly to the thoroughly abstract, permanence is something we humans covet but never attain. The closest we come—a sense of time having dropped away, whether the result of a euphoric or tragic encounter, a meditative or chemical inducement, an exalted religious or artistic experience—can provide life's most formative experiences.

Decades ago, together with eight other teenagers, I was on a survival course in the deep woods of Vermont. Late one night after we were all asleep in our tents, the course leaders bellowed for us to get up and dress quickly. We were heading out on an impromptu night hike. Holding hands and walking single file through the blackness, we slowly negotiated dense forest, thick brush, and, of particular delight, a waist-deep mud swamp. Wet, freezing, and covered in muck, we were finally led to a nearby clearing where, we were informed, the nine of us would be left for the night with nothing but three sleeping bags. Realizing the futility of our protests, however intensely delivered, we zipped the sleeping bags together, stripped down, and huddled closely under the makeshift duvet. Many cursed, others vowed to quit the course early, a few cried. But then there was the most wondrous sight. A brilliant aurora borealis filled the night sky. I had never seen anything like it. The swirling gossamer strands of light, the stunning colors bleeding one into another, all set against a backdrop of seemingly endless, uncountable stars. Suddenly, I was in a different place. The hike, the swamp, the cold, the near-naked huddling—it was all now part of a primordial throwback. Man, nature, universe. While I wore earth, I was enveloped in the dancing lights. Abandoned by the last of our communal heat, I was absorbed by the distant stars. I lost track of how long I stared at the sky before drifting off to sleep, whether minutes or hours. Duration didn't matter. For a brief moment, time had dissolved.

Episodes with this timeless quality are rare.

And they are fleeting. Time, for the most part, is a constant companion. Impermanence underlies experience. We revere the absolute but are bound to the transitory. Even those features of the cosmos that may present as enduring—the expanse of space, the distant galaxies, the stuff of matter—all lie within the reach of time. As we will explore in this chapter and the next, however stable it may appear, the universe and all it contains is mutable and precarious.

Evolution, Entropy, and the Future

Underneath reality's steadfast façade, science has revealed a relentless drama of churning particles in which it is tempting to cast evolution and entropy as embattled characters perpetually fighting for control. The tale envisions that evolution builds structure while entropy destroys it. It makes for a tidy story, but the hitch, as we have seen in earlier chapters, is that it isn't quite true. Like many simplified sketches, there is some truth to it. Evolution **is** instrumental in building structure. Entropy **does** tend to degrade structure. But entropy and evolution need not pull in opposite directions. The entropic two-step allows structure to flourish here, so long as entropy is expelled there. Life, among the premier achievements of evolution, embodies this mechanism, consuming high-quality energy, using

it to maintain and enhance its orderly arrangements, and expelling high-entropy waste to the
environment. Playing out across billions of years,
the cooperative exchange between entropy and
evolution has resulted in particulate arrangements
that are exquisite, including a life and a mind able
to produce the Ninth Symphony and vastly more
lives and minds able to experience it as sublime.

As we pivot from the journey that has taken us
from the big bang to Beethoven and turn toward
the future, will evolution and entropy continue to
be decisive factors guiding change? For Darwinian
evolution, you might think the answer is no.[1] The
dependence of reproductive success on genetic
makeup is the reason Darwinian selection has long
steered the evolutionary ship. A consequential
difference of recent times is the intervention of
modern medicine and the protections provided by
civilization more generally. Genotypes that might
have found life on the ancient African savanna challenging can do just fine in today's New York City. In
many parts of the world, your genetic profile is no
longer the dominant factor determining whether
you die as a child or issue abundant progeny as an
adult. Of course, by leveling sections of the genetic
playing field, modern advances adjust previous
selection pressures and thus exert their own variety
of evolutionary influence. Researchers also point to
numerous pressures including dietary choices (e.g.,
diets rich in milk products favor digestive systems

in which the production of lactase is prolonged beyond childhood), environmental conditions (e.g., living at high altitude gives an advantage to adaptations for surviving with less available oxygen), and mate preferences (e.g., average heights in some countries may be evolving toward statures deemed more appealing by those who are reproductively active) that drive trends in the gene pool.[2] But the greatest impact of all may come from the newfound ability to directly edit genetic profiles. Rapidly advancing techniques have the capacity to augment the mechanisms of genetic variation, random mutation and sexual mixing, to include volitional design. Should a researcher discover a genetic reconfiguration that extends human life to two hundred years with side effects being cyan skin, ten-foot stature, and a ravenous blue-centric libido, evolution will be on full display as a self-selected group of long-lived, Na'vi-like humans spreads rapidly. With the potential to wholly refashion life and perhaps design a version of sentience—whether biological, artificial, or some variety of hybrid—whose powers may dwarf our current abilities, it is anyone's guess where this will all lead.

For entropy, the answer to the question of future relevance is certainly yes. Many chapters ago we found that the second law of thermodynamics is a general consequence of applying statistical reasoning to the underlying physical laws. Might future discoveries revise the laws we now consider fundamental? Almost certainly. Will entropy and the second law maintain their place of explanatory prominence? Almost certainly too. During the transition from the classical to the radically different quantum framework, the mathematics describing entropy and the second law required an update, but because these concepts emerge from the most basic probabilistic reasoning, they continue to apply all the same. We anticipate the same will hold regardless of future developments in our understanding of physical law. It is not that we are unable to imagine physical laws that would result in entropy and the second law being irrelevant, but the laws would need to be so contrary to the features of reality inherent in all we know and all we have measured that most physicists dismiss the possibility out of hand.

In envisioning the future, greater uncertainty surrounds the control that we or some forthcoming intelligence will be able to exert over our surroundings. Might intelligent life direct the long-term fate of stars, galaxies, and even the cosmos as a whole? Might such intelligence willfully shift entropy on voluminous scales, effectively driving entropy down in enormous swaths of space, a cosmic-scale version of the entropic two-step? Might such intelligence even have the capacity to design and create entire new universes? However far-fetched these activities may sound, they fall within the realm of possibility. The dilemma for us is that their impact on the future is utterly beyond our ability to predict. Even in a lawful world, one that lacks traditional free will, the broad behavioral repertoire of intelligence—the version of freedom intelligence acquires—makes certain varieties of prediction essentially impossible. Future thought will no doubt acquire incomparable computational methods and technologies, but I suspect that predicting long-term developments that are intimately dependent on life and intelligence will remain beyond reach.

How, then, to proceed?

We will assume that the laws of physics as currently known, operating in the undirected manner they presumably have since the big bang, will be the dominant influence guiding the cosmic unfolding. We will not consider the possibility that the laws themselves or even the numerical "constants" of nature can change. Nor will we consider the possibility that these laws or constants are already slowly shifting, modifications that might currently be too small to leave a mark but might exist and might accumulate over vast timescales into substantial change.[3] We will also not consider the possibility that the dominion over which future intelligence will exert structural control will swell to the scales of galaxies and beyond. Granted, that's a lot of "nots" and "nors." But in the absence of any evidence to guide us, investigating these possibilities would amount to shooting in the dark. If these assumptions cut against your expectations for the future, you can view the account in this and the next chapter as reflecting cosmological developments that would otherwise happen in the absence of such change or intelligent intervention. My suspicion is that the clarity brought by future discoveries as well as the influences exerted by future intelligence, while surely relevant to details of the account that follows, will not require a wholesale rewriting of the cosmic unfolding we will survey.[4] A bold assumption, perhaps, but it is the most expeditious route forward and one that we will now boldly pursue.[5]

As the following pages will make evident, the very fact that we can piece together a cogent if tentative account that delineates cosmic unfolding exponentially far into the future is an extraordinary achievement, one shaped by the hands of many and as emblematic of the human longing for coherence as our species' most cherished stories, myths, religions, and artistic creations.

An Empire of Time

How should we organize our thinking about the future? Human intuition is reasonably well suited for grasping the timescales of common experience, but in analyzing key cosmological epochs of the future we will enter temporal realms so vast that even our best analogies can provide no more than a hint of the durations involved. Still, there is no better way than analogies based on familiar scales to provide mental toeholds for such an unfamiliar climb, so let's imagine that the timeline of the universe extends up the Empire State Building, with each floor representing a duration ten times that of the previous. The first floor represents ten years since the big bang, the second floor one hundred years, the third floor one thousand years, and so on. As the numbers make evident, durations grow rapidly as we climb from floor to floor—simple to describe, but easy to misconstrue. Walking, say, from floor 12 to floor 13 amounts to considering the universe from a trillion years after the big bang to ten trillion years after the bang. In ascending that single floor, nine trillion years elapse, dwarfing the entire duration represented by all previous floors. The same pattern holds as we continue to climb higher: the duration represented by each subsequent floor is far larger, exponentially larger, than the duration represented by the floors below.

With the span of a human life being roughly a hundred years, durable empires lasting about a thousand years, hardy species hanging on for millions of years, the ever-higher floors of the Empire State Building represent durations of a thoroughly distinct, seemingly aeonian sort. When we reach the Empire State Building's

observation deck on floor 86, we will be 10^{86} years—
100,000,000,000,000,000,000,000,000,000,000,
000,000,000,000,000,000,000,000,000, 000,000,
000,000,000,000,000,000,000—from the big
bang, a staggering timescale that towers over any
duration of any relevance to any human endeavor.
And yet, notwithstanding all the zeroes, when we
subsequently step to the building's uppermost landing, reaching floor 102, the duration represented by
the observation deck will, by comparison, amount
to far less than the thickness of paint coating that
final tread.

Today, it is about 13.8 billion years since the big
bang, which means that all of the developments
discussed in the previous chapters are sprinkled
between the ground floor of the Empire State
Building and just a few steps above floor 10. From
here we head exponentially far into the future.

Let's climb.

The Black Sun

Our early ancestors, even without understanding
that the sun bathes earth in a continual wash of
low-entropy energy essential to life, recognized
the central importance of the sky's watchful eye, a
burning presence overseeing the comings and goings of daily existence. As the sun set, they realized
it would rise once again, having induced the world's

most conspicuous and reliable pattern. But just as reliably, that rhythm will one day end.

For almost five billion years, the sun has supported its tremendous mass against the crushing force of gravity through the energy produced by the fusion of hydrogen nuclei in its core. That energy powers a frenzied environment of fast-moving particles that exert a strong outward pressure. And much like the pressure produced by an air pump that props up a child's inflatable bounce house, the pressure produced by fusion in the sun's core props up the sun, keeping it from collapsing under its own enormous weight. This standoff between gravity pulling inward and particles pushing outward will hold firm for about another five billion years. But then the balance will be upended. Even though the sun will still be chock-full of hydrogen nuclei, hardly any will be in the core. Hydrogen fusion produces helium, nuclei that are heavier and denser than hydrogen, and so just as sand poured into a pond displaces water as it fills the pond's bottom, helium displaces hydrogen as it fills the sun's center.

That's a big deal.

The center of the sun is where you find its hottest temperatures, currently about fifteen million degrees, well in excess of the ten million degrees required to fuse hydrogen into helium. But to fuse helium nuclei requires a temperature of about one hundred million degrees. Because the sun's temperature is nowhere near that threshold, as helium displaces hydrogen in the core, fusion's fuel supply will dwindle. The outward pressure from fusion's production of energy in the core will subside, and consequently the inward pull of gravity will gain the upper hand. The sun will begin to implode. As its spectacular heft collapses inward, the sun's temperature will skyrocket. The intense heat and pressure, still shy of the conditions necessary for helium to start burning, will spark a new round of fusion within a thin shell of hydrogen nuclei surrounding the helium core. And with such extreme conditions, hydrogen fusion will proceed at an extraordinary pace, producing a more intense outward push than the sun has ever experienced, not only halting the implosion but thrusting the sun to swell tremendously.

The fate of the inner planets hangs in the balance between two factors. How large will the sun grow? And as it does, how much mass will the sun shed? The latter question is relevant because with its nuclear engine running in overdrive, copious particles in the sun's outer layer will be blown steadily into space. A lower-mass sun, in turn, results in a diminished overall gravitational pull, causing the planets to migrate into more distant orbits. The future of any given planet depends on whether its receding trajectory can outrun the swelling sun.

Computer simulations incorporating detailed

solar models conclude that Mercury will lose the race and be swallowed by the distended sun, vaporizing quickly. Mars, orbiting at a larger distance, enjoys a head start and will be safe. Venus is likely done for, yet some simulations conclude that the swelling sun may fall just shy of reaching its receding orbit and, if so, that of earth too.[6] But even if earth is spared, conditions here will change profoundly. Earth's surface temperature will soar into the thousands of degrees, hot enough to dry out the oceans, eject the atmosphere, and flood the surface with molten lava. Unpleasant conditions, to be sure, but the giant red sun spilling across the sky would be a sight to behold. It's virtually certain, however, that it's a sight no one will ever see. If our descendants continue to thrive (having successfully dodged self-destruction, lethal pathogens, environmental disasters, deadly asteroids, and alien invasions, among other potential catastrophes), and if they aim to continue doing so, they will have long abandoned earth in search of a more hospitable home.

As the hydrogen nuclei surrounding the sun's helium core continue to fuse, the additional helium they produce will rain down, forcing the core to contract yet farther and propelling its temperature yet higher. In turn, the higher temperature will accelerate the cycle, increasing the rate of hydrogen fusion in the surrounding shell, intensifying the storm of helium pummeling the core, and driving the temperature higher still. Roughly five and a half billion years from now, the core temperature will finally be hot enough to support the nuclear burning of helium, producing carbon and oxygen. After a spectacular but momentary eruption that marks the transition to helium fusion being the sun's dominant energy source, the sun will shrink back down in size and settle into a less frenzied configuration.

But the newfound stability will be relatively short lived. In about one hundred million years, much as heavier helium displaced lighter hydrogen, heavier carbon and oxygen will do the same to lighter helium, taking over the solar core and forcing helium into surrounding layers. Nuclear burning of the new core constituents, carbon and oxygen, requires even higher temperatures, a minimum of six hundred million degrees. As the sun's core temperature is far less than this, nuclear fusion will once again grind to a halt, the inward pull of gravity will once again dominate, the sun will once again contract, and the core temperature will once again increase.

In the previous phase of this cycle, the increasing temperature sparked the onset of fusion in a shell of hydrogen surrounding the quiescent core of helium. Now the increasing temperature sparks fusion in a shell of helium surrounding a quiescent core of carbon and oxygen. But in this go-round the temperature in the core will never reach the value required for nuclear burning to be reignited there. The sun's mass is too low to provide the necessary temperature-propelling crush that, in larger stars, would ignite the fusion of carbon and oxygen into yet heavier and more complex nuclei. Instead, as the helium shell burns, showering the core with freshly made carbon and oxygen, the core will continue to contract until a quantum process—it's called the **Pauli exclusion principle**—halts the implosion.[7]

In 1925, Austrian physicist Wolfgang Pauli, a famously caustic quantum pioneer ("I don't mind your thinking slowly; I mind your publishing faster than you think"[8]), realized that quantum mechanics sets a limit to how closely two electrons can be squeezed together (more precisely, quantum mechanics excludes any two identical matter particles from occupying an identical quantum state, but the rough description will suffice). Shortly thereafter, the collective insights of a number of researchers showed that Pauli's result, notwithstanding its focus on minute particles, was the key to understanding the sun's fate, as well as the fate of all similarly sized stars. As the sun contracts, electrons in the core will be packed ever more tightly, ensuring that sooner or later the electron density will reach the limit specified by Pauli's result. When further contraction would violate Pauli's principle, a powerful quantum repulsion kicks in, the electrons stand their ground, demand their personal space, and refuse to be packed any more closely. The sun's contraction stops.[9]

Far from the core, the outer shells of the sun will

continue to expand and cool, ultimately drifting off into space, leaving behind an astoundingly dense ball of carbon and oxygen, called a white dwarf star, which will continue to glow for a handful of billions of years more. Without the required temperature for further nuclear fusion, thermal energy will slowly dissipate into space and, like the final glow of a burning ember, the remnant sun will cool and dim, ultimately transitioning into a dark frozen orb. A few steps above the tenth floor, the sun will fade to black.

It is a gentle end. All the more so when compared with a cataclysmic denouement that may be awaiting the entire universe as we continue our climb to the next floor.

The Big Rip

Toss an apple upward and the relentless tug of earth's gravity ensures that its speed steadily slows. It is a pedestrian exercise with deep cosmological significance. Ever since Edwin Hubble's observations in the 1920s, we have known that space is expanding: the galaxies are rushing away from one another.[10] But much as with the tossed apple, the gravitational pull of each galaxy on every other must, surely, be slowing the cosmic exodus. Space is expanding, but the rate of expansion must be decreasing. In the 1990s, motivated by this expectation, two teams of astronomers set out to measure the rate of cosmic slowdown. After nearly a decade of pursuit they announced their results—and rocked the scientific world.[11] The expectations were wrong. Through painstaking observations of distant supernova explosions, powerful beacons that can be seen and measured clear across the cosmos, they discovered that the expansion is not slowing down. It is speeding up. And it's not as though the shift into cosmic overdrive happened yesterday. Researchers, falling off their chairs, were confronted with astronomical observations establishing that the expansion has been picking up speed for the past five billion years.

The widely held expectation of a slowing rate of expansion had been widely held because it makes sense. To propose a quickening expansion of space is, at first blush, as absurd as predicting that a gently tossed apple will leave your hand and rocket skyward. If you saw such a bizarre thing you'd look for a hidden force, an overlooked influence responsible for pushing the apple upward. Similarly, when the data provided overwhelming evidence that the spatial expansion is speeding up, researchers picked themselves off the floor, grabbed fistfuls of chalk, and sought the cause.

The leading explanation invokes a pivotal feature of Einstein's general relativity that we encountered in our discussion of inflationary cosmology back in chapter 3.[12] Recall that according to both Newton and Einstein, clumps of matter like planets and stars exert familiar attractive gravity, but in Einstein's approach gravity's repertoire broadens. If a region of space is not host to a clump but is instead uniformly filled with an energy field—my image of choice, introduced earlier, is steam uniformly filling a sauna—the resulting gravitational force is repulsive. In inflationary cosmology, researchers envision that such energy is carried by an exotic species of field (the inflaton field), and the theory proposes that its powerful repulsive gravity drove the big bang. Although that event was nearly fourteen billion years ago, we can follow an analogous approach to explain the accelerated expansion of space we currently observe.

If we imagine that all of space is uniformly filled with another energy field—we call it **dark energy** because it doesn't generate light, but **invisible energy** would be just as apt—we can give an account of why the galaxies are all hurriedly departing. Being clumps of matter, the galaxies exert attractive gravity, mutually pulling inward and thus slowing the cosmic exodus. Being spread uniformly, the dark energy exerts repulsive gravity, pushing outward and thus quickening the cosmic exodus. To explain the accelerated expansion the astronomers observe, dark energy's push simply needs to exceed the galaxies' collective pull. And not by much. Compared with the blistering outward swelling of space during the big bang, today's expansion is gentle, and so a diminutive dark energy is all that is needed. Indeed, in a typical cubic meter of space, the amount of dark energy required to power the observed galactic speedup would keep a hundred-watt bulb running for about five trillionths of a second—almost comically tiny.[13] But space contains a lot of cubic meters. The repulsive push contributed by each and every one combines to yield an outward force able to drive the accelerated expansion measured by the astronomers.

The case for dark energy is compelling but circumstantial. No one has found a way to clutch hold of dark energy, establish its existence, and directly examine its properties. Nevertheless, because it so adeptly accounts for the observations, dark energy has become the de facto explanation for the accelerated expansion of space. Less clear, however, is the long-term behavior of dark energy. And to forecast the far future, thinking through the possibilities is essential. The simplest behavior consistent with all observations is that the value of the dark energy does not change over the course of cosmic time.[14] But simplicity, while favored conceptually, has no fundamental claim on truth. The mathematical description of dark energy allows for it to weaken, putting the brakes on accelerated expansion, or strengthen, giving additional gas to accelerated expansion. Looking out from the eleventh floor, the latter situation—repulsive gravity that grows more forceful—is the most inauspicious possibility; if realized, we are hurtling toward a violent reckoning that physicists call the **big rip.**

An increasingly powerful repulsive push of gravity would, in time, triumph over all forces that bind, with the result that everything would be torn apart. Your body is held intact by the electromagnetic force, binding together your atomic and molecular constituents, and also by the strong nuclear force, binding together the protons and neutrons inside of your body's atomic nuclei. Because these forces are far stronger than today's outward push of expanding space, your body holds firm. If you are widening, it is not because space is expanding. But if the strength of the repulsive push grows ever larger, the space inside your body will ultimately expand with such a powerful outward thrust that it will overcome the electromagnetic and nuclear forces holding you together. You will swell and ultimately burst to pieces, as will everything else.

The details depend on the rate at which repulsive gravity increases, but in one representative example worked out by physicists Robert Caldwell, Marc Kamionkowski, and Nevin Weinberg, about twenty billion years from now repulsive gravity will drive apart clusters of galaxies, about a billion years later the stars constituting the Milky Way will be flung apart like sparkles in a fireworks display, about sixty million years after that earth and the other planets in the solar system will be thrust away from the sun, a few months later the repulsive gravitational force between molecules will cause stars and planets to explode, and with the passage of just thirty minutes more the repulsion between particles constituting individual atoms will have grown so strong that even they will be blasted apart.[15] The final state of the universe depends on the currently unknown quantum nature of space and time. In loose terms that for now lack mathematical rigor, it is possible that repulsive gravity will shred the very fabric of spacetime itself. Reality started with a bang, and sometime before we reach the eleventh floor, one hundred billion years since the big bang, it may end with a rip.

While current observations allow for a dark energy that grows stronger, I, and many other physicists, consider this an unlikely possibility. When studying the equations, I'm left with a feeling that yes, the math works, barely, but no, the equations are not natural or convincing. It is a judgment based on decades of experience, not a mathematical proof, so surely it could be wrong. Still, it provides more than enough motivation for being optimistic and assuming that the big rip will not render the subsequent floors of the Empire State Building irrelevant. With that, we continue our journey up the timeline.

We don't need to climb far before we encounter the next pivotal event.

The Cliffs of Space

If the strength of the repulsive gravitational force does not increase, but remains constant, we can all breathe easy; being blasted apart by expanding space will no longer be a concern. But because repulsive gravity will continue to drive distant galaxies to race away ever more quickly, it will still have a profound long-term consequence: in about a trillion years the recessional speed of the distant galaxies will reach and then exceed the speed of light—seeming to violate the most famous rule in Einstein's universe. Closer scrutiny makes clear that, in actuality, the rule holds firm: Einstein's dictum that nothing can exceed the speed of light solely refers to the speed of objects moving **through** space. Galaxies hardly move through space at all. They're not endowed with rocket engines. Much as specks of white paint stuck to a black swatch of spandex move apart when the spandex stretches, galaxies are, for the most part, stuck to the fabric of space and move apart because space swells. The more distant one galaxy is from another, the more intervening space there is between them to swell, and so the faster the galaxies will separate. Einstein's law imposes no limit on the speed of such recession.

Despite that, light's speed limit remains immensely significant. The light each galaxy emits **does** travel through space. And much as a kayaker will be stymied if she's paddling upstream at a speed that's less than that of the stream itself, the light emitted by a galaxy that is sprinting away at superluminal speed will fight a losing battle as it tries to reach us. Traversing space at light speed, the light cannot overcome the faster-than-light-speed increase in the distance to earth. As a result, when future astronomers look past nearby stars and focus their telescopes on the deepest parts of the night sky, all they will see is velvety black darkness. The distant galaxies will have slipped beyond the bounds of what astronomers call our **cosmic horizon.** It will be as if the distant galaxies have dropped off a cliff at the edge of space.

I've focused on distant galaxies because those that are relatively nearby, a cluster of about thirty galaxies known as the Local Group, will continue to be our cosmic companions. Indeed, by the eleventh floor, the Local Group, dominated by the Milky Way and Andromeda galaxies, will likely have merged, an anticipated future union astronomers have christened **Milkomeda** (I would have lobbied for **Andromilky**). The stars of Milkomeda will all be close enough for their mutual gravitational pulls to withstand the expansion of space and keep the stellar collection intact. But our severed contact with the more distant galaxies will be a profound loss. It was through careful observations of distant galaxies that Edwin Hubble first realized that space is expanding, a discovery confirmed and refined by a century of subsequent observations. Without access to the distant galaxies, we will lose a primary diagnostic tool for tracing spatial expansion. The very data that guided us toward our understanding of the big bang and cosmic evolution will no longer be available.

Astronomer Avi Loeb has suggested that high-velocity stars that will continually escape the Milkomeda conglomerate and drift off into deep space might provide a substitute for distant galaxies, like tossing popcorn off a raft to trace the downstream currents. But Loeb, too, acknowledges that the relentless accelerated expansion will have a devastating impact on the ability of future astronomers to carry out precise cosmological measurements.[16] As a case in point, by the twelfth floor, about a trillion years after the bang, the all-important cosmic microwave background radiation, which guided our cosmological explorations in chapter 3, will have been so stretched and diluted by cosmic expansion (so **redshifted,** in technical jargon) that it will likely be impossible to detect.

It makes you wonder: assuming that the data we've gathered, establishing that the universe is expanding, were to somehow be preserved and delivered to the hands of astronomers a trillion years from now, would they believe it? Using their state-of-the-art equipment, a trillion years in the making, they will see a universe that on the largest of distances is black, about as eternal and unchanging as it gets. You can well imagine that they'd wave aside quaint results handed down from an ancient and primitive era—ours—and instead accept the erroneous conclusion that, overall, the universe is static.

Even in a world subject to a relentless rise in entropy, we have grown accustomed to measurements always improving, data sets always growing, understanding always refining. The accelerated expansion of space can subvert these expectations. Accelerated expansion can cause essential information to race away so quickly that it becomes inaccessible. Deep truths may silently beckon to our descendants from just beyond the horizon.

The Twilight of Stars

The first stars began to form on the eighth floor, roughly one hundred million years after the big bang, and will continue to form so long as the raw materials for making new stars remain. How long will that be? Well, the list of ingredients is short: all you need is a large enough cloud of hydrogen gas. As we've seen, gravity takes it from there, slowly squeezing the cloud, heating up its core, and igniting nuclear fusion. If you know the amount of gas the galaxy contains, and you know the rate at which such star formation depletes the gas reserves, you can estimate the duration over which star formation will continue. There are subtleties that make the accounting more complex (the rate of star formation in a galaxy can change over time; as stars burn, they return part of their gaseous composition to the galaxy, bulking up the reserves), but with refined calculations researchers have concluded that by the fourteenth floor, about one hundred trillion years in the future, star formation in the vast majority of galaxies will draw to a close.

Continuing the climb upward from the fourteenth floor, there is something else we will notice too. Stars will be fading away. The more massive a star, the more its heft crushes its core and the hotter its central temperature. In turn, the hotter temperature spurs a more rapid rate of nuclear fusion and thus a more rapid burn-down of the star's nuclear reserves. While the sun will burn brightly for about ten billion years, stars that are much heavier will have exhausted their nuclear fuel well before that time. By contrast, flyweight stars, down to roughly a tenth of the sun's mass, burn more gently and so live far longer. Astronomers use the catchall name **red dwarf** to label an assortment of such low-mass stars, and according to observations they likely account for the majority of stars in the universe. Their relatively low temperatures and slow, methodical burning of hydrogen (churning currents within a red dwarf ensure that almost all of the star's storehouse of hydrogen is burned in the core) allow red dwarfs to continue shining for many trillions of years, thousands of times the sun's lifetime. But by the fourteenth floor, even a late-blooming red dwarf star will be running on fumes.

And so, as we ascend from floor 14, galaxies will resemble the burnt-out cities of a dystopian future. The once-vibrant night sky full of brilliant stars will now be populated with charred cinders. Still, because the gravitational pull of a star depends only on its mass, not on whether it is shining bright or smoldering dark, those stars that host planets will mostly continue to do so.

For one more floor.

The Twilight of Astronomical Order

Looking up at a clear night sky gives the impression that the galaxy is dense with stars. It's not. Although it seems like stars are arranged cheek by jowl on a sphere that surrounds us, because their distances from earth vary widely—a feature that's mostly lost on our feeble, closely set eyes—stars are, in reality, quite far from one another. Were you to shrink the sun down to the size of a grain of sugar and place it at the Empire State Building, you'd have to drive most of the way to Greenwich, Connecticut, to encounter Proxima Centauri, our nearest stellar neighbor. And you wouldn't need to drive swiftly to ensure that Proxima would still be hanging around Greenwich by the time you got there. At this scale, typical stellar speeds clock in at less than a millimeter per hour. Like a game of tag played by widely dispersed slugs, only rarely will stars collide or even have a near miss.

That conclusion, however, is based on familiar durations—years, centuries, millennia—and so must be reconsidered in light of the far-longer timescales we are now considering. By the fifteenth floor, we are a million billion years since the bang. And over that duration there is actually a significant chance that today's distant and slow-moving stars will have had numerous close calls. In such an encounter, what will happen?

Let's focus on earth and imagine that another star wanders by. Depending on the interloper's mass and trajectory, its gravitational pull may only mildly perturb earth's motion. A lightweight intruder that keeps a good distance won't wreak havoc. But the gravitational pull of a more massive star that passes closer could easily rip earth from its orbit, sending it hurtling across the solar system and headlong into deep space. And what's true for earth is true for most other planets orbiting most other stars in most other galaxies. As we climb up the timeline, more and more planets will be flung into space by the disruptive gravitational pull of wayward stars. Indeed, although extremely unlikely, the earth could suffer this fate before the sun burns out.

Were this to happen, earth's ever-larger distance from the sun would cause its temperature to fall continually. Upper layers of the world's oceans would freeze, as would whatever else is left on the surface. Atmospheric gases, predominantly nitrogen and oxygen, would liquefy and drip from the skies. Could life survive? On earth's surface, that would be a tall order. But as we have seen, life thrives and indeed may have originated in dark thermal vents dotting the ocean floor. Sunlight can't penetrate anywhere near such depths, and so the vents will hardly be affected by the sun's absence. Instead, a substantial part of the energy powering the vents comes from diffuse but continual nuclear reactions.[17] Earth's interior contains a storehouse of radioactive elements (mostly thorium, uranium, and potassium), and as these unstable atoms decay they emit a stream of energetic particles that heat the surroundings. So whether or not earth enjoys the warmth generated by nuclear fusion in the sun, it will continue to enjoy the warmth generated by nuclear fission in its interior. Were earth to be ejected from the solar system, it is possible that life on the ocean floor would carry on for billions of years as if nothing had happened.[18]

Such stellar bumper cars will not only disrupt solar systems, but over yet longer durations will also disrupt galaxies. In near misses between meandering stars or, rarer still, head-on collisions, the speed of the heavier star tends to decrease while that of the lighter tends to increase. (Balance a ping-pong ball on a basketball and, as the stack drops to the ground and rebounds, you'll witness the collision imparting an impressive increase in speed to the ping-pong ball.)[19] In any single encounter, such exchanges will typically be modest, but over vast durations their cumulative effect can add up to significant changes in stellar velocities. The result will be a steady inventory of stars that will be kicked to speeds so high that they escape their host galaxy. Detailed calculations reveal that as we pass the nineteenth floor and continue toward the twentieth, typical galaxies will be depleted by this process. Their stars, mostly incinerated remains, will be ejected and left to drift aimlessly through space.[20]

The ubiquitous astronomical order manifested in solar systems and galaxies will have dissolved; these structures, now pervasive, will have become patterns that the universe has retired.

Gravitational Waves and the Final Sweep

If earth is fortunate and sidesteps the swelling sun on floor 11, and if it escapes being ejected by the disruptive visit of stellar neighbors, its final fate will be determined by an utterly beautiful feature of the general theory of relativity, **gravitational waves.**

In explaining general relativity's central but abstract idea of curved spacetime, physicists often invoke a familiar metaphor: we picture planets orbiting a star as if they are marbles rolling on a taut rubber sheet deformed by a bowling ball placed at the center. But the metaphor raises a question. Why don't the planets spiral toward the star and fall in? After all, the analogous fate surely befalls the marbles.[21] The answer is that rolling marbles spiral inward because they lose energy through friction. Indeed, even without any fancy equipment you can detect evidence of this: some of the lost energy makes it to your ears, allowing you to hear the marbles rolling on the rubber sheet. Orbiting planets maintain their motion because there is virtually no friction in empty space.

Even though friction is not a factor, a planet does lose a small amount of energy on every orbit. When astronomical bodies move, they disturb the fabric of space, generating ripples that propagate outward similar to those that would ripple on the rubber sheet were you to tap it persistently. Such ripples in the fabric of space are the gravitational waves Einstein predicted in papers he published in 1916 and 1918. In the decades that followed, Einstein had mixed feelings about gravitational waves, viewing them, at best, as a mere theoretical possibility that would never be observed and, at worst, as a flat-out misinterpretation of the equations. The mathematics of general relativity is so subtle that even Einstein was sometimes perplexed. It took many people many years to develop systematic methods to overcome thorny issues that would otherwise confuse attempts to link general relativity's mathematical expressions with measurable features of the world. By the 1960s, with such methods firmly in place, physicists gained confidence that gravitational waves were an unassailable consequence of the theory. Even so, no one had any experimental or observational evidence that gravitational waves were real.

About a decade and a half later, that changed. In 1974, Russell Hulse and Joe Taylor discovered the first known binary neutron-star system—a pair of neutron stars locked in a rapid orbit.[22] Subsequent observations established that over time the neutron stars were spiraling closer, evidence that the binary system was losing energy. But where was the energy going?[23] Taylor, and his collaborators Lee Fowler and Peter McCulloch, announced that the measured loss in orbital energy was in remarkable agreement with general relativity's prediction for the energy the orbiting neutron stars should be pumping into gravitational waves.[24] Although the gravitational waves produced were too feeble to be detected, these works established, albeit indirectly, that gravitational waves were real.

Three decades and a billion dollars later, the Laser Interferometer Gravitational-Wave Observatory went further, establishing the first direct detection of ripples in the fabric of space. Early in the morning on September 14, 2015, two enormous detectors, one in Louisiana and the other in Washington State, both heroically shielded from any possible disturbance save a gravitational wave, twitched. And in precisely the same way. Researchers had been preparing for this moment for nearly half a century but had finished calibrating the newly upgraded detectors barely two days earlier. The nearly immediate detection of a signal was both a surprise and a concern. Was it real? Was it the discovery of a lifetime or the handiwork of a prankster—or worse, had someone hacked the system and injected a fake signal?

After months of meticulous analysis, checking and rechecking details of the purported gravitational disturbance, the researchers announced that a gravitational wave had indeed rolled by earth. What's more, by precisely analyzing the twitch and comparing it with the results of supercomputer simulations of the gravitational waves that should be produced by various astronomical events, the researchers reverse-engineered the signal to determine the source. They concluded that 1.3 billion years ago, a time when multicellular life was just starting to coalesce on planet earth, two distant black holes were orbiting each other ever more closely and ever more quickly, closing in on the speed of light, until in a final orbital frenzy they smashed together. The collision generated a tidal wave in space, a gravitational tsunami so enormous that its power exceeded that produced by every star in every galaxy in the observable universe. The wave raced outward at the speed of light, in all directions, and so part headed toward earth, diluting in power as it spread ever more widely. About one hundred thousand years ago, as humans were migrating from the African savanna, the wave rippled through the dark matter halo surrounding the Milky Way galaxy as it continued its relentless sprint. About one hundred years ago, the wave raced past the Hyades star cluster and as it did, one member of our species, Albert Einstein, began to think about gravitational waves and wrote the first papers on the possibility. About fifty years later, as the wave dashed onward, other researchers boldly proposed that such waves might be detected and began designing and planning a device that might do so. And when the wave was but two light-days from earth, the newly upgraded version of the most advanced of these detectors was readied for operation. Two days later those two detectors shook for two hundred milliseconds, collecting data that allowed scientists to reconstruct the story I just recounted. For this achievement, team leaders Ray Weiss, Barry Barish, and Kip Thorne were awarded the 2017 Nobel Prize.

These discoveries, thrilling in their own right, are relevant here because it is on the twenty-third floor that the earth (again, assuming earth is still in orbit), having lost energy through a version of the same process—the slow but relentless production of gravitational waves—will spiral into the long-dead sun. For other planets, the story is similar, although the timescales can differ. Smaller planets more gently disturb the fabric and so have longer death spirals, as do planets whose orbits are farther from their host star. Taking earth as representative of planets that may stubbornly persist in orbit, we conclude that by the twenty-third floor such planets, resigned to their fate, will dive in for a final violent communion with their cold sun.

During their final stages, galaxies will follow an analogous sequence. At the center of most galaxies is an enormous black hole, millions or even billions of times the mass of the sun. As we climb upward from the twenty-third floor, the only stars remaining in galaxies will be burnt-out embers that, having avoided ejection, will slowly orbit the galaxy's central black hole. And much as planets slowly spiral inward as their orbital energy is funneled into gravitational waves, so too for stars around a galactic black hole. By estimating the rate of such energy transfer, researchers have concluded that by the twenty-fourth floor most stellar remains will have been consumed, falling into their galaxy's dark central abyss.[25] Should a galaxy have stragglers, burnt-out stars that are small and distant, the central black hole will offer additional assistance, relentlessly pulling on the stars, coaxing them to drift ever closer to their final demise. Taking account of both influences, central black holes will sweep most galaxies clean of stars by the thirtieth floor, 10^{30} years since the big bang, if not sooner.

By this era, a tour through the cosmos will not exactly be a riotous affair. Punctuated here and there by cold planets, burnt-out stars, and monstrous black holes, space will be dark and desolate.

The Fate of Complex Matter

In the midst of the extreme environmental transformations we've encountered, can life persist? It is a challenging question in no small part because, as emphasized at the outset of this chapter, we have no idea what life of the far future will be like. One seemingly certain characteristic is that life of any sort will need to harness suitable energy to power its life-sustaining functions—metabolic, reproductive, whatever. As stars burn down, are ejected into deep space, or spiral into omnivorous black holes, that task will become increasingly difficult. There are creative ideas, like harnessing particles of dark matter that we believe waft across space, which can produce energy as pairs collide and transform into photons.[26] But here's the thing: even if some form of life is able to tap a novel source of useful energy, as we continue our climb another challenge, more significant than all others, will likely emerge.

Matter itself may disintegrate.

At the heart of all atoms, making up all molecules, and assembled into all complex material structures from life to stars, are protons. Were protons to have a penchant for disintegrating into a spray of lighter particles (such as electrons and photons), matter would fall apart and the universe would change radically.[27] Our existence attests to the stability of protons, at least over timescales on par with the duration back to the big bang. But what about over the far-longer timescales we are now considering? For nearly half a century physicists have encountered intriguing mathematical hints that over such immense durations protons can, in fact, decay.

Back in the 1970s, physicists Howard Georgi and Sheldon Glashow developed the first **grand unified theory,** a mathematical framework that, on paper, links together the three nongravitational forces.[28] Although the strong, weak, and electromagnetic forces have vastly different properties when examined in laboratory experiments, in Georgi and Glashow's scheme these distinctions steadily diminish as the three forces are examined over smaller and smaller distances. Grand unification thus proposes that these three forces are actually different facets of a single master force, a unity in nature's workings that reveals itself only over the tiniest of scales.

Georgi and Glashow realized that with grand unification's proposed connections between forces come newfound connections between particles of matter. And such connections allow for a host of new particle transmutations, including some that would result in the decay of protons. Thankfully, the process would be slow. Their calculations showed that were you to hold a bunch of protons in the palm of your hand and wait until half disintegrate, you would have to hold them for about a thousand billion billion billion years, long enough to climb to the Empire State Building's thirtieth floor. It's a curious prediction, one that might appear to lie beyond verification. Who would have the patience to test it?

The answer emerges from a simple but clever move. Just as the odds of there being a winner in this week's lottery will be next to nothing if the state manages to sell only a handful of tickets but will vastly increase if ticket sales soar, the odds of witnessing a proton decay in a small sample is next to nothing but will vastly increase if the sample size is enlarged.[29] So fill an enormous vat with millions of gallons of purified water (every gallon provides about 10^{26} protons), surround the sample with exquisitely sensitive detectors, and stare intently, day and night, looking for the telltale sign of the decay products of a proton (which, according to the Georgi-Glashow proposal, is a particle known as a **pion,** together with an **anti-electron**).

Seeking the particulate detritus of a single decaying proton swimming in a sea of companions so numerous that their population far exceeds the grains of sand making up all beaches and all deserts on the planet might seem like a chase that would bring paroxysms of delight to wild geese everywhere. But the fact is, brilliant teams of experimental physicists have demonstrated conclusively that if a proton in the tank were to disintegrate, their detectors would sound the alarm.

I was one of Georgi's students in the mid-1980s when his unified theory was being put to the test. I was an undergraduate, studying more basic material, so didn't fully understand what was going on. But I could feel the anticipation. The unity of nature, a dream that had so driven Einstein, was about to be revealed. Then a year went by without evidence of a single proton decaying. Followed by another year. And another. The failure to observe any disintegrating protons allowed the researchers to set a lower bound on the proton's lifetime, which currently stands at about 10^{34} years.

Georgi and Glashow's proposal is magnificent. Sidelining the puzzles of quantum gravity for another day, their theory embraces the remaining three forces of nature as well as all particles of matter through a sleek, rigorous, and artful melding of mathematics and physics. It is an intellectual masterwork. And yet, in the face of their proposal, nature shrugged. Much later, I spoke with Georgi about the experience. He described the disappointing experiments as "being slapped down by nature," an experience, he added, that turned him against the whole program of unification.[30]

But the unification program continued. And continues. And a common feature of just about every approach that's been pursued—Kaluza-Klein theories, supersymmetry, supergravity, superstrings, as well as more straightforward extensions of Georgi and Glashow's own grand unification (all of which you can read about in **The Elegant Universe**)—is that protons are predicted to decay. Proposals in which the rate of such decay is close to that of Georgi and Glashow's original scheme are immediately ruled out. But many proposed unified theories predict slower rates for proton decay that are compatible with the most refined experimental limits. Typical numbers range from 10^{34} years to 10^{37} years, with some predictions being longer still.

The point is that as we have continued developing our mathematical understanding of the cosmos, proton decay has reared its head at almost every turn. It is not impossible to rig our equations to avoid proton decay, but accomplishing that often requires contorted mathematical manipulations that run counter to the theoretical accounts that past successes have proven relevant to reality. Because of this, many theorists anticipate that protons do in fact decay. This might be wrong, and in the endnotes I briefly consider the alternative.[31] But here, to be definite, I will take the proton lifetime to be about 10^{38} years.

The implication is that as we climb upward from the thirty-eighth floor, every atom that has combined into every molecule that has assembled into every structure that has ever appeared in the cosmos—rocks, water, rabbits, trees, you, me, planets, moons, stars, and so on—will disintegrate. It all falls apart. The universe will be left with isolated particulate constituents, mostly electrons, positrons, neutrinos, and photons, streaming through a cosmos that is punctuated here and there by quiescent if ravenous black holes.

On lower floors, life's dominant challenge is to harness suitable high-quality, low-entropy energy to power the processes of animate matter. The challenge from floor 38 upward is more basic. With the dissolution of atoms and molecules, the very scaffolding of life and most structure in the cosmos will have crumbled. So if life has made it this far, will it now hit the final wall? Perhaps. But, perhaps too, over the timescales we're considering—more than a billion billion billion times the current age of the universe—life will have evolved into a form that has long discarded any need for the biological architecture it currently requires. Perhaps the very categories of life and mind will be rendered coarse and clumsy by future incarnations that require new characterizations altogether.

Underlying such speculation is the assumption that life and mind are not dependent on any particular physical substrate, such as cells, bodies, and brains, but are instead collections of integrated processes. Biology has so far monopolized life's activities, but that may only reflect the vagaries of evolution by natural selection on planet earth. If some other arrangement of basic particles should faithfully execute the processes of life and mind, then that system will live and that system will think.

Our approach here is to adopt the broadest perspective and consider the possibility that even in the absence of complex atoms and molecules, some kind of thinking mind can exist. And so we ask: With our only constraint, thoroughly inflexible, being that the process of thought fully conforms to the laws of physics, can thought persist indefinitely?

The Future of Thought

To assess the future of thought may seem a classic act of hubris. Through personal experience we each know what it's like to think, but as was clear in chapter 5, the rigorous science of mind is at an early stage. For the science of motion, we progressed from Newton's laws to the radically distinct ones of Schrödinger in less than three centuries, so how can we hope to say anything relevant to the future of thought over timescales for which a billion centuries barely registers?

The question evokes one of our central themes. The universe can and must be understood from a broad range of distinct perspectives. The resulting explanations, each relevant for particular kinds of questions, must ultimately be synthesized into a coherent narrative, but you can make progress on some of these stories even with limited knowledge of many others. Newton did not have the slightest inkling about quantum physics, yet he successfully constructed an understanding of the kind of motion we encounter on everyday scales. When quantum physics came along, Newton's edifice was not dismantled. It was renovated. Quantum mechanics provided a new foundation that deepened the reach of science and gave the Newtonian structure a fresh interpretation.

It is possible that today's mathematical musings on the future of mind will prove irrelevant. After all, unless you're particularly well versed in the history of physics and philosophy, you've probably never heard of Aristotle's entelechial description of motion or Empedocles's fire-in-the-eye theory of vision. As we humans explore, we most certainly get some things—well, many things—flat-out wrong. But as with Newtonian physics, there's also a chance that such musings on mind will one day be considered part of a more sweeping chronicle. It is with this sense of optimism, rational and tempered, that we consider the far future of thought.

In 1979, Freeman Dyson wrote a visionary paper on the far future of life and mind.[32] We will closely follow his lead, incorporating updates based on more recent theoretical advances and astronomical observations. Dyson's approach, much like ours throughout these pages, takes a physicalist view of mind, deeming the act of thinking to be a physical process fully subject to physical law. And since we have a reasonably good handle on how the overall features of the universe will evolve toward the far future, we can investigate whether there will continue to be environments hospitable to thought.

Let's start by thinking about your brain. Among its other qualities, your brain is hot. It continually takes in energy, which you supply by eating and drinking and breathing; it undertakes a host of physiochemical processes that modify its detailed configuration (chemical reactions, molecular rearrangements, particle movements, and so on); and it releases waste heat to the environment. As your brain thinks (and does everything else brains do), it thus recapitulates a sequence we first encountered in chapter 2 when analyzing steam engines. Much as in that template, the heat your brain releases to the environment carries away entropy that it absorbs as well as generates through its internal workings.

If, for whatever reason, a steam engine is unable to eliminate its entropic buildup, sooner or later it will redline and fail. A similar fate will befall a brain that, for whatever reason, cannot clear away the entropic waste that its functioning continually produces. And a brain that fails is a brain that no longer thinks. Therein lies the potential challenge to the durability of brain-based thought. As the universe progresses ever further into the future, will brains maintain the capacity to jettison the waste heat they produce?

No one expects human brains to be a steady presence as we climb from today to ever-higher floors. And, certainly, by the time we've climbed high enough for atoms to start disintegrating into more basic particles, complex molecular agglomerations of any sort will become ever more rare. But the diagnostic requirement of being able to expel waste heat is so fundamental that it applies to any configuration of any kind that undertakes the process of thought. So the essential question is whether any such entity—let's call it the Thinker—regardless of how it is designed or constructed, can expel the heat that its thinking necessarily generates. If the Thinker fails to do so, it will overheat and burn up in its own entropic waste. And should the constraints imposed by physical law in an expanding universe dictate that every Thinker everywhere, sooner or later, is destined to fail in this indispensable task of entropy disposal, the future of thought itself will be imperiled.

To assess the future of thought, we thus need to understand the physics of thought. How much energy does the Thinker's thought require and how much entropy does the process of thinking generate? At what rate does the Thinker need to expel waste heat and at what rate can the universe absorb it?

Thinking Slow

Earlier, in chapter 2, I emphasized that entropy counts the number of rearrangements of a physical system's microscopic constituents—its particles—that "pretty much look the same." In analyzing the Thinker, there's a particularly useful way to restate this. If a system has low entropy, then the configuration of its particles is one among relatively few possibilities that all look the same—one among relatively few doppelgängers. Consequently, if I tell you which configuration among these possibilities the system actually realizes, I will have provided you with only a small amount of information. Like specifying one particular can of Campbell's tomato soup on a meagerly stocked grocery store shelf, I will have distinguished this particular configuration of particles from only a small number of possibilities. If a system has high entropy, then the configuration of its particles is one among a great many possibilities that all look the same—one among a great many doppelgängers. Consequently, if I tell you which configuration among these possibilities the system actually realizes, I will have provided you with heaps of information. Like specifying that can of tomato soup on a ridiculously overstocked grocery store's shelf, I will have distinguished this particular configuration of particles from an enormous number of possibilities. So for a system with low entropy, its particle configuration has low-information content; for a system with high entropy, its particle configuration has high-information content.

The link between entropy and information is important because regardless of where thinking takes places—within a human brain or within the abstract Thinker—to think is to process information. The information-entropy connection therefore tells us that information processing, the function of thought, can also be described as entropy processing. And since, as you may recall from chapter 2, the processing of entropy—shifting entropy from here to there—requires the transfer of heat, we have a comingling of three concepts: thought, entropy, and heat. Dyson leveraged the mathematical version of the links between each to quantify the heat that the Thinker needs to expel based on the number of thoughts the Thinker has. (For the mathematically inclined, the formula is in the endnotes.)[33] Many thoughts implies that a lot of heat needs to be expelled. Fewer thoughts implies that less heat needs to be expelled.

Now, to power its thinking, the Thinker must extract energy from its surroundings. And because heat is itself a form of energy, the amount of energy the Thinker takes in must be at least as large as the amount of heat the Thinker needs to expel. The input energy has higher quality (so it can be readily harnessed by the Thinker) than the output heat (which is waste and will thus be dispersed), but the Thinker can't release more than it absorbs. So Dyson's calculation specifies the minimum high-quality energy the Thinker needs to absorb from the environment, thereby quantifying the challenge: as stars burn out, solar systems unravel, galaxies disperse, matter disintegrates, and the universe expands and cools, the Thinker will face the increasingly difficult task of gathering the concentrated, high-quality, low-entropy energy it needs to continue cogitating. With provisions becoming scarce, the Thinker needs an effective strategy of resource management and waste disposal—a detailed plan, that is, for taking in low-entropy energy and flushing out high-entropy heat. Following Dyson, let's come up with one.

As a first step, let's make the reasonable assumption that the speed of the Thinker's internal processes, whatever they may be, scale with the Thinker's temperature.[34] At higher temperatures, particles move more speedily, and so the Thinker thinks more swiftly, consumes energy more rapidly, and builds up waste more quickly. At lower temperatures, all of this slows. Faced with a universe that's expanding, cooling, and winding down, the Thinker, who aspires to continue thinking for as long as possible, needs to place a premium on conservation, executing a long, slow burn instead of a quick, intense flash. We therefore advise the Thinker to follow the universe's lead: as time goes by, the Thinker should continually lower its temperature, slow down its thinking, and decrease the rate at which it consumes the universe's diminishing supply of quality energy.

Since thinking is all the Thinker does, the prospect of thinking slower is not particularly appealing. We console the Thinker. "You're thinking about this all wrong," we tell the Thinker. "Since **all** your internal processes will slow down together, your subjective experience won't change at all. You won't notice any alteration to your thinking. You might see various processes in the environment seeming to run more quickly, but your thoughts will seem to proceed with their usual alacrity." Relieved, the Thinker agrees to follow the strategy but voices one final concern. "If I follow this approach, will I be able to think new thoughts forever?"

This is the central question, and so we anticipated that the Thinker would ask it. And we're ready. The math reveals that much like a car whose mile-per-gallon consumption gets ever better the slower it drives, the Thinker's thought-per-energy consumption gets ever better the slower it thinks. That is, the Thinker's thinking becomes ever more efficient at ever lower temperatures. For this reason, the Thinker can actually think an **infinite** number of thoughts and yet only require a **finite** supply of energy (much as an infinite sum such as $1 + \frac{1}{2} + \frac{1}{4} + \ldots$ can add up to a finite number, in this case 2). We excitedly inform the Thinker of the result: "By following the plan, not only will you be able to keep on thinking forever, you will be able to do so with only a finite supply of energy!"[35]

Rejoicing, the happy Thinker is about to put the plan into action. But then we hit an unexpected snag. There is one other pesky implication of the math we have so far overlooked: somewhat as a cooler cup of coffee expels less heat to its surroundings than a hotter one, the cooler the Thinker becomes, the less able it is to release the waste heat its thinking generates. "You know very little about me," the Thinker reminds us, "so perhaps discretion is called for before spreading rumors that I have problems expelling waste." Point taken. But that, really, is the beauty of the calculation. The reasoning merely assumes that the Thinker is subject to the known laws of physics and is composed of elementary particles like electrons. The analysis is thus completely general. We do not need to know anything about the Thinker's detailed physiology or construction to conclude that as the Thinker's temperature decreases, the rate at which it can expel entropy will drop below the rate at which it produces entropy. With that realization, we have no choice but to break the news. "Although thinking at ever lower temperatures is essential for prolonging thought as well as for needing only a finite supply of energy, there will come a point when your entropy will build up more quickly than you can expel it. And from there on, if you try to think further, you will burn up in your own thoughts."[36]

Before the crestfallen Thinker can fully think this through, a member of our crack team proposes a way forward: hibernation. The Thinker needs periodically to give thinking a rest—turn off its mind and go to sleep—pausing entropy production while continuing to clear out all of its waste heat. If the break from thinking is long enough, then when the Thinker awakes it will have expelled all waste and so will no longer face the danger of burning up. And since the Thinker won't be thinking during the downtime, when it awakes it won't even notice the hiatus. Emboldened by the solution, one originally proposed by Dyson in his groundbreaking paper, we assure the Thinker that with this rhythm thought can continue forever.

But can it?

A Final Thought on Thought

Two developments in the decades since Dyson's paper are particularly relevant to the strategy. One clarifies the link between the act of thinking and the production of entropy, which leads to a modest reinterpretation of the result. The other brings to bear the accelerated expansion of space, which has the potential to undermine the conclusion entirely, placing thought squarely in the entropic cross hairs.

First, the reinterpretation. The core of Dyson's reasoning is that the act of thinking necessarily produces heat. I made this plausible by recalling that thought is linked to information, information is linked to entropy, and entropy is linked to heat. But the links are subtle, and more recent insights, largely coming from computer science, show that there are clever ways of carrying out elementary information processing—like adding one and one and getting two—without any degradation in energy.[37] With the assumption that thought and computation are cut from the same cloth, a Thinker invoking such a strategy would not generate any waste at all.

Nonetheless, related considerations from computer science show that a version of the thought-entropy-heat connection that drove our initial analysis does remain intact, it just has a slightly different flavor. The results show that if a computer **erases** any of its memory banks, waste heat is necessarily produced. (Recall that waste heat is generally produced by processes that are difficult to reverse, like shattering a glass; erasing data makes it difficult to reverse a computation and so it is not particularly surprising that erasures produce heat.)[38] Taking this into account, our advice to the Thinker needs only gentle modification. The Thinker **can** think without the need to purge heat so long as the Thinker never erases a memory. But assuming the Thinker is of finite extent, it will have a finite memory capacity that will sooner or later fill to its limit. Once it does, all the Thinker can do internally is reshuffle the fixed information it has in memory, endlessly ruminating on old thoughts—not a version of immortality many of us would choose. If the Thinker wants the creative capacity to think new thoughts, to lay down new memories, to explore new intellectual terrain, then it will have to allow for erasures, thereby producing heat and taking us right back to the situation discussed in the previous section and the hibernation strategy recommended there.

The second development is more pressing. The discovery that the expansion of space is accelerating raises a new and possibly insurmountable hurdle for endless thought.[39] If, as the data currently suggest, the accelerated expansion continues unabated, then as we encountered on floor 12, distant galaxies will disappear as if they have fallen over a cliff at the edge of space. That is, we are surrounded by a distant spherical horizon marking the boundary of what, even in principle, we can see. Everything more distant than the boundary recedes from us at greater than light speed, and so any light emitted from such distances will never reach us. Physicists call the distant boundary our **cosmological horizon.**

You can picture the distant cosmological horizon as an enormous glowing sphere, much like a spherical array of distant heat lamps that generates a background temperature in space. I'll explain why this is in the next chapter (it is closely related to the physics of black holes, which also have glowing horizons, as discovered by Stephen Hawking), but here let me stress that the temperature from the glowing cosmological horizon is completely distinct from the 2.7 kelvin microwave background temperature left over from the big bang. Over time, the microwave background temperature will continue to cool, closing in on absolute zero as space continues to expand and the microwave radiation continues to dilute in intensity. The temperature arising from the cosmological horizon behaves differently. It's constant. It's tiny—based on the measured rate of accelerated expansion, it's about 10^{-30} kelvin—but it's enduring. And in the long run, endurance matters.

Heat only flows spontaneously from things that are hotter to those that are cooler. When the Thinker's temperature is higher than that of the universe, it has the opportunity to radiate its waste heat into space. But were the Thinker's temperature to decrease below that of space, heat would flow in the other direction—from space to the Thinker—thwarting the Thinker's need to flush out its waste heat. This implies that the hibernation strategy is destined to fail. As the Thinker continues to decrease its temperature (which, remember, is what allows it to continue thinking indefinitely on a finite energy budget), sooner or later it will reach the tiny value of 10^{-30} kelvin. At that point, game over. The universe won't accept its waste. One more thought (or, more precisely, one more erasure) and the Thinker fries.

The conclusion rests on the assumption that the accelerated expansion of space will persist unchanged. No one knows if this will prove to be the case. The acceleration might increase, propelling us toward a big rip, further diminishing the prospects for life and thought. Or it might decrease. That would obviate a cosmological horizon, turn off the distant heat lamps, and allow the temperature of the universe to decrease indefinitely. As physicists Will Kinney and Katie Freese showed, this cosmological possibility would reinstate Dyson's original optimism, allowing the Thinker, diligently following the hibernation schedule, to continue thinking indefinitely far into the future.[40]

Far be it from me to diminish a lone ray of hope for the future of thought, but it's useful to recap where things stand. Our entire chain of reasoning is forged in optimism. In a universe that may lack everything from stars and planets right down to molecules and atoms, we've assumed that the Thinker can exist. While stable elementary particles—like electrons, neutrinos, and photons—will be wafting about, it takes a rose-colored imagination for the mind's eye to picture gathering them up and producing a thinking structure. Yet, to be as broadminded as possible, we've assumed that such an entity may be formed. And it is surely gratifying to learn that if the universe expands in the right way, there's at least a chance that such Thinkers can think indefinitely. All the same, it is hard to avoid the conclusion that the far future of thought is precarious.

Indeed, if the accelerated expansion does not slow, there will come a time when thought takes its final bow. Our understanding is too coarse to make a precise prediction, but putting rough numbers into the equations suggests this could happen within the next 10^{50} years. A big unknown, as we noted at the outset, is whether intelligent life will be able to intercede in the cosmic unfolding, perhaps affecting the evolution of stars and galaxies, mining unanticipated sources of high-quality energy, or even controlling the rate of spatial expansion. Because of the complexity of intelligence, it is impossible to weigh in with anything more than wild conjecture, which is why I've chosen to avoid such influences entirely. So, putting intelligent intervention to the side and diligently hewing to the second law of thermodynamics, we conclude that by the time we climb to the fiftieth floor, the universe may very well have hosted its final thought.

By most scales humans have contemplated, 10^{50} years is a spectacularly long span. It can accommodate the stretch from the big bang until today more than a billion billion billion billion times over. Yet when considered from the timescale of, say, the seventy-fifth floor, 10^{50} years is fleeting—it is much less, ridiculously less, than our experience of a time delay between turning on a table lamp and light reaching our eyes. And, of course, if the universe is eternal, any duration, however long, registers as infinitesimal. Narrated from the perspective of these longer scales, the cosmological accounting would go like this: a moment after the big bang, life arose, briefly contemplated its existence within an indifferent cosmos, and dissolved away. It is a cosmic recapitulation of Pozzo's lament as he rails against those awaiting Godot, "They give birth astride of a grave, the light gleams an instant, then it's night once more."

Some will deem this future bleak. Even with his more rudimentary mid-twentieth-century understanding, Bertrand Russell, whose assessment we encountered back in chapter 2, surely did. My view is different. To me, the future that science now envisions highlights how our moment of thought, our instant of light, is at once rare, wondrous, and precious.

10

THE TWILIGHT OF TIME

Quanta, Probability, and Eternity

Long after thought concludes, with no cogitating beings left to notice, the laws of physics will continue to do what they have always done—delineate the unfolding of reality. As they do, the laws will manifest an essential realization: quantum mechanics and eternity form a powerful union. Quantum mechanics is a particular sort of starry-eyed dreamer, allowing for a vast collection of possible futures while grounding its madcap vision by specifying the likelihood of any given outcome. Over familiar timescales we can safely ignore those outcomes whose quantum probabilities are so fantastically small that we would have to wait far longer than the current age of the universe to have a reasonable chance of encountering them. But over timescales so vast that by comparison the current age of the universe is evanescent, many possibilities we could previously brush aside now require due consideration. And if there truly is no end date for time, then any and all outcomes not strictly forbidden by the quantum laws—familiar to bizarre, likely to implausible— can rest assured that sooner or later they will be given their moment to shine.[1]

In this chapter, we will examine a handful of such rare cosmological processes, biding their time, waiting to be tapped on the shoulder and called upon to enter reality.

The Disintegration of Black Holes

In the middle of the twentieth century, with their decisive role in the final episodes of World War II, physicists enjoyed marked prominence. The dominant areas of research were nuclear and particle physics, investigations that in the words of Freeman Dyson had endowed physicists with the seemingly godlike powers to "release this energy that fuels the stars . . . to lift a million tons of rock into the sky."[2] General relativity, by contrast, was widely viewed as a niche discipline that had already lived out its glory days. Physicist John Wheeler would change that. Wheeler's contributions to nuclear and quantum physics were numerous and influential, but he had an abiding affection for the general theory of relativity. He also had an uncanny knack for inspiring others with his enthusiasm. During the following decades, Wheeler would train some of the world's most masterful physicists, who would work with him to reinstate general relativity as a vibrant field of scientific research.

Black holes were a particular fascination for Wheeler. According to general relativity, once something falls inside a black hole it can't escape. It's gone. Permanently. Thinking this through in the early 1970s, Wheeler was led to a puzzle that he mentioned to his student Jacob Bekenstein. Black holes seemed to offer a ready-made strategy for violating the second law of thermodynamics. Take a hot cup of tea, Wheeler mused, and toss it into a nearby black hole. Where does the tea's entropy go? Since the inside of a black hole is permanently inaccessible to those on the outside, the hot tea, together with its entropy, seem to have disappeared. Wheeler worried that disposal of entropy into a black hole provided a reliable means for willfully breaching the second law.

After a few months, Bekenstein came back to Wheeler with a resolution. The tea's entropy is not gone, he declared. The entropy has simply been transferred to the black hole. Much as grabbing a hot sauté pan transfers some of the pan's entropy to your hand, Bekenstein suggested that anything falling into a black hole transfers its entropy to the black hole itself.

It is a natural response, one that had also occurred to Wheeler.[3] However, it immediately slams into a problem. Entropy, as we have seen, counts the number of rearrangements of a system's constituents that leave it "pretty much looking the same." Or, more precisely, entropy counts the distinct configurations of a system's microscopic constituents that are compatible with its macroscopic state. If the tea transfers its entropy to the black hole, the entropy should show up as an increase in the number of internal rearrangements of the black hole that have no effect on the black hole's macroscopic features.

Here's the problem: In the late 1960s and early 1970s, physicists Werner Israel and Brandon Carter used the equations of general relativity to show that a black hole is fully determined by just three numbers: the black hole's mass, the black hole's angular momentum (how fast it is spinning), and the black hole's electrical charge.[4] Once you have measured these macroscopic features, you have all the information necessary to fully describe the black hole. Which means that any two black holes with the same macroscopic features—the same mass, the same angular momentum, and the same electric charge—are identical, right down to the last detail. So unlike a collection of pennies in which specifying, say, thirty-eight heads and sixty-two tails allows for billions upon billions of different configurations of the coins, and unlike a container of steam in which specifying the volume, temperature, and pressure allows for a gargantuan number of distinct configurations of the molecules, when it comes to black holes, specifying the mass, angular momentum, and the electric charge rigidly points to one and only one configuration. With no other configurations to count, no look-alikes to enumerate, it would seem that black holes do not carry any entropy at all. Toss in a cup of tea and, apparently, its entropy will vanish. When faced with a black hole, the second law of thermodynamics seems to capitulate.

Bekenstein would have none of it. Black holes, he proclaimed, **do** have entropy. What's more, when something falls in, the black hole's entropy increases in just the right way to make the world safe for the second law. To grasp the gist of Bekenstein's reasoning, note first that when something falls into a black hole, its mass is not lost. Everyone who studied and understood general relativity agreed that anything falling in shows up as an increase in the mass of the black hole itself. To visualize the process, picture a black hole's **event horizon,** the spherical surface defining the black hole's boundary, marking locations beyond which there is no coming back. The math shows that the radius of the event horizon is proportional to the mass of the black hole: less mass entails a smaller horizon, more mass a larger horizon. When you throw something in, the black hole's mass increases, and so you should picture its horizon swelling outward in response. The black hole eats and its spherical waist widens.

Following the spirit of Bekenstein's approach,[5] imagine now tossing in a particularly special probe, one carefully designed to examine how a black hole responds to entropy. To that end, we prepare a single photon whose wavelength is so long—whose possible locations are so spread out—that when it encounters the black hole, the most precise description we can give of the outcome is expressed by a single unit of information: either the photon fell into the black hole or it did not. By design, the photon's position is so nebulous that if it is captured by the black hole we cannot provide a more detailed description such as specifying that the photon entered the black hole through this or that spot on the horizon. Such a photon carries a single unit of entropy and so allows us to examine mathematically how the black hole responds when it eats a single entropy meal.

Since the photon has energy, and since energy and mass are two sides of the same Einsteinian coin (from $E = mc^2$), if the black hole consumes the photon, its mass increases slightly and its event horizon expands slightly. But the payoff is in the particulars. Bekenstein noticed a crucial pattern: by tossing in one unit of entropy, the black hole's event horizon would expand by one unit of area (a so-called **quantum unit of area** or a **Planck area,** which is about 10^{-70} square meters).[6] Toss in two units of entropy, and the surface area would grow by two units of area. And so on. The surface area of the black hole's event horizon thus seems to keep track of the entropy the black hole has ingested. Bekenstein elevated the pattern to a proposal: **the total entropy of a black hole is given by the total area of its event horizon** (measured in Planck units). This was the new idea Bekenstein delivered to Wheeler.

Bekenstein could not explain the surprising link between the entropy of a black hole and its outer surface, its event horizon; the link is unexpected because the entropy of an ordinary object, like the cup of tea, is contained in its interior, its volume. Nor could Bekenstein explain how his proposal related to the conventional framework in which entropy should enumerate the possible rearrangements of a black hole's microscopic ingredients (an issue that would lie mostly dormant until the mid-1990s, when string theory would provide insight). But as an accounting device, his proposal offered a quantitative way to rescue the second law of thermodynamics. The fix is immediate: when tracking total entropy, you need to tally not only the contributions of matter and radiation but also the contributions from black holes. Throwing your tea into a black hole reduces entropy at your breakfast table, but if you calculate the increase in the surface area of the black hole's event horizon, you'll realize that the entropic decrease you enjoy at home is offset by the entropic increase in the black hole itself. By providing an algorithm for including black holes in the entropy accounting, Bekenstein bucked up the second law, allowing it once again to walk with its head held high.

When Stephen Hawking heard about Bekenstein's proposal, he considered it ludicrous. Many other physicists had a similar take. Fully determined by just three numbers and consisting of mostly empty space (everything that falls into a black hole is drawn relentlessly toward its central singularity), black holes had acquired an aura of utter simplicity. The view, coarsely put, was that black holes cannot carry disorder because there is nothing within them to be disordered. Leading the charge against Bekenstein's proposal, Hawking launched his own calculations using a delicate comingling of the mathematical methods of general relativity and quantum mechanics, which he anticipated would quickly reveal a fallacy in Bekenstein's reasoning. Instead, the calculations led Hawking to a conclusion so shocking that it took him some time to believe it. Hawking's analysis not only confirmed Bekenstein's, but also revealed complementary surprises: black holes have a **temperature** and black holes **glow.** They radiate. Black holes are black in name only. Or, said more precisely, black holes are black only if you ignore quantum physics.

Briefly, here is the essence of Hawking's reasoning.

According to quantum mechanics, any tiny region of space will always harbor quantum activity. Even if the region appears empty, seemingly containing no energy at all, quantum theory shows that its energy content actually rapidly fluctuates up and down, yielding zero energy only on **average.** These are the same type of quantum fluctuations that gave rise to the temperature variations in the cosmic microwave background radiation that we encountered in chapter 3. Through $E = mc^2$, such quantum energy fluctuations can also show up as quantum mass fluctuations—particles and their antiparticle partners popping into existence in otherwise empty space. This is happening right now in front of your eyes, yet however intently you stare you'll see no evidence of it. The reason is that quantum mechanics also dictates that such particle-antiparticle pairs quickly find each other, annihilate and fade back into empty space. We do detect indirect signatures of these ephemeral machinations because it is only when we include them in our calculations that we achieve the stunning agreement between predictions and measurements that has justifiably made quantum mechanics the centerpiece of fundamental physics.[7]

Hawking revisited these quantum processes but now imagined them taking place just outside the event horizon of a black hole. When a particle-antiparticle pair pops into this environment, sometimes the two particles will annihilate quickly, just as they would anywhere else. But, and this is the point, Hawking realized that on occasion they will not annihilate. Sometimes one member of the pair will get sucked into the black hole. The surviving particle, now bereft of a partner with which to annihilate (and tasked with conserving total momentum), turns tail and rushes outward. With this happening repeatedly in every tiny region of space all along the surface of the black hole's spherical horizon, the black hole will appear to radiate particles in all directions, what we now call **Hawking radiation.**

What's more, according to the calculations, each such particle that falls into the black hole has **negative** energy (perhaps not surprising, given that the partner particle escaping the hole has positive energy, and total energy must be conserved). As the black hole consumes these negative mass particles, it's as if it is eating negative calories, resulting in its mass going down, not up. Viewed from the outside, the black hole thus appears to steadily shrink as it radiates particles. Were it not that the source of the radiation is exotic—a black hole immersed in the quantum bath of fluctuating particles inherent in empty space—the process would appear thoroughly pedestrian, like a glowing chunk of charcoal radiating photons as it slowly wastes away.[8]

Just as a growing black hole, whether consuming hot tea or turbulent stars, fully conforms to the second law of thermodynamics, so too for a shrinking black hole. The decrease in the area of a shrinking black hole's event horizon means its own entropy decreases, but the radiation the black hole emits, streaming outward and spreading across an ever-wider spatial expanse, transfers a more-than-compensating cache of entropy to the environment. The choreography is familiar: as black holes radiate, they dance the entropic two-step.

Hawking's result made all of this mathematically precise. Among much else, he discovered a precise formula for the temperature of a glowing black hole. I'll give a qualitative explanation of his result in the next section (and for the mathematically inclined, the formula is in the endnotes[9]), but the feature most relevant to us here is that the temperature is **inversely** proportional to the mass of the black hole. Much as mature Great Danes are large and mild while shih tzu puppies are small and manic, large black holes are calm and cool while small black holes are frenzied and hot. Some numbers, courtesy of Hawking's formula, make this explicit. For a large black hole, like the one at the center of our galaxy with four million times the mass of the sun, Hawking's formula pegs its temperature at the tiny value of a hundredth of a trillionth of a degree above absolute zero (10^{-14} kelvin). For a smaller black hole, with the mass of the sun, the temperature is higher, but far from balmy, just shy of a tenth of millionth of a degree (10^{-7} kelvin). A tiny black hole, with the mass, say, of an orange, would be blazing with a temperature of about a trillion trillon degrees (10^{24} kelvin).

A black hole whose mass is larger than the moon's has a temperature that is lower than that of the 2.7 degree microwave background radiation currently suffusing the cosmos. Handy for erudite cocktail party chatter, this is a numerical factoid of cosmological significance. Because heat spontaneously flows from higher to lower temperatures, heat will flow from the frigid microwave-filled environment surrounding such a black hole to the yet more frigid black hole itself. Although the black hole emits Hawking radiation, on balance it will take in more energy than it releases, slowly increasing its heft. Because even the smallest black holes so far discovered by astronomical observations are much more massive than the moon, they are all in the process of plumping up. However, as the universe continues to expand, the microwave background radiation will continue to dilute and its temperature will continue to cool. In the far future when the background temperature of space drops below that of any given black hole, the energy seesaw will pivot, the black hole will emit more than it receives, and it will start shrinking as a result.

In the fullness of time, black holes will waste away too.

There are many questions about black holes that remain at the forefront of contemporary research, and one of considerable importance to our discussion here concerns the final moments of a black hole's existence. As a black hole radiates, its mass decreases and, in turn, its temperature increases. What happens when the black hole is almost gone, when its mass nears zero and its temperature soars toward infinity? Does it explode? Does it fizzle? Something else? We don't know. Even so, the quantitative understanding of Hawking radiation allowed physicist Don Page to determine the rate at which a given black hole shrinks and thus the time it will take to reach its final moment—whatever the details of that moment may be.[10] Taking the mass of the sun as representative of those black holes that form from a dying star, Page's result shows that by about floor 68 of the Empire State Building, 10^{68} years after the bang, such black holes will have radiated away.

The Disintegration of Extreme Black Holes

The black holes believed to inhabit the center of most if not all galaxies have gargantuan masses. As astronomical surveys have progressed, each record holder has been unseated by the next, with champion masses heading toward one hundred billion times that of the sun. A black hole of that mass has an event horizon so large that it would stretch from the sun past the orbit of Neptune and a fair way toward the Oort cloud. Even if you're a bit rusty on Oort and his distant cloud, just know that it takes sunlight well over one hundred hours to reach it, so we are talking about a black hole with a monstrous span. But as I'll now explain, the enormous size of these black holes belies their placid demeanor.

According to general relativity, the recipe for building a black hole is dead simple: gather any amount of mass and form it into a ball of a sufficiently small size.[11] Of course, even a passing familiarity with black holes leads you to expect that "sufficiently small" means **really** small, **spectacularly** small, **ludicrously** small. And in some cases your expectation is right on the mark. To turn a grapefruit into a black hole, you'd need to squeeze it down to about 10^{-25} centimeters across; to turn the earth into a black hole you'd need to squeeze it down to about two centimeters across; and for the sun, you'd need to squeeze it to about six kilometers across. Each example requires a fantastic crushing of matter, contributing to the widespread intuition that to form a black hole you need stupendous densities. But were you to continue cataloging examples well beyond the mass of the sun, focusing on the formation of ever-larger black holes, you'd come upon a pattern you might find surprising.

As the amount of matter used to create a black hole increases, the required density to which that matter must be crushed **decreases.** If you'll indulge one, well two, mathy sentences, the reason is immediately apparent: Because the radius of a black hole's event horizon scales with its mass, its volume scales as mass cubed, and so the average density—mass per volume—**drops** with the mass squared. Increase the mass by a factor of two and the density drops by a factor of four; increase the mass by a factor of a thousand and the density drops by a factor of a million. Math to the side, the qualitative point is that in forming a black hole, the larger the mass, the less that mass needs to be crushed. To build a black hole like the one in the center of the Milky Way, whose mass is about four million times that of the sun, you need matter whose density is about one hundred times that of lead, so you've still got some serious crushing ahead of you. To build one with mass one hundred million times that of the sun, the necessary density drops all the way to that of water. And to build one that's four billion times the mass of the sun, the density you need is on par with that of the air you're now breathing. Gather together four billion times the mass of the sun in air, and unlike the case with a grapefruit, or the earth, or the sun, to create a black hole you would not need to squeeze the air at all. Gravity acting on the air would form a black hole on its own.

I'm not advocating bags of air as realistic raw material for creating supermassive black holes, but the fact that a black hole weighing four billion times as much as the sun would have an average density of air is remarkable, and a telling illustration of how the properties of black holes can differ from popular conceptions.[12] Gigantic when assessed by their mass and size, such black holes are dainty when assessed by their average densities, rendering them decidedly gentle giants. In this sense, larger black holes are less extreme than smaller black holes, a realization that gives an intuitive explanation of Hawking's finding that the more massive the black hole, the lower its temperature and the more subdued its glow.

The longevity of large black holes thus benefits from two related factors: they have more mass to radiate and, with their lower temperatures, they radiate that mass more slowly. Plugging numbers into the equations, we find that a black hole whose mass is about one hundred billion times that of the sun will wither away at so leisurely a pace that only as we reach the top floor of the Empire State Building, floor 102, will such a black hole spew its last burst of radiation and finally, truly, fade to black.[13]

An End of Time

Gazing on the universe from the 102nd floor, we will not see much beyond a diffuse mist of particles wafting through space. Occasionally, the attraction between an electron and its antiparticle, the positron, will draw them ever closer along inward spiraling trajectories until they annihilate in a tiny flash, a pinprick of light momentarily penetrating the blackness. If the dark energy has drained away and the rapid expansion of space has diminished, it is possible that particles may accumulate into even larger black holes that will radiate yet more slowly, yielding yet longer lifetimes. But if the dark energy persists, particles will be driven apart increasingly quickly by the accelerated expansion, ensuring that they will rarely if ever encounter one another. Curiously, the conditions have an affinity with those just after the big bang, when space was also populated with separate particles. The difference is that in the early universe, the particles were so dense that gravity easily coaxed them into structures like stars and planets, while in the later universe, the particles will be so widely dispersed and the quickening expansion of space so unrelenting that such clumping will be extraordinarily unlikely. It is a cosmic version of dust to dust, with the early dust primed to dance the entropic two-step, being driven by gravity into orderly astronomical structures, while the later dust, spread so thinly, will be content to drift quietly through the void.

Physicists sometimes liken this future era to the end of time. Not that time stops. But when the action amounts to no more than an isolated particle moving from this spot in the vast reaches of space to that spot, it's reasonable to conclude that the universe has finally transitioned to oblivion. Still, our willingness in this chapter to consider yet longer durations raises to relevance processes so improbable that they would otherwise be summarily dismissed. Barely conceivable though they are, these rare events may punctuate oblivion with infrequent but far-reaching possibilities.

The Disintegration of Emptiness

At a press conference on July 4, 2012, held at CERN, the European Center for Nuclear Research, spokesperson Joe Incandela announced the discovery of the long-sought Higgs particle. I was watching the live feed at the Aspen Center for Physics in a room packed with colleagues. It was about two a.m. Everyone erupted into wild cheers. The camera cut to Peter Higgs, removing his glasses and wiping his eyes. Higgs had proposed the particle bearing his name nearly fifty years earlier, had successfully fought the resistance unfamiliar ideas sometimes encounter, and had waited a lifetime to learn that he was right.

While on a long walk in the outskirts of Edinburgh, a young Peter Higgs solved a puzzle that had been frustrating researchers around the world. The mathematics for describing the strong, weak, and electromagnetic forces, as well as the particles of matter these forces influence, was rapidly coming together. Working shoulder to shoulder, theorists and experimenters were writing a quantum mechanical manual laying out the workings of the microworld. But there was one glaring omission. The equations couldn't explain how the fundamental particles acquired mass. Why is it that if you were to push on fundamental particles (like electrons or quarks), you would feel the particles resisting your effort? This resistance reflects the particle's mass but the equations seemed to tell a different story: according to the math, the particles should be massless and thus should offer no resistance at all. Needless to say, the mismatch between reality and the mathematics was driving the physicists batty.

The reason the math seemed to countenance only massless particles is a touch technical, but it comes down to symmetry. Much as a cue ball looks the same as you turn it this way and that, the equations describing fundamental particles look the same as you swap this mathematical term with that one. In each case, the insensitivity to change—

of orientation for the cue ball and of mathematical rearrangement for the equations—reflects a high degree of underlying symmetry. For the cue ball, the symmetry ensures that it rolls smoothly. For the equations, the symmetry ensures that the mathematical analysis unfurls smoothly. As researchers in particle physics had realized, without the symmetry the equations would be inconsistent, yielding nonsense similar to the result of dividing one by zero. Hence the puzzle: the analysis revealed that the same mathematical symmetry that ensures healthy equations also requires massless particles (perhaps not surprising, as zero is itself a highly symmetric number, holding firm to its value when multiplied or divided by any other number).

That's where Higgs came in. He argued that, intrinsically speaking, particles **are** massless, just as the pristine symmetric equations required. However, Higgs continued, when thrust into the world, the particles acquire mass through an environmental influence. Higgs envisioned that space is filled with an invisible substance, now called the **Higgs field,** and that particles pushed through the field experience a drag force somewhat like that experienced by a Wiffle ball flying through air. Even though a Wiffle ball weighs next to nothing, if you hold it outside the window of a car revving up to ever-higher speeds, your hand and arm will get quite a workout: the Wiffle ball feels massive because it is plowing through the resistance exerted by the air. Similarly, Higgs proposed, when you push on a particle it feels massive because it is plowing through the resistance exerted by the Higgs field. The more hefty a particle the more it resists your push, which according to Higgs means the particle experiences a stronger resistance from his space-permeating field.[14]

If you're not already familiar with the notion of the Higgs field but have diligently read the previous chapters, the idea may not sound particularly exotic. Modern physics has grown accustomed to the idea of invisible substances suffusing space, latter-day versions of the ancient ether. From the inflaton field that may have driven the big bang to the dark energy that may be responsible for the accelerated expansion of the universe now measured, physicists of the past few decades have not been shy about proposing that space is filled with invisible stuff. But in the 1960s, the idea was radical. Higgs was suggesting that if space were truly empty in the conventional and intuitive sense, particles would have no mass at all. He thus concluded that space must not be empty, and the peculiar substance it harbors must be just right for imbuing particles with their evident mass.

The first paper in which Higgs made the case for this new proposal was dismissed out of hand. "I was told I was talking nonsense," Higgs recalled of the reaction.[15] But those who carefully studied the idea realized its merits and it slowly gained currency.

Ultimately, it was embraced fully. I first encountered the Higgs proposal in a graduate course in the 1980s, and it was presented with such certitude that for a while I didn't realize the proposal had yet to be confirmed experimentally.

The strategy for testing the proposal is as easy to describe as it is challenging to carry out. When two particles, say two protons, slam together at high speed, the collision should jiggle the surrounding Higgs field. On occasion, this would theoretically knock free a tiny droplet of the field, which would show up as a new type of elementary particle— a **Higgs particle**—what Nobel laureate Frank Wilczek calls a "chip off the old vacuum." A sighting of this particle would thus provide the theory's smoking gun, a goal that inspired more than thirty years of research, by more than three thousand scientists, from over three dozen countries, using the world's most powerful particle accelerator, with a price tag exceeding fifteen billion dollars. The conclusion of that odyssey, announced in that Independence Day press conference, was signaled by a tiny bump in an otherwise smooth graph produced by data collected at the Large Hadron Collider—experimental confirmation that the Higgs particle was in hand.

It is a wonderful episode in the annals of human discovery, deepening our understanding of the properties of particles and bolstering our confidence in the capacity of mathematics to reveal hidden aspects of reality. The relevance of the Higgs field for our journey on the cosmic timeline comes from a related but distinct consideration—at some point in the future the value of the Higgs field may change. And much as the drag experienced by a Wiffle ball would change if the density of air it encountered was different, the masses of the fundamental particles would change if the value of the Higgs field they encountered was different. For all but the most minuscule of shifts, such a change would almost certainly destroy reality as we know it. Atoms and molecules and the structures they build depend intimately on the properties of their particulate constituents. The sun shines because of the physics and chemistry of hydrogen and helium, which depend on the properties of protons, neutrons, electrons, neutrinos, and photons. Cells do what cells do mostly because of the physics and chemistry of the molecular constituents, which again depend on the properties of the fundamental particles. If you change the masses of the fundamental particles you change how they behave, and so you change more or less everything.

A wealth of laboratory experiments and astronomical observations have established that for most if not all of the past 13.8 billion years, the masses of the fundamental particles have been constant and thus the value of the Higgs field has been stable. Yet, even if there is only a minute probability that in the future the Higgs field can jump to a different value, that probability will be amplified into a near certainty by the enormous durations we are now considering.

The relevant physics for a Higgs jump is called **quantum tunneling,** a process best grasped by first considering it in a simpler setting. Place a small marble in an empty champagne flute, and if no one disturbs it, you would expect the marble to remain there. After all, the marble is hemmed in by barriers on all sides and doesn't have enough energy to climb the walls of glass and escape through the top. Nor does it have enough energy to penetrate directly through the glass. Similarly, if you place an electron in a trap shaped like a tiny champagne flute, hemming in its position with barriers on all sides, you would expect that it too would remain in place. Indeed, most of the time the electron does. But sometimes it doesn't. Sometimes the electron disappears from the trap and rematerializes outside it.

Surprising as such a Houdini-like move may be for us, in quantum mechanics it is business as usual. Using Schrödinger's equation, we can calculate the probability that an electron will be found in this or that location, such as on the inside or on the outside of the fluted trap. The math shows that the more formidable the trap—the taller and thicker the sides—the smaller the likelihood that the electron will escape. But, and this is key, for the probability to be zero, the trap would need to be infinitely wide or infinitely high, and in the real world that just doesn't happen. And a nonzero probability, however small, means that by waiting long enough, sooner or later the electron **will** make it to the other side. Observations confirm that it does. Such a transit through a barrier is what we mean by "quantum tunneling."

I've described quantum tunneling in terms of a particle penetrating a barrier, changing its location from here to there, but it can also involve a field penetrating a barrier, changing its value from this to that. Such a process, involving the Higgs field, may determine the long-term fate of the universe.

In the units physicists conventionally use, the current value of the Higgs field is 246.[16] Why 246? No one knows. But the drag force mustered by a Higgs field with this value (together with the precise manner in which each particle interacts with it) successfully explains the masses of the fundamental particles. But why has the Higgs value been stable for billions of years? The answer, we believe, is that the Higgs value, like the marble in the flute or the electron in the trap, is hemmed in on all sides by formidable barriers: if the Higgs field was to try migrating from 246 to a larger or smaller number, the barrier would forcefully drive it back to its original value, much like the marble would be driven back to the bottom of the flute should someone momentarily shake the glass. And were it not for quantum considerations, the Higgs value would permanently remain at 246. But as Sidney Coleman discovered in the mid-1970s, quantum tunneling changes the story.[17]

Just as quantum mechanics allows an electron occasionally to tunnel out of a trap, so too does it allow for the value of the Higgs field to tunnel through a barrier. Were this to occur, the Higgs field would not change its value across all of space simultaneously. Instead, in some tiny region singled out by the random nature of quantum events, the Higgs would make its move, tunneling through the barrier to a different value. Then, much as a marble that tunnels through a champagne flute will drop to a lower height, the Higgs field's value would drop to a lower energy. The lure of lower energy would then coax the Higgs field at nearby locations to make the transition too, a domino-like effect that would yield an ever-growing sphere within which the Higgs value would have changed.

Inside this sphere, the new Higgs value would cause particle masses to change, so the familiar features of physics, chemistry, and biology would no longer hold. Outside the sphere, where the Higgs value had yet to shift, particles would retain their usual properties, and so all would seem normal. Coleman's analysis revealed that the boundary of the sphere, marking the transition from old Higgs value to new, would spread outward at very nearly the speed of light.[18] Which means that for those of us on the outside it would be virtually impossible to see the wall of doom approaching. By the time we saw it, it would be upon us. One moment it would be life as usual. The next moment we would cease to be. Might new structures and perhaps new forms of life ultimately emerge in this realm populated by particles with unfamiliar properties? Possibly. But these questions are currently beyond our ability to answer.

Physicists cannot pinpoint when the Higgs might make such a jump. The timescale depends on particle and force properties that have yet to be determined with adequate precision. Moreover, as a quantum process, it can only be predicted probabilistically. Current data suggest that the Higgs is likely to tunnel to a different value somewhere between 10^{102} and 10^{359} years from now—somewhere between floors 102 and 359 (a range that would even challenge the reach of the Burj Khalifa).[19]

Because the Higgs field redefines what we mean by emptiness—the emptiest of empty space anywhere in the observable universe contains the Higgs field with value 246—quantum tunneling of the Higgs field's value reveals an instability of empty space itself. Wait long enough, and even empty space will change. While the timescale for such change, such disintegration, gives little cause for anxiety, note that there is a chance that the tunneling event could happen today. Or tomorrow. That is the burden of living in a quantum universe in which future events are governed by probability. Just as you might drop a few hundred pennies and they all land heads—possible but unlikely—we might be on the verge of getting slammed by a wall of shifted Higgs field trailing a new variety of empty space in its wake. Possible, but unlikely too.

That this probability is minuscule would seem a good thing. Being swept away by a light-speed wall of doom, while swift and painless, is something most of us would rather avoid. However, as we turn our attention to even longer timescales, we will encounter quantum processes that are not only bizarre but have the capacity to undermine everything we hold to be true about reality. In response, some physicists have cultivated a fondness for theories in which the universe will end well before we would have to face the implosion of rational thinking itself.

Boltzmann Brains

As we have ascended the timeline, we have witnessed the second law of thermodynamics in action. From the big bang to the formation of stars, the dawn of life, the processes of mind, the depletion of galaxies, and on through the disintegration of black holes, entropy has been relentlessly on the rise. This consistent growth can obscure the fact that the second law's decree is probabilistic.

Entropy **can** decrease. The particles of air currently spread throughout your room **can** all simultaneously coalesce into a ball hovering near the ceiling, leaving you gasping for breath. It's just so unlikely, and the timescale for it to happen is just so enormous, that we acknowledge the possibility but wisely get on with our lives. However, since we are now taking the long view, let's throw off our temporal provincialism and consider some fairly mind-blowing entropy-decreasing possibilities.

Imagine that you've been reading this book for the past hour, sitting in your favorite chair, and now and then sipping tea from your favorite mug. If asked how this cozy arrangement came about, you'd say you bought the mug in New Mexico from a local potter, that you inherited the chair from your fraternal grandmother, and that you've always been interested in the workings of the universe, which led you to this book. If encouraged to provide more details, you'd talk about your upbringing, your siblings, your parents, and so on. If pushed yet harder to reach back in time and provide a more complete account, you might ultimately talk about the very material we've covered in earlier chapters.

All of this is based on a curious fact: everything you know reflects thoughts, memories, and sensations that **currently** reside in your brain. The purchase of the mug has long since passed. What remains is a configuration of particles inside your head that holds the memory. The same is true for your memories of inheriting your grandmother's chair, of being curious about the universe, and of having read about various concepts in this book. From a staunchly physicalist perspective, all of that is in your head right now because of the particular arrangement of the particles that are in your head right now. Which means that if a random spray of particles flitting through the void of a structureless, high-entropy universe should, by chance, spontaneously dip to a lower-entropy configuration that just happens to match that of the particles currently constituting your brain, that collection of particles would have the same memories, thoughts, and sensations that you do. Whether in honor or reproach, I don't know which, such hypothetical, free-floating, untethered minds formed by the rare but possible spontaneous coming together of particles into a special, highly ordered configuration have become known as **Boltzmann brains.**[20]

Alone in the frigid darkness of space, a Boltzmann brain would not think many thoughts before it expired. However, a spontaneous coming together of particles could also yield accessories that would prolong its functioning: the housing of a head and body, a supply of food and water, an appropriate star and planet, to mention a few. Indeed, a spontaneous coming together of particles (and fields) could yield today's entire universe or recreate the conditions that set off the big bang, allowing a universe much like ours to unfold anew.[21] Admittedly, when it comes to a spontaneous drop in entropy, the odds overwhelmingly favor drops that are smaller: fewer particles coming together in structures that are more tolerant of imprecise arrangements. And by overwhelmingly favored, I mean **overwhelmingly** favored. Exponentially favored. And since we have a particular interest in the far future of thought, a solitary Boltzmann brain is the minimal and hence most likely random formation of particles that can briefly cerebrate and thus wonder how in the world it came to be.[22]

What makes this more than the beginnings of a B-grade sci-fi plot is that as we look to the far future, the conditions appear ripe for these bizarre-sounding processes to actually happen. An essential ingredient is the accelerated expansion of space. Earlier we noted that such expansion results in a cosmological horizon—a distant surrounding sphere marking the boundary beyond which objects recede from us faster than the speed of light, cutting off any possibility of contact or influence. Now, much as Hawking showed that quantum mechanics implies that a black hole horizon has a temperature and emits radiation, Hawking and his collaborator Gary Gibbons used similar reasoning to show that a cosmological horizon has a temperature and emits radiation as well. Our analysis in the previous chapter focusing on the future of thought relied on this very fact, concluding that the tiny temperature of our cosmological horizon, about

10^{-30} kelvin, may well be enough to cause future Thinkers, desperately trying to continue thinking indefinitely, to ultimately burn up in their own thoughts. As we will now see, over the course of far-longer timescales similar considerations offer the future of thought the potential for a curious revival.

In the far future, the radiation emitted by the cosmological horizon will provide a dim but consistent source of particles (predominantly massless particles, photons, and gravitons) that will meander through the region of space the horizon surrounds. On occasion, collections of these particles will collide and, via $E = mc^2$, transmute their energy of motion into the production of a smaller number of more massive particles like electrons, quarks, protons, neutrons, and their antiparticles. By resulting in fewer particles and less motion, these processes decrease entropy, but wait long enough and such unlikely things will happen. And will continue to happen. On rarer occasions still, some of the protons, neutrons, and electrons so produced will move in just the right way to join into this or that atomic species. The enormous duration required for such rare processes explains why they are irrelevant in the synthesis of atomic nuclei after the big bang or within stars, but now, with unlimited time on our hands, such processes matter. Over an even longer temporal expanse, the atoms will randomly join into an array of ever-more-complex configurations, ensuring that every now and then on the road to eternity a collection will coalesce into this or that macroscopic structure—bobbleheads to Bentleys. In the absence of thinking beings, all of these will come and go without notice. But every so often the randomly formed macroscopic structure will be a brain. Long extinct, thought will make a momentary comeback.

What is the timescale for such resurrection? With a rough calculation (which math enthusiasts can find in the endnotes[23]) we can estimate that there's a reasonable chance that a Boltzmann brain will form within $10^{10^{68}}$ years. That's a long time. Whereas we could write out the duration represented by the peak of the Empire State Building, 10^{102} years, a one followed by 102 zeroes, on about a line and a half, to write out $10^{10^{68}}$—which is a 1 followed by 10^{68} zeroes—we could replace every character on every page of every book that has ever been printed and even then we would not make a dent. Even so, it is not as though anyone would be hanging around, glancing at their watch, waiting for the entropic drop to get a move on and produce a brain. The universe could persist for nearly an eternity in a run-of-the-mill, disordered, high-entropy state, and no one would complain.

Which raises an interesting, somewhat personal concern. Where did your brain come from? The question sounds silly, but humor me. In answering, you naturally follow your memories and knowledge to explain that you were born with your brain, and that your inception is part of a sequence we can trace back through your ancestral lineage, through the evolutionary record of life, through the formation of the earth, the sun, and so on, all the way back to the big bang. On the face of it, this seems to make good sense. Most of us would give a version of the same response. But as the previous chapters made clear, the window of time during which brains can form in the manner you've recounted is limited—generously, the span is likely between the Empire State Building's tenth and fortieth floors. The window of time for brains to form in the Boltzmannian manner is incomparably longer—it may well be unlimited.[24] As time continues to roll onward, Boltzmann brains will, rarely but reliably, continue to coalesce, and so the total number of such brains that come and go will grow ever larger. A survey of a long-enough stretch on the timeline would thus reveal that the total population of Boltzmann brains far exceeds the total population of traditional ones. The same is true even if we focus on only those Boltzmann brains whose particulate configurations imprint the erroneous belief that they arose in the traditional biological manner. Once again, however rare a process, over arbitrarily long durations it will happen arbitrarily many times.

If you then ask yourself for the most likely way that you acquired the beliefs, memories, knowledge, and understanding that you currently hold,

the dispassionate answer based on sheer population size is clear: your brain just spontaneously
formed from particles in the void, with all of its
memories and other neuropsychological qualities
imprinted through the particular configuration of
the particles. The story you told of how you came
to be is touching but false. Your memories and the
various chains of reasoning that have led to your
knowledge and your beliefs are all fictitious. You do
not have a past. You have just come into existence
as a disembodied brain endowed with thoughts
and memories of things that never happened.[25]

Beyond its utter strangeness, this scenario comes
with a devastating conclusion, the very reason
I have focused on spontaneously forming brains
and not the myriad other inanimate objects that
randomly coalescing particles can also realize. If a
brain, yours or mine or anyone's, can't trust that
its memories and beliefs are an accurate reflection
of events that happened, then no brain can trust
the supposed measurements and observations and
calculations that constitute the basis of scientific
understanding.[26] I have memories of learning general relativity and quantum mechanics, I can think
through the chain of reasoning that supports these
theories, I can recall looking at the data and observations these theories so impressively explain, and
so on. But if I can't trust that these thoughts were
imprinted by the actual events to which I attribute
them, I can't trust that the theories are anything

more than mental figments, and so I can't trust any conclusions to which the theories point. For the kicker, among such conclusions, now rendered untrustworthy, is the likelihood that I'm a spontaneously created brain floating in the void. The deep skepticism that emerges from the possibility of spontaneous brain formation forces us to be skeptical of the very reasoning that led us to entertain the possibility in the first place.

In short, rare spontaneous drops in entropy, which are entailed by the laws of physics, can shake our confidence in the laws themselves and all they supposedly entail. By considering the laws operating over arbitrarily long durations, we are plunged into a skeptical nightmare, rattling our trust in everything. Not a happy place to be. So how can we regain confidence in the foundations of rational thought that have facilitated our vigorous climb up the Empire State Building and beyond? Physicists have developed a number of strategies.

Some conclude that Boltzmann brains are much ado about nothing. Sure, this perspective acknowledges, Boltzmann brains can form. But ease your mind. You are definitely not one of them. Here's how to prove it: Look out on the world and take in all you see. If you are a Boltzmann brain, the odds are overwhelming that a moment later you won't exist. A brain that can last longer is a brain that's part of a larger and more ordered support system and thus requires a yet rarer fluctuation to even lower entropy, making its formation that much more unlikely. So if your second glance at the world seems much like your first, your confidence that you are not a Boltzmann brain increases. Indeed, according to this perspective, every next moment of a similar sort makes your argument stronger and your confidence greater.

Notice, though, that the argument assumes each of the moments in such a sequence is, in the conventional sense, real. If right now you have a memory of looking out at the world a dozen times during the past minute, repeatedly assuring yourself that you are not a Boltzmann brain, that memory reflects the state of your brain right now and is thus compatible with your brain having turned on just now imprinted with those very memories. By taking the scenario fully to heart, you realize that the empirical observations you used to argue that you are not a Boltzmann brain may themselves be part of the fiction. I may have memories of say-ing to myself "I think, therefore I am," but viewed from any given moment, an accurate accounting requires that I say instead, "I think I thought, therefore I think I was." In reality, the memory of such thoughts does not ensure that the thoughts ever happened.

A more convincing approach is to challenge the underlying scenario itself: Central to the argument for Boltzmann brains is the existence of a distant cos-mological horizon continuously radiating particles, the raw materials for building complex structures, including minds. Over the long haul, if the dark energy filling space were to dissipate away, then accelerated expansion would draw to a close and the cosmological horizon would withdraw. Without a distant surrounding surface radiating particles, the temperature of space would close in on zero, and with that the chance of spontaneously forming complex macroscopic structures would close in on zero too. There is as yet no evidence for a weakening (or a strengthening) of dark energy, but future observational missions will study the possibility with greater precision. A conservative assessment is that the jury is still out.[27]

More radical still are approaches in which the universe, or at least the universe as we know it, simply will not exist arbitrarily far into the future. In the absence of the fantastically long durations we have been considering, the likelihood of Boltzmann brains forming becomes so ridiculously tiny that we can safely ignore the process entirely. If the universe were to end long before the time-scale that would make the production of Boltzmann brains likely, we could set our skepticism aside and comfortably revert to our previous account of the origin and development of our brains, including our memories, knowledge, and beliefs.[28]

How might such a speedy end to the universe come about?

Is the End Near?

Earlier, we considered the possibility that the Higgs field might take a quantum leap to a new value, resulting in a sudden change of particle properties that would rewrite many of the basic processes of physics, chemistry, and biology. The universe would carry on but almost certainly without us. If this disjuncture should happen long before the timescales necessary for Boltzmann brains to form—as the data on the Higgs field currently suggest—ordinary brains would dominate the population, and we would sidestep the skeptical morass.[29]

A yet more emphatic resolution would emerge from a quantum leap in which the value of the dark energy would suddenly change. Currently, the accelerated expansion of the cosmos is driven by a positive dark energy suffusing every region of space. But just as positive dark energy yields an outward-thrusting repulsive gravity, negative dark energy yields an inward pulling attractive gravity. Consequently, a quantum tunneling event in which the dark energy leaped to a negative value would mark a transition from the universe's swelling outward to its collapsing inward. Such an about-face would result in everything—matter, energy, space, time—being squeezed to extraordinary density and temperature, a kind of reverse big bang that physicists call the **big crunch**.[30] Much as there is uncertainty regarding what happened at time zero, setting off the bang, there is uncertainty regarding what would happen at the final moment, the crunch itself. What's evident, though, is that were the crunch to happen in far less time than $10^{10^{68}}$ years, the peculiar implications of Boltzmann brains would again be rendered moot.

In one final approach, interesting beyond considerations of Boltzmann brains, physicist Paul Steinhardt and collaborators Neil Turok and Anna Ijjas imagine parlaying such a potential universe-ending crunch into a more upbeat universe-producing bounce.[31] According to this theory, regions of space like ours go through phases of expansion followed by contraction, with the cycles repeating indefinitely. The big bang becomes the big bounce—a rebound from the previous period of contraction. The idea itself is not entirely new. Shortly after Einstein completed the general theory of relativity, a cyclic version of cosmology was proposed by Alexander Friedmann and subsequently developed by Richard Tolman.[32] Tolman's aim, in particular, was to dodge the question of how the universe began. If the cycles extend infinitely far to the past, there was no beginning. The universe always existed. Tolman found, however, that the second law of thermodynamics thwarts this vision. The continual buildup of entropy from one cycle to the next implies that the universe we currently inhabit could be preceded only by a finite number of cycles, thus requiring a beginning after all. In their new version of the cyclic approach, Steinhardt and Ijjas argue that they can surmount this problem. They have established that during each cycle a given region of space stretches far more than it contracts, ensuring the entropy it contains is thoroughly diluted. Cycle upon cycle, the total entropy across the entirety of space increases, as per the second law of thermodynamics. But in any finite region, such as the one giving rise to our observable realm, the entropic buildup that stymied Tolman is no longer a concern. Expansion dilutes away all matter and radiation, while the subsequent contraction harnesses the power of gravity to replenish just enough high-quality energy to start the cycle anew.

The duration of each cycle is determined by the value of the dark energy which, based on today's measurements, sets the duration on the order of hundreds of billions of years. As this is far less than the typical time required for Boltzmann brains to form, cyclic cosmology provides another potential solution for preserving rationality. While there would be ample time during a given cycle to produce brains in the ordinary manner, the cycle would conclude well before there would be time to produce brains in the Boltzmannian manner. With reasonable confidence we could all then declare that our memories were laid down by events that really happened.

Looking to the future, the cyclic approach suggests

that our climb up the Empire State Building would be cut short, ending somewhere in the vicinity of the eleventh or twelfth floor, when the contracting phase of space would result in a bounce that concludes our cycle and initiates the next. The linearity of the skyscraper metaphor would also need an updating to a spiral shape (a soaring version of the Guggenheim Museum comes to mind), with each lap representing a cosmological cycle. Moreover, since the cycles might persist indefinitely into the past as well as the future, we would need to envision the structure extending infinitely far in both directions. Reality as we know it would be part of a single lap around the cosmological track.

In recent years, cyclic cosmology has emerged as a main competitor to the inflationary theory. Although both can explain cosmological observations, including the all-important temperature variations in the microwave background radiation, the inflationary theory continues to dominate cosmological research. In part this reflects the uphill battle to interest physicists in an alternative to a theory that over the course of four decades has propelled cosmology into a mature and precise science. That ours is called the golden age of cosmology is largely attributable to the inflationary theory. Of course, truth in science is not determined by polls or popularity. It is determined by experiments, observations, and evidence. And the inflationary and cyclic theories do make one significantly different observational prediction, which may one day figure prominently in adjudicating between them: The burst of inflationary expansion at the big bang would likely have so vigorously disturbed the fabric of space that the gravitational waves produced might still be detectable. The more gentle expansion of the cyclic model results in gravitational waves too mild to be observed. In the not-too-distant future, observations may thus have the capacity to tip the balance between the two cosmological approaches.[33]

Among researchers, inflation remains the foremost cosmological theory, which is why we have focused upon it in earlier chapters. Even so, it remains thoroughly exciting to imagine future observations deepening our knowledge of the cosmos and rendering our era but one of many, perhaps infinitely many, moments of incomplete understanding. While this would impact our discussion of the earliest stages of the universe as well as its unfolding past floor 12 or so, the core considerations of entropy and evolution that have guided us throughout the bulk of our journey would persist all the same. Most impactful of all, were the cyclic theory confirmed, we would learn that the most ubiquitous of all patterns—birth, death, and rebirth—is recapitulated over cosmological scales. It is an enticing template. Thinkers as far back as the ancient Hindus, Egyptians, and Babylonians imagined that instead of a beginning, middle, and end, the universe, like days and seasons, might go through a sequence of dovetailing cycles. In the not too distant future, data collected by gravitational wave observatories may reveal whether this pattern is embraced by the cosmos itself.[34]

Thought and the Multiverse

Would a journey at arbitrary speed into the depths of space reach an end? Might it go on forever? Or perhaps circle back on itself in a cosmic Magellanic journey? No one knows. Within the inflationary theory, the most intently studied mathematical formulations imply that space is endless, explaining in part why researchers have paid most attention to this possibility. For the far future of thought, endless space provides a particularly outlandish consequence, so let's follow the dominant inflationary perspective and assume that space is infinite.[35]

The vast majority of infinite space would be beyond our ability to see. Light emitted from a distant location is visible to our telescopes only if there has been ample time for it to have traversed the space between us. Using the maximal travel time possible—the duration back to the big bang, 13.8 billion years—we can calculate that the maximum distance we can see in any direction is about 45 billion light-years (you might have thought the limit would be 13.8 billion light-years, but because space expands while the light is in transit, the span is larger). If you grew up on a planet more distant from earth than that, there is no way we could have as yet communicated or directly influenced each other. So assuming space is infinite, you can picture it as a patchwork of widely separated 90-billion-light-year regions, with each region having evolved independently of the others.[36] Physicists like to think of each such region as its own independent universe, with the entire collection of such regions being a **multiverse.** Accordingly an infinite spatial expanse gives rise to a multiverse containing infinitely many universes.

In studying these universes, physicists Jaume Garriga and Alex Vilenkin[37] established a pivotal feature. If you were to watch a series of films that showed the cosmological unfolding in each, the films could not all be different. Because each of the regions has a finite size, and each contains a large but finite amount of energy, there are only finitely many distinct histories that can possibly play out. Intuitively, you might think otherwise. You might expect that there would be infinitely many variations because, given any history, you can always modify it by nudging this particle that way or that particle this way. But here's the thing: if your nudges are too small, they'll fall below the sensitivity limit of quantum uncertainty, and will thus be meaningless; if your nudges are too large, the particles won't remain within the region or their energies will exceed the maximum available.

Constrained on both small scales and large, there are only finitely many variations, and so only finitely many different films are possible.

Now, with infinitely many regions and finitely many films, there simply aren't enough different films to go around. We are guaranteed that the films will repeat; indeed, we are guaranteed that they will repeat infinitely many times. We are also guaranteed that each film will be used. The quantum jitters that result in one history being different from another are random and hence they sample every possible configuration. No history is left behind. The infinite collection of universes thus realizes every possible history, and each such history is realized infinitely often.

This entails a peculiar conclusion: the reality that you and I and everyone else experiences is happening out there in other regions—in other universes—over and over again. Modify that reality in any manner that is not strictly forbidden by the laws of physics (you can't violate the conservation of energy or electric charge, for example) and it is also out there, over and over again. It tickles the mind to fathom realms where alternate realities play out—Lee Harvey Oswald misfires, Claus von Stauffenberg succeeds, James Earl Ray doesn't. Quantum aficionados will recognize a similarity to the so-called Many Worlds interpretation of quantum physics, which envisions that every possible outcome allowed by the quantum laws takes place in its own separate universe. Physicists have debated for more than half a century whether this approach to quantum mechanics is mathematically sensible and whether, if it is, the other universes are real or merely useful mathematical fictions. The essential difference in the cosmological theory we are now recounting is that the other worlds—the other regions—are not a matter of interpretation. If space is infinite, the other regions **are** out there.

From all that we've explored in this and previous chapters, it is reasonable to conclude that here in our region, in our universe, our days, and those of thinking beings more generally, are numbered. The number may be large, but somewhere along the climb up the Empire State Building, or perhaps beyond, life and mind will more than likely reach their end. Against this backdrop, Garriga and Vilenkin offer a curious sort of optimism. They note that because every history plays out across the infinite collection of universes, some will necessarily enjoy rare but fortuitous drops in entropy that keep particular stars and planets intact, or yield new environments containing sources of high-quality energy, or any of a wide array of unlikely developments that will allow life and thought to persist far longer than otherwise expected. Indeed, as Garriga and Vilenkin argue, if you select **any** finite duration, however long, there will be universes among the infinite collection in which unlikely processes swim against the entropic stream to keep life alive for at least that duration. And so, among the infinity of universes, some will host life and mind arbitrarily far into the future.

It is hard to know how the inhabitants of such regions would explain their good fortune in managing to survive. Or even if they'd be aware of their good fortune. Perhaps they'd have worked out the same understanding of physics as we have and would recognize that random fluctuations can result in rare and fortuitous outcomes. But that very knowledge would at the same time make clear that what they're experiencing, while possible, is extraordinarily unlikely. From this realization they might go on to conclude that they need to rework their understanding of physics. Think about it. Although the probabilistic laws of quantum physics allow for the possibility that I can walk through a solid wall, if I did, and did so repeatedly, we'd want to revamp our understanding of quantum physics. Not because I would have contravened the quantum laws. I would not have. It is simply that if supposedly unlikely events happen, and happen often, we're apt to seek better explanations according to which the events are not so unlikely after all. Of course, it is also possible that the inhabitants of such lucky realms would not be focused on explanations at all and would simply go with the flow and happily live on indefinitely.

As the odds are next to nothing that we inhabit such a region or that we are sufficiently close to one to make our escape there, perhaps as our own end comes into view we will gather what we've learned, discovered, and created and pack it into a capsule that we'll launch in the hope that it might someday reach one of the more fortunate realms. If we are not part of a lineage that extends to eternity, perhaps we can transmit the essence of our accomplishments to those who are. Perhaps, however indirectly, we can leave a trace on eternity. Garriga and Vilenkin study a version of this scenario, and together with insights from philosopher David Deutsch, conclude that the plan is hopeless. Across the infinity of universes and the vastness of timescales, random quantum fluctuations will produce far more fake capsules than our descendants will be able to produce real ones, ensuring that any reliable imprint of who we are and what we have accomplished will be lost in the quantum noise.

Life and thought here in our universe, in what we have long considered **the** universe, will likely draw to a close. Perhaps there is consolation in knowing that somewhere in the vast reaches of infinite space, well beyond the boundary of our realm, life and thought may persist, conceivably indefinitely. Still, even though we can contemplate eternity, and even though we can reach for eternity, apparently we cannot touch eternity.

THE NOBILITY
OF BEING

Mind, Matter, and Meaning

T he guide at Pilanesberg National Park, rifle slung low across his back, was double-checking that those accompanying him on foot would respond appropriately should an elephant or a hippo or a lion come too close for comfort. "You . . . stay . . . still," he said, emphasizing each word as he slowly panned across the group. "Run from a lion? You spend the rest of your life trying to win the race." Gently laughing, we all murmured "yes" and "of course" and "absolutely." Just then I glanced down at the sleeve of my loose-fitting shirt. Identifying precisely what was clinging to my cuff was of little concern. To me it was a tarantula. And it was making its way upward. I freaked. My arm flew back and forth, knocking glasses off the breakfast table. I jumped from my chair, and plates that had survived my initial flailing were now falling too. In the mayhem, the tarantula, or whatever creepy thing it actually was, detached. By the time I regained composure, the little nickel-size creature was on the ground, slowly crawling away. "Ah," the guide said, smiling, when all had settled down, "the universe has spoken for our physicist friend. You travel in the Jeep." And I did.

The universe had not spoken for me. The attack was random and its timing blind chance. Were I a disinterested party, I would offer my standard riposte, noted earlier, that in the absence of such an event there would be no surprise that such a coincidence had **not** happened. But the truth is, for a brief moment, the embarrassing episode felt significant. I was already uneasy about a safari on foot, was wondering if I should back out, and then I was delivered a tailor-made reminder that this particular risk is not in the best interest of someone who, when lost in thought, can be startled nearly to death by an unanticipated hello. Rationally, I know this kind of talk is silliness. The universe is not keeping tabs on what I do or the dangers I face. Still, while the atavistic instincts inflamed by the tarantula attack were gradually subsiding, rational thought was a step or two from regaining full command.

Sensitivity to pattern is, in part, how we've

prevailed. We look for connections. We take note of coincidences. We mark regularities. We assign significance. But only some of these assignments result from considered analyses delineating demonstrable features of reality. Many emerge from an emotional preference for imposing a semblance of order on the chaos of experience.

Order and Significance

I often speak as if our mathematical equations are out there in the world, relentlessly controlling all physical processes, quarks to the cosmos. That may be the case. Perhaps we will one day establish that mathematics is fundamentally stitched into the tapestry of reality. When you work with the equations day in and day out it surely feels that way. However, I am more confident in asserting that nature is lawful—that the universe is made of ingredients whose behaviors follow a lawful progression—the very basis of the journey we have taken in this book. The equations at the core of modern physics represent our most precise statement of the laws. Through diligent experiment and observation we have established that these equations provide a spectacularly accurate account of the world. But there is no guarantee that they are expressed in nature's intrinsic lexicon. Although I consider it unlikely, I allow for the possibility that in the future, when we proudly show alien visitors our equations, they will politely smile, tell us that they too started with math but then discovered the **real** language of reality.

Historically, the physical intuition of our ancestors was informed by the patterns evident in familiar encounters, from falling rocks to snapping branches to rushing streams; there is manifest survival value in having an innate sense of everyday mechanics. In time, we employed our cognitive capacities to go beyond such survival-promoting intuitions, illuminating and codifying patterns in realms spanning from the microworld of individual particles to the macroworld of clustered galaxies, many of which have little or no adaptive value. By shaping our intuition and developing our cognitive skills, evolution initiated our education in physics but our more comprehensive understanding has emerged from the force of human curiosity expressed through the language of mathematics. The resulting equations articulated in this language are of profound utility in exploring the deep structure of reality, but they may nevertheless be constructs of the human mind.

I hold to a version of this perspective when we shift focus to qualities that guide our evaluation of human experience. Right and wrong, good and evil, destiny and purpose, value and meaning are all profoundly useful concepts, but I am not among those who believe that moral judgments and assignments of significance transcend the human mind. We invent these qualities. Not from whole cloth. Our Darwinian-selected minds are predisposed to be attracted to or repulsed by or scared of various ideas and behaviors. Worldwide, care for the young scores high, while incest is abhorrent. Fairness in day-to-day dealings is widely valued, as is loyalty to family and compatriots. As our ancestors gathered in groups, the interplay of these and numerous other predispositions with on-the-ground encounters created feedback loops: Behavior of individuals influenced the effectiveness of group living, leading to the gradual articulation of communal codes of conduct. In turn, such behavioral codes contributed differing degrees of survival value to those who followed them.[1] Much as natural selection shaped our intuition for basic physics, it also had a hand in shaping our innate sense of morality and value.

Even among those who concur with the belief that moral codes are not imposed from on high or floating in an abstract realm of truth, there is a healthy debate regarding the role of human cognition in determining how these early sensibilities developed. Some suggest that, similar to the developmental pattern for physics, evolution imprinted a rudimentary moral sense, but our cognitive powers have allowed us to leap beyond that innate base to fashion independent attitudes and beliefs.[2] Others suggest that we are adept at using our cognitive dexterity to explain our moral commitments, but these accounts are just-so stories, rationalizations of judgments anchored in our evolutionary past.[3]

A point worthy of reemphasis is that none of these positions relies on a traditional conception of free will. In describing human behavior, we invoke an amalgam of factors, from instinct and memory to perception and societal expectation. Yet, as argued earlier, this type of high-level account—lying at the core of how we humans make sense of the world—emerges from a complex chain of processes ultimately resting on the dynamics of nature's fundamental constituents. We all are collections of particles, beneficiaries of innumerable evolutionary battles that have unshackled our behaviors and given us the capacity to delay entropic decay. But such triumphs grant us no freely willed powers over physical progression; the unfolding does not await our wishes, judgments, and moral appraisals. Or, put more precisely, our wishes, judgments, and moral appraisals are simply part of the world's physical progression, as dictated by nature's dispassionate laws.

Our description of that progression invokes impersonal mathematical rules that lay out in symbols how the universe will develop from one moment to the next. And for much of the past, prior to the emergence of collections of particles capable of reflecting on reality, this story was the full story. Familiar as we now are with the essential details, we can recount our most refined if provisional version of that story—swiftly, briefly, and, for ease of language, with an anthropomorphic tinge.

Some 13.8 billion years ago, within ferociously swelling space, the energy contained in a tiny but ordered cloud of inflaton field disintegrated, shutting off repulsive gravity, filling space with a bath of particles, and seeding the synthesis of the simplest atomic nuclei. Where quantum uncertainty rendered the density of the bath slightly higher, the gravitational pull was slightly stronger, enticing particles to fall together in ever-growing clumps, forming stars, planets, moons, and other heavenly bodies. Fusion within stars, as well as rare but powerful stellar collisions, melded simple nuclei into more complex atomic species, which, upon raining down on at least one planet in the making, were coaxed by molecular Darwinism to assemble into arrangements capable of self-replication. Random variations of the arrangements that happened to abet molecular fecundity spread widely. And among these were molecular pathways for extracting, storing, and dispersing information and energy—the rudimentary processes of life—which, through the long haul of Darwinian evolution, became increasingly refined. In time, complex, self-directed, living beings emerged.

Particles and fields. Physical laws and initial conditions. To the depth of reality we have so far plumbed, there is no evidence for anything else.

Particles and fields are the elementary ingredients. The physical laws prompted by the initial conditions dictate progression. Because reality is quantum mechanical, the pronouncements of the laws are probabilistic, but even so the probabilities are rigidly determined by mathematics. Particles and fields do what they do without concern for meaning or value or significance. Even when their indifferent mathematical progression yields life, physical laws maintain complete control. Life has no capacity to intercede or overrule or influence the laws.

What life can do is facilitate groups of particles to act in concert and manifest collective behaviors that, compared to the inanimate world, are novel. The particles constituting marigolds and marbles adhere fully to nature's laws, yet marigolds grow larger and follow the sun while marbles don't. Through the force of selection, evolution takes a hand in shaping life's behavioral repertoire, favoring activities that advance survival and reproduction. Among these, ultimately, is thought. The capacity to form memories, analyze situations, and extrapolate from experience provides potent artillery in the arms race for survival. Powering a string of victories across tens of thousands of generations, thought gradually refines, resulting in thinking species that acquire various degrees of self-awareness. The wills of such beings are not free in the traditional sense of stepping outside the unfolding dictated by physical law, but their highly organized structure allows for a wealth of responses—from inner emotions to external behaviors—that, at least so far, are unavailable to collections of particles lacking life or mind.

Add in language, and one such self-aware species rises above the needs of the moment to see itself as part of an unfolding from past to future. With that, winning the battle is no longer the only concern. We are no longer satisfied to merely survive. We want to know why survival is significant. We seek context. We search for relevance. We assign value. We judge behavior. We pursue meaning.

And so we develop explanations of how the universe came to be and how it might end. We tell and retell stories of minds making their way through worlds, real and fanciful. We imagine realms populated by departed ancestors or semi-powerful or all-powerful beings that reduce death to a stepping-stone in an ongoing existence. We paint and carve and etch and sing and dance to touch these other realms, or to pay homage to them, or simply to imprint the future with something that attests to our brief time in the sun. Perhaps these passions take hold and become part of what it means to be human because they enhance survival. Stories prepare the mind for responding to the unexpected; art develops imagination and innovation; music sharpens sensitivity to pattern; religion binds adherents into strong coalitions. Or perhaps the explanation is less lofty: some or all of the activities may emerge and persist because they leverage or tag along with other behaviors and responses that have played a more direct role in advancing survival. But even with their evolutionary origin still fodder for debate, these aspects of human behavior manifest a widespread need to step beyond the mere eking out of transitory survival. They reveal a pervasive longing to be part of something larger, something lasting. Value and meaning, decidedly absent from the bedrock of reality, become intrinsic to a restless urge that elevates us above indifferent nature.

Mortality and Significance

Whereas Gottfried Leibniz wondered why there is something rather than nothing, the deeply personal dilemma is that self-aware somethings, like us, subsequently dissolve into nothing. To acquire a temporal perspective is to realize that the vibrant activity animating one's own mind will one day cease.

Against the backdrop of that awareness, the previous chapters have explored the full expanse of time from our best understanding of its beginning to the closest our mathematical theories can take us to its end. Will our understanding continue to develop? Of course. Will details, some minor and others significant, be enhanced or replaced? No doubt.

But the rhythm of birth and death, emergence and disintegration, creation and destruction that we've witnessed playing out along the timeline will persist. The entropic two-step and the evolutionary forces of selection enrich the pathway from order to disorder with prodigious structure, but whether stars or black holes, planets or people, molecules or atoms, things ultimately fall apart. Longevity varies widely. Yet the fact that we will all die, and the fact that the human species will die, and the fact that life and mind, at least in this universe, are virtually certain to die are expected, run-of-the mill, long-term outcomes of physical law. The only novelty is that we notice.

A frequent if fraught expectation, lightly entertained by many and intensely pursued by some, is that we would be entirely better off if death would bow out of human proceedings altogether. From ancient myth to modern fiction, thinkers have pondered the possibility. Perhaps it's telling that in these excursions things don't always turn out so well. The immortals in Jonathan Swift's land of Luggnagg continue to age and are declared legally dead at eighty as they drift into irrelevance. Having endured for more than three hundred years, Karel Čapek's heroine Elina Makropulos allows the formula for a life-extending elixir to go up in flames rather than continue on in a state of profound boredom. Living in an endless world absent death, writes the protagonist in Jorge Luis Borges's "The

Immortal," "no one is anyone, one single immortal man is all men . . . I am god, I am hero, I am philosopher, I am demon and I am world, which is a tedious way of saying that I do not exist."[4]

Philosophers have treaded in these waters too, offering systematic assessments of life in a world without death. Some, like Bernard Williams, who was inspired by Karl Janacek's operatic adaptation of Čapek's play, reach similarly gloomy conclusions.[5] Williams argues that with endless time each of us would satiate every objective that drives us onward, leaving us listless in the face of a mind-numbingly monotonous eternity. Others, like Aaron Smuts, inspired in part by Borges's story, contend that immortality would drain the decisions that shape a human life—how to spend one's time and with whom—of the consequences essential to their significance. Make the wrong choice? No problem. You've got eternity to make it right. The satisfaction of achievement would also fall victim to immortality. Those with limited abilities would reach their potential and then experience eternal frustration; those with abilities capable of deepening without limit would be guaranteed to improve continually, deflating the sense of accomplishment that comes from outperforming expectations.[6]

Notwithstanding these concerns, I suspect that we are sufficiently resourceful—and endowed with endless time we would become all the more so—to grow into thoroughly well-adjusted immortals.

Our needs and capacities would likely transform beyond recognition, rendering assessments based on what keeps us engaged and motivated in the here and now of little or no relevance. Should everlasting joie de vivre require a different flavor of joie, we would find it or invent it or develop it. This is no more than a hunch, of course, but to conclude that we would necessarily grow bored suggests an unduly parochial vision of the immortal mind.

While science will continue to extend life spans, our trek to the far future suggests that immortality will forever remain beyond reach. Despite that, thinking about life that never ends clarifies the relevance of life that does. The imagined fate of value and significance in an immortal world makes clear that in a mortal one understanding a great many of our decisions, choices, experiences, and reactions requires seeing them in the context of limited opportunity and finite duration. Not that we spring to our feet each morning wailing "Carpe diem!" but the deep-seated knowledge that there are just so many mornings when we will rise at all instills an intuitive calculus of value, one that would be very different in a world with unlimited do-overs. The explanations we give for the subjects we study, the trades we learn, the work we pursue, the risks we take, the partners we join, the families we build, the objectives we set, the concerns we entertain—all reflect the recognition that our opportunities are scarce because our time is limited.

We each respond to that recognition in our own way, but there are common qualities that run through the human sense of value. Among these is a surprisingly strong yet often unspoken need for a future populated by descendants who will carry on after we are gone.

Descendants

Many years ago I was asked to participate in a post-performance talkback with the audience of an off-Broadway show in which a collection of characters realize that earth will shortly be destroyed by an asteroid. My fellow discussant was my brother; the producers anticipated that commentary on the end of the world from siblings whose lives had followed divergent but relevant paths—one immersed in science and the other religion—would be a crowd-pleaser. Frankly, I didn't think much about the issues before the event, and in those days I was a good deal more susceptible to the energy of an audience. The more my brother veered toward ethereal realms, the more blunt I became. "Earth is a pedestrian planet orbiting an unremarkable star in the suburbs of an ordinary galaxy. If we're taken out by an asteroid, the universe won't so much as blink. In the grand scheme of things, it just won't matter." The starkness was welcomed by some, I presume those who identified as no-nonsense skeptics bravely facing up to the realities of existence. But for others, regrettably, my remarks came off as smug. Well, at least one audience member felt that way: an elderly woman who chided me for running roughshod over what she described as an essential need we all have for the species to continue. "Which news would affect you more," she asked, "being told you have a year to live or that in a year earth will be destroyed?"

At the time I said something facile about it depending on whether either outcome would entail physical pain, but later, as I mulled the question over, I found it unexpectedly illuminating. A terminal prognosis affects people in different ways—focusing attention, providing perspective, stoking regret, fueling panic, delivering composure, inspiring epiphany. I anticipated that my own reaction would lie somewhere among these. But the prospect that earth and all of humankind would be wiped out triggered a different kind of reaction. The news would make everything seem rather pointless. Whereas my own impending end would heighten intensity, endowing with significance moments that might have otherwise receded into the daily humdrum, contemplating the end of the entire species seemed to do the opposite, yielding a sense of futility. Would I still get up in the morning and want to pursue research in physics? Maybe for the comfort of doing something familiar, but with no one left to build on today's discoveries, the pull of advancing knowledge would weaken. Would I finish the book I was writing? Maybe for the satisfaction of tying up loose ends, but with no one to read the finished work motivation would run thin. Would I still send my kids to school? Maybe for the calm offered by routine, but with no future what would they be preparing for?

I found the contrast with how I would react to learning the date of my own demise surprising. While one realization seemed to intensify awareness of life's value, the other seemed to drain it away. In the years since, this realization has helped shape my thinking about the future. I had long since had my youthful epiphany regarding the capacity of mathematics and physics to transcend time; I was already convinced of the existential significance of the future. But my image of that future was abstract. It was a land of equations and theorems and laws, not a place populated with rocks and trees and people. I am not a Platonist but, still, I implicitly envisioned mathematics and physics transcending not only time, but also the usual trappings of material reality. The doomsday scenario refined my thinking, making it patently evident that our equations and theorems and laws, even if they tap into fundamental truths, have no intrinsic value. They are, after all, a collection of lines and squiggles drawn on blackboards and printed in journals and textbooks. Their value derives from those who understand and appreciate them. Their worth derives from the minds they inhabit.

This refinement in thinking went far beyond the role of equations. By leading me to imagine a future bereft of anyone to receive all that we value, absent anyone to add their own iconic imprint and pass it on to future generations, the doomsday scenario revealed how hollow that future would feel. While immortality of the individual may sap significance, immortality of the species seems necessary to secure it.

I can't be sure how widespread this reaction to news of an impending end would be, but I suspect it would be common. Philosopher Samuel Scheffler recently initiated scholarly investigation of the issue, exploring a variant of the question posed to me decades ago. How would you respond, Scheffler asks, if you learned that thirty days after your own death everyone remaining would be obliterated? It's a more revealing version of the scenario as it excises one's own premature mortality and so shines a tighter spotlight on the role of descendants in anchoring value. Scheffler's carefully reasoned conclusion resonates with my own informal musings:

> Our concerns and commitments, our values and judgments of importance, our sense of what matters and what is worth doing—all these things are formed and sustained against a background in which it is taken for granted that

human life is itself a thriving, ongoing enter-
prise . . . We need humanity to have a future
for the very idea that things **matter** to retain a
secure place in our conceptual repertoire.[7]

Other philosophers have weighed in too, providing opinions that delineate a wider range of
perspectives. Susan Wolf suggests that recognition
of our shared fate might elevate the care for others
to newfound heights, but even so, she concurs that
our vision of a future populated by humans is essential to the value we ascribe to our undertakings.[8]
Harry Frankfurt offers a different view, suggesting
that many things we value would be unaffected by
the doomsday scenario, most prominently artistic
pursuits and scientific research. The intrinsic grati-
fication of these activities, he believes, would be
enough for many to keep at it. I've already given my
contrarian view regarding scientific research, which
serves to emphasize a related point, obvious but
telling: people will respond to the news in different
ways.[9] The best we can do is envision dominant
trends. For me, and many others too, to engage
in creative pursuits and scholarly undertakings is
to feel part of a long, rich, and ongoing human
dialogue. Even if a given physics paper I write does
not set the world on fire, the paper nevertheless
makes me feel part of the conversation. Yet, if I
know that I am the last to speak, and if I know that

there will be no one in the future to reflect on what I say, I'm left wondering why I should bother.

In Scheffler's scenario, as well as in the question I was asked years earlier, the doomsdays are hypothetical but the timescales for the world's destruction are easily grasped. In this book, the doomsdays we've explored are genuine but their timescales make them extraordinarily remote. Does this change of scale, a colossal change at that, affect the conclusions? It's an issue that both Scheffler and Wolf consider, entertainingly framed by the wonderful scene in **Annie Hall** in which nine-year-old Alvy Singer has concluded that there's no point in doing homework given that in a few billion years the expanding universe will break apart and destroy everything. Alvy's shrink, let alone his mother, considers Alvy's concern ludicrous. Audiences laugh because they regard Alvy's worry as farcical. Scheffler shares these intuitions yet notes that he does not have a fundamental justification for why we think it reasonable to have an existential crisis in the face of imminent destruction but silly to do so when such destruction is far in the future. He chalks it up to the difficulty we have grasping timescales that are vastly beyond the range of human experience. Wolf agrees, noting that if the immediate demise of humanity would render life meaningless, then the same should be true even if the end is far off. Indeed, as she notes, on cosmic timescales the delay of a few billion years is not long at all.

I agree. Forcefully so.

As we've seen repeatedly, the notion of a duration being long or short has no absolute meaning. Long or short is a matter of perspective. The time represented by the observation deck of the Empire State Building, floor 86, is enormous by everyday standards, but comparing that duration to the time represented by floor 100 is like comparing the blink of an eye to ten thousand centuries. Our familiar human perspective leads us to judgments that while relevant are also parochial. Because of this, I view the scenario of imminent demise as no more than a tool that employs artificial urgency to catalyze an authentic response. The intuition we glean remains relevant to an end that awaits our descendants in the far future; that future, viewed from a larger context, **is** a moment away.

While it is indeed challenging to internalize timescales that are significantly beyond anything we experience, the journey we've taken in this book has populated the cosmic timeline with landmarks that serve to make the abstract concrete. I can't say that I have an innate sense of the timescales marked out along the metaphor of the Empire State Building in the same way that I sense the timescales of daily life or those of my generation or even a few generations, but the sequence of transformative events we have explored provides handholds for grasping the future. There is no need to chant, and a lotus position is optional, but if you find a quiet place and let your mind slowly and freely float along the cosmic timeline, moving through and then past our epoch, past the era of distant receding galaxies, past the era of stately solar systems, past the era of graceful swirling galaxies, past the era of burnt-out stars and wandering planets, past the era of glowing and disintegrating black holes, and onward to a cold, dark, nearly empty but potentially limitless expanse—in which the evidence that we once existed amounts to an isolated particle located here instead of there or another isolated particle moving this way instead of that—and if you are at all like me and let that reality fully settle in, the fact that we've traveled fantastically far into the future hardly diminishes the shuddering yet awestruck feeling that wells up inside. Indeed, in one essential way, the enormous sweep of time only adds weight to the nearly unbearable lightness of being; compared to the timescale we've reached, the epoch of life and mind is infinitesimal. By today's scales, its entire span, from the earliest microbes to the final thought, would be less than the duration required for light to traverse an atomic nucleus. The entire duration of human activity—whether we annihilate ourselves in the next few centuries, are wiped out by a natural disaster in the next few millennia, or somehow find a way to carry on until the death of the sun, the end of the Milky Way, or even the demise of complex matter—would be more fleeting still.

We are ephemeral. We are evanescent.

Yet our moment is rare and extraordinary, a recognition that allows us to make life's impermanence and the scarcity of self-reflective awareness the basis for value and a foundation for gratitude. While we may long for a perdurable legacy, the clarity we gain from exploring the cosmic timeline reveals that this is out of reach. But that very same clarity underscores how utterly wondrous it is that a small collection of the universe's particles can rise up, examine themselves and the reality they inhabit, determine just how transitory they are, and with a flitting burst of activity create beauty, establish connection, and illuminate mystery.

Meaning

Most of us deal quietly with the need to lift ourselves beyond the everyday. Most of us allow civilization to shield us from the realization that we are part of a world that, when we're gone, will hum along, barely missing a beat. We focus our energy on what we can control. We build community. We participate. We care. We laugh. We cherish. We comfort. We grieve. We love. We celebrate. We consecrate. We regret. We thrill to achievement, sometimes our own, sometimes of those we respect or idolize.

Through it all, we grow accustomed to looking out to the world to find something to excite or soothe, to hold our attention or whisk us to someplace new. Yet the scientific journey we've taken suggests strongly that the universe does not exist to provide an arena for life and mind to flourish. Life and mind are simply a couple of things that happen to happen. Until they don't. I used to imagine that by studying the universe, by peeling it apart figuratively and literally, we would answer enough of the how questions to catch a glimpse of the answers to the whys. But the more we learn, the more that stance seems to face in the wrong direction. Looking for the universe to hug us, its transient conscious squatters, is understandable, but that's just not what the universe does.

Even so, to see our moment in context is to realize that our existence is astonishing. Rerun the big bang but slightly shift this particle's position or that field's value, and for virtually any fiddling the new cosmic unfolding will not include you or me or the human species or planet earth or anything else we value deeply. If a super intelligence were to look at the new universe as a whole, much as we look at a collection of tossed pennies as a whole or the air we're now breathing as a whole, it would conclude that the new universe pretty much looks the same as the original. For us, it would be vastly different. There wouldn't be an "us" to notice. By shifting our attention away from fine details, entropy has provided an essential organizing principle for grasping the large-scale trends in how things transform. But whereas we generally don't care if this penny is heads or that tails, or if one particular oxygen molecule happens to be here or there, there are certain fine details that we do care about. Profoundly so. We exist because our specific particulate arrangements won the battle against an astounding assortment of other arrangements all vying to be realized. By the grace of random chance, funneled through nature's laws, we are here.

It is a realization that echoes across each stage of human and cosmic development. Think of what Richard Dawkins described as the nearly infinite collection of potential people, would-be carriers of the nearly infinite collection of base pair sequences in DNA, none of whom will ever be born. Or think of the moments constituting cosmic history, from the big bang through your birth and on to today, filled with quantum processes whose relentless probabilistic progression at each of a nearly limitless collection of junctures could have yielded that outcome instead of this, resulting in an equally sensible universe but one that would not include you or me.[10] And yet, with this astronomical number of possibilities, against astonishing odds, your sequence of base pairs and mine, your molecular combination and mine now exist. How spectacularly unlikely. How thrillingly magnificent.

And the gift is greater still: our particular molecular

combinations, our specific chemical and biological and neurological arrangements, give us the enviable powers that have occupied much of our attention in earlier chapters. Whereas most life, miraculous in its own right, is tethered to the immediate, we can step outside of time. We can think about the past, we can imagine the future. We can take in the universe, we can process it, we can explore it with mind and body, with reason and emotion. From our lonely corner of the cosmos we have used creativity and imagination to shape words and images and structures and sounds to express our longings and frustrations, our confusions and revelations, our failures and triumphs. We have used ingenuity and perseverance to touch the very limits of outer and inner space, determining fundamental laws that govern how stars shine and light travels, how time elapses and space expands—laws that allow us to peer back to the briefest moment after the universe began and then shift our gaze and contemplate its end.

Accompanying these breathtaking insights are deep and persistent questions. Why is there something rather than nothing? What sparked the onset of life? How did conscious awareness emerge? We have explored a range of speculations, but definitive answers remain elusive. Perhaps our brains, well adapted for survival on planet earth, are just not structured for resolving these mysteries. Or perhaps, as our intelligence continues to evolve, our engagement with reality will acquire a wholly different character, with the result that to-day's towering questions become irrelevant. While either is possible, the fact that the world as we now understand it, remaining mysteries and all, holds together with such a tight mathematical and logi-cal coherence, and the fact that we have been able to decipher so much of that coherence, suggests to me that neither is the case. We are not lacking the brainpower. We are not staring at Plato's wall, unaware of a radically different kind of truth, just beyond reach, with the power to suddenly provide startling new clarity.

As we hurtle toward a cold and barren cosmos, we must accept that there is no grand design. Particles are not endowed with purpose. There is no final answer hovering in the depths of space awaiting discovery. Instead, certain special collections of particles can think and feel and reflect, and within these subjective worlds they can create purpose. And so, in our quest to fathom the human condi-tion, the only direction to look is inward. That is the noble direction to look. It is a direction that forgoes ready-made answers and turns to the highly personal journey of constructing our own mean-ing. It is a direction that leads to the very heart of creative expression and the source of our most resonant narratives. Science is a powerful, exquisite tool for grasping an external reality. But within that rubric, within that understanding, everything else is the human species contemplating itself, grasping what it needs to carry on, and telling a story that reverberates into the darkness, a story carved of sound and etched into silence, a story that, at its best, stirs the soul.

Acknowledgments

I am grateful to the many people who provided invaluable feedback while I was writing **Until the End of Time.** For thoughtfully reading the manuscript, sometimes more than once, and offering perspectives, criticisms, and suggestions that substantially enhanced the presentation, I owe great thanks to Raphael Gunner, Ken Vineberg, Tracy Day, Michael Douglas, Saakshi Dulani, Richard Easther, Joshua Greene, Wendy Greene, Raphael Kasper, Eric Lupfer, Markus Pössel, Bob Shaye, and Doron Weber. For carefully reading and responding to particular sections or chapters, and/or answering queries, I thank David Albert, Andreas Albrecht, Barry Barish, Michael Bassett, Jesse Bering, Brian Boyd, Pascal Boyer, Vicki Carstens, David Chalmers, Judith Cox, Dean Eliott, Jeremy England, Stuart Firestein, Michael Graziano, Sandra Kaufmann, Will Kinney, Andrei Linde, Avi Loeb, Samir Mathur, Peter de Menocal,

Brian Metzger, Ali Mousami, Phil Nelson, Maulik Parikh, Steven Pinker, Adam Riess, Benjamin Smith, Sheldon Solomon, Paul Steinhardt, Giulio Tononi, John Valley, and Alex Vilenkin. I thank the entire team at Knopf, including copy editor Amy Ryan, assistant editor Andrew Weber, designer Chip Kidd, production editor Rita Madrigal, and my editor, Edward Kastenmeier, who offered many deeply insightful suggestions and, together with my agent, Eric Simonoff, thoroughly supported the project at every stage of development. Finally, my heartfelt gratitude for the steadfast love and support of my family: my mother, Rita Greene; my siblings, Wendy Greene, Susan Greene, and Joshua Greene; my children, Alec Day Greene and Sophia Day Greene; and my wonderful wife and dearest friend, Tracy Day.

Notes

Preface

1. The quote is from an early mentor of mine, a graduate student in the mathematics department of Columbia University in the 1970s, Neil Bellinson, who generously gave his time and unique talent to teach mathematics to a young student—me—who had nothing to offer save a passion for learning. We were discussing a paper on human motivation I was writing for a psychology course at Harvard taught by David Buss, now at the University of Texas at Austin.

2. Oswald Spengler, **Decline of the West** (New York: Alfred A. Knopf, 1986), 7.

3. Ibid., 166.

4. Otto Rank, **Art and Artist: Creative Urge and Personality Development,** trans. Charles Francis Atkinson (New York: Alfred A. Knopf, 1932), 39.

5. Sartre articulates this perspective through the

reflections of the condemned character Pablo Ibbieta in his wonderful short story "The Wall." Jean-Paul Sartre, **The Wall and Other Stories,** trans. Lloyd Alexander (New York: New Directions Publishing, 1975), 12.

Chapter 1: The Lure of Eternity

1. William James, **The Varieties of Religious Experience: A Study in Human Nature** (New York: Longmans, Green, and Co., 1905), 140.

2. Ernest Becker, **The Denial of Death** (New York: Free Press, 1973), 31. Becker credited Otto Rank with being his dominant influence.

3. Ralph Waldo Emerson, **The Conduct of Life** (Boston and New York: Houghton Mifflin Company, 1922), note 38, 424.

4. E. O. Wilson invokes the word "consilience" to describe his vision of a coming together of disparate knowledge to yield deeper understanding. E. O. Wilson, **Consilience: The Unity of Knowledge** (New York: Vintage Books, 1999).

5. In later chapters I will discuss evidence suggesting a pervasive influence of humankind's emerging awareness of mortality, but as there is little to no incontrovertible data attesting to the ancient human mind-set, the conclusion is not universally accepted. For an alternate perspective, which argues that death anxiety is a modern affliction, see for example Philippe Ariès, **The Hour of Our Death,** trans. Helen

Weaver (New York: Alfred A. Knopf, 1981). Becker's perspective, building on the insights of Otto Rank, is that death anxiety is deeply ingrained in the species.

6. Vladimir Nabokov, **Speak, Memory: An Autobiography Revisited** (New York: Alfred A. Knopf, 1999), 9.

7. Robert Nozick, "Philosophy and the Meaning of Life," in **Life, Death, and Meaning: Key Philosophical Readings on the Big Questions,** ed. David Benatar (Lanham, MD: The Rowman & Littlefield Publishing Group, 2010), 73–74.

8. Emily Dickinson, **The Poems of Emily Dickinson,** reading ed., ed. R. W. Franklin (Cambridge, MA: The Belknap Press of Harvard University Press, 1999), 307.

9. Henry David Thoreau, **The Journal, 1837–1861** (New York: New York Review Books Classics, 2009), 563.

10. Franz Kafka, **The Blue Octavo Notebooks,** trans. Ernst Kaiser and Eithne Wilkens, ed. Max Brod (Cambridge, MA: Exact Change, 1991), 91.

Chapter 2: The Language of Time

1. The broadcast, on BBC's Third Programme, January 28, 1948, at 9:45 p.m., was of a debate that took place in the previous year. https://genome.ch.bbc.co.uk/35b8e9bdcf 60458c976b882d80d9937f.

2.   Bertrand Russell, **Why I Am Not a Christian** (New York: Simon & Schuster, 1957), 32–33.

3.   This is of course a highly simplified description of a steam engine, modeled on the so-called **Carnot cycle,** which involves four steps: (1) Steam in a canister absorbs heat from a source (generally described as a heat reservoir) as it pushes against a piston, doing work at a constant temperature. (2) The canister is disconnected from the heat source and allowed to continue to push the piston, now performing work as the steam's temperature drops (but its entropy is constant, since there is no heat flow). (3) The canister is then connected to a second heat reservoir, at a lower temperature than the first, and work is done at this constant lower temperature to slide the piston back toward its original position, expelling waste heat in the process. (4) Finally, the canister is disconnected from the cooler reservoir as work continues to be exerted on the piston, completing its journey back to its original position, as the temperature of the steam is brought up to its original value as well. The cycle then begins again. In an actual steam engine—as opposed to a theoretical one we analyze mathematically—these steps, or ones that are comparable, are accomplished in a variety of ways dictated by issues of engineering and practicality.

4.   Sadi Carnot, **Reflections on the Motive Power of Fire** (Mineola, NY: Dover Publications, Inc., 1960).

5.   Modeling a baseball as a single massive particle with no internal structure is a gross approximation of the baseball itself. However, the application of Newton's laws to this approximate model of the baseball yields the **exact** classical motion of the baseball's center of mass. For the center of mass motion, Newton's third law ensures that all internal forces cancel each other out and so the center of mass motion depends solely on the external forces applied.

6.   One study (B. Hansen, N. Mygind, "How often do normal persons sneeze and blow the nose?" **Rhinology** 40, no. 1 [Mar. 2002]: 10–12) concluded that on average, people sneeze about once per day. As there are about 7 billion people on earth, that yields 7 billion sneezes worldwide per day. Since there are about 86,000 seconds in a day, we find about 80,000 sneezes per second worldwide.

7.   The description I have given is fine as a broadbrush summary, but there are more exotic physical systems in which to ensure that reverse-run sequences are allowed by the laws of physics we must subject the system to two other manipulations beyond the reversal of time: we must also reverse the charges of all particles (so-called **charge conjugation**) and also reverse the roles of left- and right-handedness (so-called **parity reversal**). The laws of physics, as currently understood, necessarily respect the conjunction of all three of these reversals, something known as the **CPT theorem** (with C standing for

charge conjugation, P for parity reversal, and T for time reversal).

8. For two tails, the calculation is $(100 \times 99)/2 = 4{,}950$; for three tails, $(100 \times 99 \times 98)/3! = 161{,}700$; for four tails, $(100 \times 99 \times 98 \times 97)/4! = 3{,}921{,}225$; for five tails $(100 \times 99 \times 98 \times 97 \times 96)/5! = 75{,}287{,}520$; for 50 tails $(100!/(50!)^2) = 100{,}891{,}344{,}545{,}564{,}193{,}334{,}812{,}497{,}256$.

9. More precisely, entropy is the **logarithm** of the number of members in a given group, an essential mathematical distinction ensuring that entropy has sensible physical properties (for example, when two systems are brought together, their entropies add), but one that for our qualitative discussion we can safely ignore. In parts of chapter 10, we will implicitly use the more precise definition, but for now we are fine.

10. In this example, for pedagogical ease, we will consider only the steam—molecules of H_2O—that is floating in your bathroom. We ignore the role of air and any other substances that are present. For simplicity, we also ignore the internal structure of the water molecules and treat them as structureless point particles. When we refer to the steam's temperature, bear in mind that liquid water transitions to steam at 100° C, but once formed the temperature of steam can be raised higher still.

11. Physically, temperature is proportional to the average kinetic energy of the particles, and so is calculated mathematically by averaging the square of each particle's velocity. For our

purposes, thinking of temperature in terms of average speed—the magnitude of velocity—is adequate.

12. More precisely, the first law of thermodynamics is a version of the law of energy conservation that (i) recognizes heat as a form of energy and (ii) takes account of the work done by or on a given system. Conservation of energy thus states that the change in internal energy of a system arises from the difference between the net heat it absorbs and the net work it does. The particularly well-informed reader may note that when we consider energy and its conservation in a global setting—across the entirety of the universe—subtleties emerge. We will not need to explore these, so we can safely assume the straightforward statement that energy is conserved.

13. Much as in the example of steam in your bathroom, in which I ignored molecules of air, for simplicity I am not explicitly considering collisions between the hot molecules released by the baking bread and the cooler molecules of air wafting through your kitchen and through the rest of your house. Such collisions would, on average, increase the speed of the air molecules and decrease the speed of those released by the baking bread, ultimately bringing both types of molecules to the same temperature. The decreased temperature of the bread molecules would act to lower their entropy, but the increased temperature of the air molecules would

result in a more than compensating entropic increase, so the combined entropy of both groups would indeed increase. In the simplified version I've described, you can think of the average speed of the molecules released by the baking bread as remaining constant as they spread out; their temperature would thus remain fixed, and so the increase in their entropy would be due to their filling a larger volume.

14. For the mathematically informed reader, there is a key technical assumption underlying this discussion (as well as most treatments of statistical mechanics in textbooks and in the research literature). Given any macrostate, there **are** compatible microstates that will evolve toward lower-entropy configurations. For example, consider the time-reversed version of any unfolding that yielded a given microstate starting from an earlier lower-entropy configuration. Such a "time-reversed" microstate would evolve toward lower entropy. Generally, we categorize such microstates as "rare," or "highly tuned." Mathematically, such categorization requires the specification of a **measure** on the space of configurations. In familiar situations, using the uniform measure on such a space does indeed render entropy-decreasing initial conditions "rare"—that is, of small measure. However, according to a measure that was chosen to peak around such entropy-decreasing initial configurations, they would, by design, not be rare. As far as we know, the choice of measure is an empirical one; for

the kinds of systems we encounter in everyday life, the uniform measure yields predictions that agree with observations, and so is the measure we invoke. But it is important to note that the choice of measure is justified by experiment and observation. When we consider exotic situations (such as the early universe) for which we lack analogous data leading us to a particular choice of measure, we need to acknowledge that our intuitions about "rare" or "generic" do not have the same empirical basis.

15. There are a few relevant points, glossed over in this paragraph, that affect the meaning of a "maximum entropy" state when applied to the universe. First, in this chapter we are not taking into consideration the role of gravity. In chapter 3, we will. And as we will see, gravity has a profound impact on the nature of high-entropy particle configurations. In fact, while it won't be our focus, in a given finite volume of space the maximum entropy configuration is a black hole—an object deeply dependent on gravity— that completely fills the spatial volume (for details, see, for example, my book **The Fabric of the Cosmos,** chapter 6 and chapter 16). Second, if we consider arbitrarily large regions of space—even infinitely large—the highest entropy configurations of a given amount of matter and energy are those in which the constituent particles (matter and/or radiation) are uniformly distributed over an ever-larger volume. Indeed, black holes, as we will discuss

in chapter 10, ultimately evaporate (through a process discovered by Stephen Hawking), yielding higher-entropy configurations in which particles are increasingly spread out. Third, for the purpose of this section, the only fact we need is that the entropy currently present in any given volume of space is not at its maximum value. If that volume contained, say, the room you are now inhabiting, entropy would increase if all the particles making up you, your furniture, and any other of the room's material structures were to collapse into a small black hole, which would subsequently evaporate yielding particles that would spread through an even larger volume of space. The very existence of interesting material structures—stars, planets, life, and so on—therefore implies that entropy is lower than what it potentially could be. It is such special, comparatively low-entropy configurations that call out for an explanation of how they arose. In the next chapter, we will take up this challenge.

16. For the particularly diligent reader, there is one additional detail worth spelling out. When the steam pushes on the piston, it expends some of the energy it absorbed from the fuel, but in the process the steam does not relinquish any of its entropy to the piston (assuming that the piston has the same temperature as the steam). After all, whether the piston is **here** or, having been pushed, is a short distance from **here** has no impact on its internal order or disorder; its entropy is unchanged. And with no entropy transferred

to the piston, the entropy remains fully within the steam itself. This means that as the piston is reset to its original position, ready for the next thrust, the steam must somehow expel all the excess entropy it is harboring. This is accomplished, as emphasized in the chapter, by the steam engine expelling heat to its surroundings.

17. Bertrand Russell, **Why I Am Not a Christian** (New York: Simon & Schuster, 1957), 107.

Chapter 3: Origins and Entropy

1. Georges Lemaître, **"Recontres avec Einstein," Revue des questions scientifiques** 129 (1958): 129–32.

2. The full story of Einstein's conversion to an expanding universe involved two factors. First, Arthur Eddington showed mathematically that Einstein's earlier proposal of a static universe suffered from a technical flaw: The solution was unstable, meaning that if the expanse of space were nudged to expand slightly, then it would continue expanding; if nudged to contract slightly, it would continue contracting. Second, the observational case, as discussed in this chapter, made it increasingly clear that space is not static. The combination of both realizations convinced Einstein to drop the notion of a static universe (although some have argued that the theoretical considerations may have had the most significant influence). For details of this history, see Harry Nussbaumer, "Einstein's

conversion from his static to an expanding universe," **European Physics Journal—History** 39 (2014): 37–62.

3.   Alan H. Guth, "Inflationary universe: A possible solution to the horizon and flatness problems," **Physical Review D** 23 (1981): 347. The technical term for the "cosmic fuel" is a **scalar** field. Unlike the more familiar electric and magnetic fields that provide a vector at each location in space (the magnitude and the direction of the electric or magnetic field at the location), a scalar field provides only a single number at each location in space (numbers from which the field's energy and pressure can be determined). Note that Guth's paper, and many subsequent treatments, emphasize the role of inflation in addressing a collection of cosmological issues that had previously stymied researchers—the monopole problem, the horizon problem, and the flatness problem being the most prominent. For an accessible and illuminating discussion of these issues, see Alan Guth, **The Inflationary Universe** (New York: Basic Books, 1998). Following Guth, I like to motivate inflation by raising the more intuitive problem of identifying the outward push that drove the big bang's spatial expansion.

4.   The cooling I refer to takes place after the inflationary burst has concluded and the universe has entered a phase of less rapid but still significant spatial expansion. For simplicity, I have left out some intermediate steps in the cosmological

unfolding. The early universe cooled because much of the energy it contained was carried by electromagnetic waves, and such waves stretch as space expands. This elongation of the electromagnetic waves—the so-called redshifting of the radiation—decreases their energy and lowers their overall temperature. Note, though, that even though the temperature is cooling, overall entropy is increasing due to the expanding volume of space.

5. There is a minority perspective that does attribute the fog to an inherent quantum limitation on the precision of measurements and not to a fundamentally blurry reality. In this approach—usually called "Bohmian mechanics," after physicist David Bohm, but sometimes referred to as the "de Broglie–Bohm theory," including attribution to Nobel laureate Louis de Broglie—particles retain sharp and definite trajectories. The trajectories are different from those predicted by classical physics (there is an additional quantum force that acts on particles as they move), but to use the language in the chapter, such trajectories could be drawn with a sharp quill. The uncertainty and fuzziness of the more traditional formulation of quantum mechanics shows up as statistical uncertainty regarding the initial conditions of any given particle. The difference between the two perspectives, while essential to the picture of reality each theory paints, has virtually no impact on quantitative predictions.

6.   Inflationary cosmology is a framework of theories—as opposed to a specific theory—based on the premise that during an early phase of its development the universe underwent a brief period of rapid accelerated expansion. The precise manner in which this phase arose and the precise details of its unfolding vary from one mathematical formulation to another. The simplest versions are in tension with ever-more-precise observational data, which has shifted focus to somewhat more complex versions of the inflationary theory. Detractors argue that the more complex versions are less convincing and that, moreover, these versions demonstrate that the inflationary paradigm is too flexible for data to ever rule it out. Proponents argue that all we are witnessing is the natural progression of science: we continually adjust our theories to bring them in line with the most precise information provided by observational measurements and mathematical concerns. More generally, and in more technical terms, a statement widely embraced by cosmologists is that the universe experienced a phase during which the size of the comoving horizon decreased. What is less clear is whether that phase is correctly described by inflationary cosmology, in which the dynamics is driven by the uniform energy suffusing space supplied by a scalar field (see note 3 of this chapter), as I have described, or whether such a phase may have arisen through a different mechanism (such as bouncing cosmologies, brane inflation, colliding

brane worlds, variable speed of light theories, among others that physicists have proposed). In chapter 10, we will briefly discuss the possibility of a bouncing cosmology, as developed by Paul Steinhardt, Neil Turok, and various of their collaborators, in which the universe undergoes numerous cycles of cosmological evolution.

7. For the particularly diligent reader, let me address an important point shadowing the discussion. If all you know about a given physical system is that it has less than the maximum available entropy, then the second law of thermodynamics allows you to draw not one but two conclusions: the most likely evolution of the system toward the future will increase its entropy **and** the most likely evolution of the system toward the past will also increase its entropy. Such is the burden of time symmetric laws—equations that operate in exactly the same way whether evolving today's state toward the future or toward the past. The challenge is that the higher-entropy past to which such considerations lead is incompatible with the lower-entropy past attested to by memory and records. (We remember partially melted ice cubes as previously being less melted, thus having lower entropy, not more melted, which would be higher entropy.) More pointedly, a high-entropy past would undermine our confidence in the very laws of physics because such a past would not include the experiments and observations that support the laws themselves. To avoid such a loss of confidence in our

understanding we must **enforce** a low-entropy past. Generally, we do so by introducing a new assumption, one named the **past hypothesis** by philosopher David Albert, which declares that entropy is anchored at a low value near the big bang and has on average been growing larger ever since. This is the approach we have implicitly taken in this chapter. In chapter 10, we will explicitly analyze the unlikely but conceivable possibility of a low-entropy state emerging from a previous high-entropy configuration. For background and more details, see chapter 7 of **The Fabric of the Cosmos.**

8. Mathematical descriptions of entropy make this precise: within any region, there are many more ways for the value of a field to vary (higher here, lower there, much lower way over there, and so on) than there are ways for it to be uniform (same value at every location), and thus the required conditions have low entropy. However, there is a hidden technical assumption that is important to call out. For ease, I will use classical language, but the considerations have a direct translation to quantum physics. In the microworld, no configuration of particles or fields is fundamentally singled out over any other and so we generally deem each to be as likely as any other. But this is an assumption that relies on what philosophers call the **principle of indifference.** With no a priori evidence distinguishing one microscopic configuration from another, we assign them equal probabilities of

being realized. When we shift our focus to the macroworld, the likelihood of one macrostate versus another is then determined by the ratio of the number of microstates that yield each. If there are twice as many microstates that yield a particular macrostate compared to those that yield another, that macrostate is twice as likely to occur.

Notice, though, that fundamentally, the justification for the principle of indifference must be empirically based. Indeed, common experience confirms the validity of a multitude of uses, implicit though they may be, of the principle of indifference. Take our example of tossed pennies. By assuming that each "microstate" of the coins (a state specified by listing each coin's disposition, such as coin 1 is heads, coin 2 is tails, coin 3 is tails, and so on) is as likely as any other, we conclude that those "macroscopic" arrangements (states specified only by giving the overall number of heads and tails, not the disposition of individual coins) that can be realized by many microstates are more likely. When we toss the coins, this assumption is empirically confirmed by the rarity of those outcomes that can be realized by only a small number of microstates (such as all heads) and the ubiquity of those that can be realized by a large number of microstates (such as half heads and half tails).

The relevance to our cosmological discussion is that when we say that a uniform patch of inflaton field is "unlikely," we are similarly invoking the principle of indifference. We are implicitly assuming that each possible microscopic configuration of the field (the field's precise value at every location) is as likely as any other so, again, the likelihood of a given macroscopic configuration is proportional to the number of microstates that realize it. However, in contrast to the case of tossed pennies, we have no empirical evidence to support this assumption. The fact that it seems reasonable is based on our experience in the everyday macroscopic world where the principle of indifference is supported by observation. But for the cosmological unfolding, we are privy to only a single run of the experiment. A hard-nosed empirical approach would conclude that however special some configurations may seem based on the principle of indifference, if they lead to the universe we observe, then they are singled out and, as a class, deserve to be called not just "likely" but "definite" (subject to the usual provisional nature of all scientific explanations). Mathematically, such a shift in what we call likely and unlikely is known as a change in the measure over configuration space (see chapter 2, note 14). The initial measure, assigning equal probabilities to each possible configuration, is called a "flat" measure. Observations can thus motivate the introduction of a "non-flat" measure that singles out certain classes of configurations as more probable.

Physicists are generally unsatisfied with such an approach. Introducing a measure over a space of configurations to ensure the greatest weight is given to those that lead to the world as we know it strikes physicists as "unnatural." Physicists seek a fundamental, first-principles, mathematical structure that will yield such a measure as output as opposed to including it as part of the input. Important issues are whether this is asking for too much and whether success would simply shift the question one step further back to the implicit assumptions underlying any first-principles approach. These are not nitpicking concerns. Much of the past thirty years of theoretical work in particle physics has been aimed at addressing issues of fine-tuning in our most refined theories (fine-tuning of the Higgs field in the standard model of particle physics; fine-tuning required to address the horizon and flatness problems in standard big bang cosmology). To be sure, such research has led to profound insights into both particle physics and cosmology, but might there come a point when we simply have to accept certain features of the world as given, without a deeper explanation? I like to think that the answer is no, as do a great many of my colleagues. But there is no guarantee that this will be the case.

9.   Andrei Linde, personal communication, July 15, 2019. Linde's preferred approach is for the inflationary phase to be initiated by a quantum-tunneling event from a realm of all possible geometries and fields, one in which the very concepts of time and temperature may not

yet have meaning. By judiciously using aspects of quantum formalism, Linde has argued that the quantum creation of conditions leading to inflationary expansion may well be a common process in the early universe that suffers from no quantum suppression.

10.   It is natural to think that the more powerful a telescope (the larger the dish, the greater the size of the mirror, and so on), the farther the objects are that it will be able to resolve. But there is a limit. If an object is so distant that any light it has emitted since its birth would not have yet had sufficient time to reach us, then regardless of the equipment we use, we will be unable to see it. We say that such objects lie beyond our **cosmic horizon,** a concept that will play a particularly important part in our discussion of the far future in chapters 9 and 10. In inflationary cosmology, space expands so rapidly that surrounding regions are indeed driven beyond our cosmic horizon.

11.   Based on indirect evidence (the motion of stars and galaxies), there is wide consensus that space is suffused with particles of dark matter—particles that exert a gravitational force but which do not absorb or produce light. But because searches for dark matter particles have so far come up empty-handed, some researchers have suggested alternatives to dark matter in which observations are explained through modifications of the gravitational force law. With the continued failure of numerous ongoing experiments to directly

detect particles of dark matter, the alternative theories are attracting increased attention.

12. The direction of heat flow, from hotter substances or environments to cooler ones, is a direct consequence of the second law of thermodynamics. When hot coffee cools to room temperature, transferring some of its heat to the molecules of air in the room, the air slightly heats up and so its entropy increases. The increase in the entropy of the air exceeds the decrease in entropy of the cooling coffee, ensuring that overall entropy increases. Mathematically, the change in entropy of a system is given by the change in its heat divided by its temperature ($\Delta S = \frac{\Delta Q}{T}$, where S denotes entropy, Q denotes heat, and T denotes temperature). When heat flows from a hotter system to a cooler one, the magnitude of the change in heat for each system is the same, but as the equation shows, the decrease in entropy of the hotter system is less than the increase in temperature of the cooler one (due to the factor of T in the denominator), and so the net change will yield an overall increase in entropy.

13. From the standpoint of energy conservation, as the molecules move outward, their gravitational potential energy increases, and so their kinetic energy decreases.

14. For the mathematically inclined and physically trained reader, you can understand this with a back-of-the-envelope calculation using classical statistical mechanics, in which entropy is

proportional to phase space volume. Assume the shrinking gas cloud satisfies the (famous) virial theorem, which relates the average kinetic energy of the particles, K, to their average potential energy, U, via $K = -U/2$. Then, because gravitational potential energy is proportional to $1/R$, with R the radius of the cloud, we see that K is also proportional to $1/R$. Further, since kinetic energy is proportional to the square of particle speeds, we learn that the average speed of the particles is proportional to $1/\sqrt{R}$. The phase space volume accessible to the particles in the cloud is thus proportional to $R^3(1/\sqrt{R})^3$ where the first factor represents the spatial volume accessible to the particles and the second factor represents the momentum space volume accessible to the particles. We see that the decrease in spatial volume dominates over the increase in momentum space volume, yielding an overall decrease in entropy as the cloud shrinks. Note too that the virial theorem ensures that as the cloud shrinks, the decrease in potential energy exceeds the increase in kinetic energy (due to the factor of "2" in the theorem relating K and U), so not only does the entropy of the shrinking part of the cloud decrease, its energy decreases too. That energy is radiated to the surrounding shell, whose energy increases, as does its entropy.

Chapter 4: Information and Vitality

1. Letter from F. H. C. Crick to E. Schrödinger, 12 August 1953.

2. J. D. Watson and F. H. C. Crick, "Molecular Structure of Nucleic Acids: A Structure for Deoxyribose Nucleic Acid," **Nature** 171 (1953): 737–38. The central figure in the discovery was chemist and crystallographer Rosalind Franklin, whose "photograph 51" was provided without her knowledge to Watson and Crick by Wilkins. It was this photograph that was instrumental in Watson and Crick completing the double helix model of DNA. Franklin died in 1958, four years before the Nobel Prize for unraveling the structure of DNA was awarded—and the Nobel cannot be awarded posthumously. Had Franklin still been alive, it is unclear how the Nobel committee would have acted. See, for example, Brenda Maddox, **Rosalind Franklin: The Dark Lady of DNA** (New York: Harper Perennial, 2003).

3. Maurice Wilkins, **The Third Man of the Double Helix** (Oxford: Oxford University Press, 2003), 84.

4. Erwin Schrödinger, **What Is Life?** (Cambridge: Cambridge University Press, 2012), 3.

5. **Time** magazine, Vol. 41, Issue 14 (5 April 1943): 42.

6. Erwin Schrödinger, **What Is Life?** (Cambridge: Cambridge University Press, 2012), 87.

7. K. G. Wilson, "Critical phenomena in 3.99

dimensions," **Physica** 73 (1974): 119. See Ken
Wilson's Nobel Prize lecture for a semitechnical
discussion and references therein: https://www
.nobelprize.org.

8. In various guises, the notion of nested stories,
sometimes described as "levels of understanding" or "levels of explanation," has been invoked
by scholars from a broad range of scientific
disciplines. Psychologists speak of explaining
behavior at a biological level (invoking physiochemical causes), a cognitive level (invoking
higher-level brain functions), and a cultural
level (invoking social influences); some cognitive
scientists (going back to neuroscientist David
Marr) organize understanding of information
processing systems in terms of a computational
level, an algorithmic level, and a physical level.
Common to many of the hierarchical schemas
espoused by philosophers and physicists is a
commitment to **naturalism**—a term often used
but difficult to define precisely. Most who use it
would agree that naturalism rejects explanations
that invoke supernatural entities and instead
relies solely on qualities of the natural world. Of
course, to make this position precise, we need to
specify discernible limits on what constitutes the
natural world, a task that is easier said than done.
Tables and trees lie squarely within its domain,
but what about the number five or Fermat's
Last Theorem? How about the emotion of joy
or the sensation of red? How about the ideals of
inalienable freedom and human dignity?

Over the years, questions like these have inspired many variations on the theme of naturalism. One extreme position holds that the only legitimate knowledge of the world comes from the concepts and analyses of science—a position sometimes labeled "scientism." Here, too, the perspective requires its proponents to define the terms with precision: What constitutes science? Clearly, if science is taken to mean conclusions based on observations, experience, and rational thinking, the boundaries of science extend well beyond the disciplines we typically find represented in university science departments. As you can imagine, this results in claims of a significant overreach of science.

Less extreme positions thread a naturalistic commitment through various organizing principles. Philosopher Barry Stroud has advocated what he calls "expansive or open-minded naturalism" in which the explanatory boundaries are not set in stone from the outset. Instead, expansive naturalism reserves the freedom to build layers of understanding that invoke everything from nature's material ingredients to psychological qualities to abstract mathematical statements, as called for to explain observations, experiences, and analyses (Barry Stroud, "The Charm of Naturalism," **Proceedings and Addresses of the American Philosophical Association** 70, no. 2 [November 1996], 43–55). Philosopher John Dupré has advocated "pluralistic naturalism," which argues that the dream of a unity within science is a dangerous myth, and instead our explanations must emerge from "diverse and overlapping projects of inquiry" that span the traditional sciences and beyond to include, among other disciplines, history, philosophy, and the arts (John Dupré, "The Miracle of Monism," in **Naturalism in Question,** ed. Mario de Caro and David Macarthur [Cambridge, MA: Harvard University Press, 2004], 36–58). Stephen Hawking and Leonard Mlodinow introduced the notion of "model dependent realism," which describes reality in terms of a collection of distinct stories, each based on a different model or theoretical framework for explaining observations, whether in the microworld of particles or the macroworld of everyday happenings (Stephen Hawking and Leonard Mlodinow, **The Grand Design** [New York: Bantam Books, 2010]). Physicist Sean Carroll has invoked "poetic naturalism," to refer to explanations that extend scientific naturalism to include language and concepts catered to different domains of interest (Sean Carroll, **The Big Picture** [New York: Dutton, 2016]). And as pointed out in chapter 1, note 4, E. O. Wilson uses the term "consilience" to express a coming together of knowledge from widely disparate disciplines to provide a depth of understanding that would otherwise be unattainable.

I am not much for jargon, but were I to label my own view, the one that will guide our discussion across this book, I would call it "nested naturalism." Nested naturalism, as will become clear in this and subsequent chapters, is committed to the value and the universal applicability of reductionism. It takes as a given that there is a fundamental unity in the workings of the world, and posits that such unity will be found by pursuing the reductionist program to whatever depth it leads. Everything that takes place in the world admits a description in terms of nature's fundamental constituents following the dictates of nature's fundamental laws. Nested naturalism also emphasizes, though, that such a description has limited explanatory power. There are many other levels of understanding that wrap around the reductionist account much as the outer parts of a nest wrap around its innermost structure. And depending on the questions being pursued, these other explanatory stories can provide accounts that are far more insightful than the one provided by reductionism. All of the accounts must be mutually consistent, but new and useful concepts can emerge at higher levels that do not admit lower-level correlates. For example, when studying many water molecules, the concept of a water wave is both sensible and useful. When studying a single water molecule, it is not. Similarly, in exploring the rich and varied stories of human experience, nested naturalism freely invokes accounts at whatever levels of structure prove most illuminating, all the while ensuring that the accounts fit into a coherent description.

9. Throughout, all references to "life" implicitly

mean "life as we know it, on planet earth" and so I will not provide this qualification.

10. One significant hurdle in forming atoms with large atomic weights is that there are no stable nuclei that contain either five or eight nucleons. As nuclei build up by sequentially adding protons and neutrons (hydrogen and helium nuclei), the instability at steps five and eight creates a bottleneck that hinders big bang nucleosynthesis.

11. The ratios I have given provide the relative abundances by mass. Since each helium nucleus is roughly four times the mass of each hydrogen nucleus, a count of the number of hydrogen atoms compared to the number of helium atoms yields a different ratio, roughly 92 percent hydrogen and 8 percent helium.

12. For a full history see Helge Kragh, "Naming the Big Bang," **Historical Studies in the Natural Sciences** 44, no. 1 (February 2014): 3. Kragh has suggested that although Hoyle favored his own cosmological theory (the steady state model, in which the universe always existed), his use of the term "big bang" may not have been meant derisively. Instead, Hoyle may have used "big bang" as a colorful way of distinguishing his own theory from this particular competitor.

13. S. E. Woosley, A. Heger, and T. A. Weaver, "The evolution and explosion of massive stars," **Reviews of Modern Physics** 74 (2002): 1015.

14. One study analyzed hundreds of thousands of possible trajectories and concluded that almost

all would have required the sun to be ejected at such a high speed that either it would lose its protoplanetary disk or if planets had already formed, they would be dispersed (Bárbara Pichardo, Edmundo Moreno, Christine Allen, et al., "The Sun was not born in M67," **The Astronomical Journal** 143, no. 3 [2012]: 73). Another study, which makes a different assumption for the location where Messier 67 itself was formed, concluded that a slower ejection speed might be adequate to launch the sun on its way, and with this slower speed, the planets or the protoplanetary disk would be preserved (Timmi G. Jørgensen and Ross P. Church, "Stellar escapers from M67 can reach solar-like Galactic orbits," arxiv.org, arXiv:1905.09586).

15. A. J. Cavosie, J. W. Valley, S. A. Wilde, "The Oldest Terrestrial Mineral Record: Thirty Years of Research on Hadean Zircon from Jack Hills, Western Australia," in **Earth's Oldest Rocks,** ed. M. J. Van Kranendonk (New York: Elsevier, 2018), 255–78. The most recent data is consistent with the original study described in John W. Valley, William H. Peck, Elizabeth M. King, and Simon A. Wilde, "A Cool Early Earth," **Geology** 30 (2002): 351–54; John Valley, personal communication, 30 July 2019.

16. Werner Heisenberg, **Physics and Philosophy: The Revolution in Modern Science** (London: Penguin Books, 1958), 16.

17. Max Born, **"Zur Quantenmechanik der Stoßvorgänge," Zeitschrift für Physik** 37,

no. 12 (1926): 863. In the initial version of this paper, Born associated quantum wave functions directly with probabilities, but in a footnote added later, he corrected the relationship to involve the norm squared of the wave function.

18. Wolfgang Pauli's exclusion principle, which we will discuss in chapter 9, is also essential to determining the allowed quantum orbitals of electrons around a nucleus. The exclusion principle establishes that no two electrons (more generally, no two matter particles of the same species) can occupy the same quantum state. Consequently, the individual quantum orbitals determined by Schrödinger's equation can each accommodate at most one electron (or, including the spin degree of freedom, two electrons). Many of these orbitals have the same energy, which in our analogy corresponds to seats that are located on the same level in the quantum theater. But once each of these seats is taken—once each quantum orbital is occupied—that level cannot accommodate any additional electrons.

19. If you recall your middle school chemistry, you will realize that I have made a modest simplification. In a more precise description, I would note that (because of quantum mechanics) atoms organize their tiers into a variety of subtiers, whose angular momenta have different values. Sometimes a higher tier, with less angular momentum, has less energy than a lower tier with more angular momentum. If so, electrons will

populate that subtier of the higher tier before completing the lower tier.

20. More precisely, stability is achieved when an atom's outer subshell (its valence shell) is full. You may recall from high school the "rule of eight," which notes that atoms usually need eight electrons in their valence shell and so will either donate, receive, or share electrons with other atoms to reach that number.

21. Albert Szent-Györgyi, "Biology and Pathology of Water," **Perspectives in Biology and Medicine** 14, no. 2 (1971): 239.

22. Our focus in this chapter is on plants and animals, which are constituted of eukaryotic cells (cells that contain a nucleus). Researchers thus say that the lineages converge on the "last eukaryotic common ancestor," or LECA. More generally, if we also consider bacteria and archaea, the lineages converge further back on the "last universal common ancestor," or LUCA.

23. A. Auton, L. Brooks, R. Durbin, et al., "A global reference for human genetic variation," **Nature** 526, no. 7571 (October 2015): 68.

24. Scientists have developed various measures for comparing DNA overlap between species. One approach compares base pairs for those genes the species share (which is the origin of the roughly 1 percent genetic difference quoted between humans and chimps), while another compares entire genomes (which yields a human-chimp genetic difference that is somewhat larger).

25. More precisely, researchers describe the code explained in the next paragraph as "nearly" universal, reflecting the fact that in particular special cases variations have been discovered. Nevertheless, even these modest modifications all share the same basic coding structure as the one described in the chapter.

26. With three-letter codes and four distinct letters, there are sixty-four possible combinations. But since these sequences are coding for only twenty amino acids, a number of different sequences can and do code for the same amino acid. Historically, among the first papers unraveling this genetic code were F. H. C. Crick, Leslie Barnett, S. Brenner, and R. J. Watts-Tobin, "General nature of the genetic code for proteins," **Nature** 192 (1961): 1227–32; J. Heinrich Matthaei, Oliver W. Jones, Robert G. Martin, and Marshall W. Nirenberg, "Characteristics and Composition of Coding Units," **Proceedings of the National Academy of Sciences** 48, no. 4 (1962): 666–77. By the mid-1960s, through the efforts of a number of researchers, most notably Marshall Nirenberg, Robert Holley, and Har Gobind Khorana, the code was completed, for which these three leaders of the efforts were awarded the 1968 Nobel Prize.

27. The precise definition of a gene is still subject to debate. In addition to the protein coding information, a gene comprises auxiliary sequences (that need not be contiguous with

the coding region) that can impact the precise way a cell uses the coding data (for example, enhancing or suppressing the rate of production of a given protein, among other regulatory functions).

28. The key insight, proton-based electric currents powering ATP synthesis, was proposed by British biochemist Peter Mitchell, for which he won the 1978 Nobel Prize. (P. Mitchell, "Coupling of phosphorylation to electron and hydrogen transfer by a chemiosmotic type of mechanism," **Nature** 191 [1961]: 144–48.) Although various details of Mitchell's proposal required subsequent refinement, the Nobel Prize was awarded for his insights into "biological energy transfer." Mitchell was an unusual scientist. Fed up with the various inane qualities of the academic world (I can sympathize), he established an independent charitable company, Glynn Research, where he, various colleagues, and a staff of up to ten carried out biochemical research. Details of his fascinating life are recounted in John Prebble and Bruce Weber, **Wandering in the Gardens of the Mind: Peter Mitchell and the Making of Glynn** (Oxford: Oxford University Press, 2003). For details of the modern understanding of energy extraction and transport within cells see, for example, Bruce Alberts et al., **Molecular Biology of the Cell,** 5th ed. (New York: Garland Science, 2007), chapter 14. The informed reader will note one qualification to

the universality of this process: the extraction of energy via **fermentation** (an energy extraction process that does not use oxygen).

29. Charles Darwin, **The Origin of Species** (New York: Pocket Books, 2008).

30. In my analogy I am imagining a business iterating their product through random trial and error. However, there are other ways in which trial and error can be integrated in a more effective way. For instance, in developing various computational algorithms, computer scientists start with one algorithm, randomly modify it, discard those modifications that decrease the algorithm's speed, and then further modify those that remain (the modified algorithms that increase speed). Iterating this procedure, we have a natural-selection-inspired approach that samples a wide range of possibilities, leading toward faster computational procedures. Of course, studying modified algorithms on a computer is much less costly than trying out a randomly modified product in the market place. Blind trial and error can thus be a useful strategy in various tasks so long as the cost in both time and resources to iterate round upon round of random modification is small (or if modifications can be tested in a massively parallel manner).

31. Eric T. Parker, Henderson J. Cleaves, Jason P. Dworkin, et al., "Primordial synthesis of amines and amino acids in a 1958 Miller H_2S-rich spark discharge experiment," **Proceedings of**

the National Academy of Sciences 108, no. 14
(April 2011): 5526.

32.   Cell walls can naturally form from common
chemicals, like fatty acids, that have one end
that seeks water and another end that avoids
it. This relation to water can coax such molecules
to form double-wide barriers, with the water-
loving ends of the molecules on the outside
and the water-averting ends holding the two
walls together—a cell wall. For a discussion in
the context of the RNA World scenario, see
G. F. Joyce and J. W. Szostak, "Protocells and
RNA Self-Replication," **Cold Spring Harbor
Perspectives in Biology** 10, no. 9 (2018).

33.   Various researchers, including chemist Svante
Arrhenius, astronomer Fred Hoyle, astrobi-
ologist Chandra Wickramasinghe, and physicist
Paul Davies, among others, have suggested that
some of the falling rocks may have themselves
carried particularly hardy seeds of life, ready-
made molecules that could replicate and catalyze
reactions. Intriguing as this is, raising the possibility that life-bearing space rocks may have
landed on a great many planets throughout the
cosmos, the proposal doesn't advance our understanding of life's origin, as it shifts the question
to the origin of the seeds.

34.   David Deamer, **Assembling Life: How Can
Life Begin on Earth and Other Habitable
Planets?** (Oxford: Oxford University Press,
2018).

35.   A. G. Cairns-Smith, **Seven Clues to the Origin**

of Life (Cambridge: Cambridge University Press, 1990).

36. W. Martin and M. J. Russell, "On the origin of biochemistry at an alkaline hydrothermal vent," **Philosophical Transactions of the Royal Society B** 367 (2007): 1187.

37. Erwin Schrödinger, **What Is Life?** (Cambridge: Cambridge University Press, 2012), 67.

38. The energy carried by the incoming photons is more concentrated (their wavelengths are shorter, lying in the visible part of the spectrum, and their number is fewer) and is thus of higher quality; the energy carried by outgoing photons is more dilute (their wavelengths are longer, lying in the infrared part of the spectrum, and their numbers are greater) and is thus of lower quality. The utility of solar power thus derives not just from the voluminous quantity of energy supplied by the sun but from the sun's energy being high quality, carrying much lower entropy than the heat released back into space by the earth. As noted in the chapter, for every photon earth receives from the sun it radiates a couple dozen back into space. To estimate this number, note that photons from the sun are emitted from an environment whose temperature is about 6000 K (the surface temperature of the sun), while those released by earth are emitted from an environment whose temperature is about 285 K (surface temperature of the earth). A photon's energy is proportional to such temperatures (considering photons to

be an ideal gas of particles), and hence the ratio of photons absorbed by earth from the sun and then re-released is given by the ratio of the two temperatures, 6000 K/285 K, which is about 21 photons, or roughly two dozen.

39. Erwin Schrödinger, **What Is Life?** (Cambridge: Cambridge University Press, 2012), 1.

40. Albert Einstein, **Autobiographical Notes** (La Salle, IL: Open Court Publishing, 1979), 3. For a beautiful modern treatment of thermodynamic principles in the context of living systems, providing insightful examples illustrating many of the essential concepts we are invoking, see Philip Nelson, **Biological Physics: Energy, Information, Life** (New York: W. H. Freeman and Co., 2014).

41. J. L. England, "Statistical physics of self-replication," **Journal of Chemical Physics** 139 (2013): 121923. Nikolay Perunov, Robert A. Marsland, and Jeremy L. England, "Statistical Physics of Adaptation," **Physical Review X** 6 (June 2016): 021036-1; Tal Kachman, Jeremy A. Owen, and Jeremy L. England, "Self-Organized Resonance During Search of a Diverse Chemical Space," **Physical Review Letters** 119, no. 3 (2017): 038001-1. See also G. E. Crooks, "Entropy production fluctuation theorem and the nonequilibrium work relation for free energy differences," **Physical Review E** 60 (1999): 2721; and C. Jarzynski, "Nonequilibrium equality for free energy differences," **Physical Review Letters** 78 (1997): 2690.

42.   England also points out that because the physical structure of a living entity is not just momentarily ordered but maintains its order over long periods of time—even for a period after it dies—a significant part of the waste energy life produces may be a by-product of building such stable structures. For life, then, it may be that a dominant contribution to the entropic two-step is linked to structure formation in addition to the ongoing preservation of homeostasis. Also note that while living systems need to take in high-quality energy, they need that energy to be in a form that does not disrupt the system's internal organization. For a mechanical illustration, a wineglass can be driven to vibrate by a tone that has the right frequency, but if too much energy is transferred, the glass can shatter. To avoid an analogous outcome, some degrees of freedom in a dissipative system may cluster in configurations that avoid resonance with the energy impinging from the environment. Life involves an appropriate balance between these extremes.

Chapter 5: Particles and Consciousness

1.   Albert Camus, **The Myth of Sisyphus,** trans. Justin O'Brien (London: Hamish Hamilton, 1955), 18.

2.   Ambrose Bierce, **The Devil's Dictionary** (Mount Vernon, NY: The Peter Pauper Press, 1958), 14.

3. Will Durant, **The Life of Greece,** vol. 2 of **The Story of Civilization** (New York: Simon & Schuster, 2011), 8181–82, Kindle.

4. As I make frequent mention of mathematical equations articulating the laws of physics, it is worth briefly writing down our most refined version of these equations. Even if you don't grasp the symbols, it may still be of interest to see the general "look" of the mathematics.

   Einstein's field equations from the general theory of relativity are: $R_{\mu\nu} - \frac{1}{2} g_{\mu\nu}R + \Lambda g_{\mu\nu} = \frac{8\pi G}{c^4} T_{\mu\nu}$, where the left-hand side describes the curvature of spacetime, as well as the cosmological constant, Λ, and the right-hand side describes the mass and energy that is the source of the curvature (the source of the gravitational field). In this expression (and in those that follow) the Greek indices run from 0 to 3, representing the four spacetime coordinates.

   Maxwell's equations from electromagnetism are $\partial^{\alpha}F_{\alpha\beta} = \mu_0 J_\beta$ and $\partial_{[\alpha}F_{\varrho\sigma]} = 0$, where the left-hand side of these equations describes the electric and magnetic fields, and the right-hand side of the first equation describes the electric charges giving rise to these fields.

   The equations for the strong and weak nuclear forces are generalizations of Maxwell's equations. The essential new feature is that whereas in Maxwell's theory we can write the "field strength" $F_{\alpha\beta} = \partial_\alpha A_\beta - \partial_\beta A_\alpha$ in terms of A_α, which is known as the "vector potential," for the nuclear forces there are a collection of field strengths $F^a_{\alpha\beta}$ as well as a collection of vector potentials A^a_α, which are related by $F^a_{\alpha\beta} = \partial_\alpha A^a_\beta - \partial_\beta A^a_\alpha + gf^{abc}A^b_\alpha A^c_\beta$. The Latin indices run over the generators of the Lie algebras su(2) for the weak nuclear force and the su(3) for the strong nuclear force, and f^{abc} are the structure constants of these algebras.

Schrödinger's equation of quantum mechanics is $i\hbar\frac{\partial\psi}{\partial\tau} = H\psi$, where H is the Hamiltonian and ψ is the wave function, whose (properly normalized) norm squared provides quantum mechanical probabilities. The melding of quantum mechanics and the electromagnetic, weak and strong nuclear forces, including as well the known particles of matter and the Higgs particle, constitutes the Standard Model of Particle Physics. Typically, the Standard Model is expressed in an equivalent but distinct formalism known as the path integral (an approach pioneered by physicist Richard Feynman). The melding of quantum mechanics and general relativity is an ongoing topic of advanced research.

5. Augustine, **Confessions,** trans. F. J. Sheed (Indianapolis, IN: Hackett Publishing, 2006), 197.

6. Thomas Aquinas, **Questiones Disputatae de Veritate,** questions 10–20, trans. James V. McGlynn, S.J. (Chicago: Henry Regnery Company, 1953). https://dhspriory.org/thomas/QDdeVer10.htm#8.

7. William Shakespeare, **Measure for Measure,**

ed. J. M. Nosworthy (London: Penguin Books, 1995), 84.

8. Gottfried Leibniz, letter to Christian Goldbach, 17 April 1712.

9. Otto Loewi, "An Autobiographical Sketch," **Perspectives in Biology and Medicine** 4, no. 1 (Autumn 1960): 3–25. Loewi incorrectly noted that the dream took place Easter Sunday 1920, although the year was 1921.

10. For an in-depth history, see Henri Ellenberger, **The Discovery of the Unconscious** (New York: Basic Books, 1970).

11. Peter Halligan and John Marshall, "Blindsight and insight in visuo-spatial neglect," **Nature** 336, no. 6201 (December 22–29, 1988): 766–67.

12. The culprit was James Vicary, who in 1957 claimed that subliminal flashes encouraging audiences to eat popcorn and drink Coca-Cola resulted in significant sales increases in both. Later, Vicary admitted that the claims were without merit.

13. Researchers have established the capacity of a wide variety of subliminal stimuli to influence conscious activities. In this paragraph I describe one example, subliminal influences on simple numerical determinations. But similar subliminal influences have been demonstrated for recognizing words—see, for example, Anthony J. Marcel, "Conscious and Unconscious Perception: Experiments on Visual Masking and Word Recognition," **Cognitive Psychology** 15

(1983): 197–237—as well as for the perception and evaluation of a broad spectrum of images and objects.

14. L. Naccache and S. Dehaene, "The Priming Method: Imaging Unconscious Repetition Priming Reveals an Abstract Representation of Number in the Parietal Lobes," **Cerebral Cortex** 11, no. 10 (2001): 966–74; L. Naccache and S. Dehaene, "Unconscious Semantic Priming Extends to Novel Unseen Stimuli," **Cognition** 80, no. 3 (2001): 215–29. Note that in these experiments, the initial stimulus is rendered subliminal via a **masking** procedure in which geometrical shapes are flashed before and after the stimulus. For a review, see Stanislas Dehaene and Jean-Pierre Changeux, "Experimental and Theoretical Approaches to Conscious Processing," **Neuron** 70, no. 2 (2011): 200–27, and Stanislas Dehaene, **Consciousness and the Brain** (New York: Penguin Books, 2014).

15. Isaac Newton, letter to Henry Oldenburg, 6 February 1671. http://www.newtonproject.ox.ac.uk/view/texts/normalized/NATP00003.

16. Philosophers, psychologists, mystics, and a range of other thinkers have espoused various definitions of consciousness. Depending on context, some may be more useful than the approach we are taking, some less so. Our focus here is on the "hard problem," and for this purpose the description given in the chapter will serve us well.

17. My reference here to protons, neutrons, and

electrons is a shorthand for the state of my brain articulated in terms of nature's most refined ingredients, whatever those ingredients (particles, fields, strings, etc.) may turn out to be.

18. Thomas Nagel, "What Is It Like to Be a Bat?" **Philosophical Review** 83, no. 4 (1974): 435–50.

19. When I speak of understanding typhoons or volcanoes—or any macroscopic body—in terms of fundamental particles, I am speaking from an "in principle" perspective. As chaos theory has long emphasized, tiny differences in the initial conditions of a collection of particles will yield enormous differences in the future configuration of the particles. This is true even for small collections. In practice, this fact significantly impacts the kinds of predictions we are able to make, but it does not entail any mystery. Chaos theory provides a significant and profound set of insights, but the theory was not developed to fill a perceived gap in our understanding of the underlying physical laws. When it comes to consciousness, however, the problem raised in the chapter—how can mindless particles yield mindful sensations?—has suggested to some researchers that there is a gap of a far more fundamental nature. They have argued that the mind's sensations cannot emerge from large collections of particles, regardless of the coordinated motions those particles might follow.

20. Frank Jackson, "Epiphenomenal Qualia," **Philosophical Quarterly** 32 (1982): 127–36.

21. Daniel Dennett, **Consciousness Explained** (Boston: Little, Brown and Co., 1991), 399–401.

22. David Lewis, "What Experience Teaches," **Proceedings of the Russellian Society** 13 (1988): 29–57. Reprinted in David Lewis, **Papers in Metaphysics and Epistemology** (Cambridge: Cambridge University Press, 1999): 262–90, which builds on earlier insights in Laurence Nemirow, "Review of Nagel's Mortal Questions," **Philosophical Review** 89 (1980): 473–77.

23. Laurence Nemirow, "Physicalism and the cognitive role of acquaintance," in **Mind and Cognition,** ed. W. Lycan (Oxford: Blackwell, 1990), 490–99.

24. Frank Jackson, "Postscript on Qualia," in **Mind, Method, and Conditionals, Selected Essays** (London: Routledge, 1998), 76–79.

25. In his 1995 paper, Chalmers discusses both vitalism and electromagnetism as useful references for thinking about the hard problem. The key distinguishing feature of the hard problem, as Chalmers has defined it, is that it necessarily addresses subjective qualities of experience and thus, he argues, cannot be resolved by acquiring a more refined understanding of the brain's objective functions. In this section, I find it useful to frame the problem somewhat differently, contrasting open issues that science can resolve, at least in principle, using its currently established paradigm (which defines the arena

within which reality, as we know it, takes place) and open issues for which this paradigm may prove inadequate. In this framing, a problem is hard if to solve it we must fundamentally shift the existing approach to describing the world (in the example of electricity and magnetism, scientists had to introduce fundamentally new qualities—space-filling electric fields, magnetic fields, and electric charges). In that Chalmers argues that the hard problem cannot be solved by solely using the material ingredients at the core of our fundamental physical descriptions of reality, the framing I introduce, while different, captures an essential part of the issue. Note, too, that according to Chalmers, the very reason vitalism gradually disappeared is that the question it highlighted **was** one of objective function: How can physical ingredients carry out the objective functions of life? As science better understood the functional capacities of physical ingredients (biochemical molecules and so on), the enigma vitalism sought to address diminished. According to Chalmers, this progression will not be recapitulated with the hard problem. Physicalists do not share this intuition and thus anticipate progress in understanding brain function giving insight into subjective experience. For more details, see David Chalmers, "Facing Up to the Problem of Consciousness," **Journal of Consciousness Studies** 2, no. 3 (1995): 200–19, and David Chalmers, **The Conscious Mind: In Search of a Fundamental**

Theory (Oxford: Oxford University Press, 1997), 125.

26. There are innumerable cases in the clinical literature in which excising specific sections of the brain results in the loss of targeted brain functions. One such case is particularly close to home. After brain surgery to remove a malignant tumor, my wife, Tracy, temporarily lost the ability to name a wide variety of common nouns. As she describes it, it was as if the surgery sliced out the data bank in which her knowledge of the names of various items had been stored. She could still conjure a mental picture of such nouns, like a pair of red shoes, but she was unable to name the image in her mind.

27. Giulio Tononi, **Phi: A Voyage from the Brain to the Soul** (New York: Pantheon, 2012); Christof Koch, **Consciousness: Confessions of a Romantic Reductionist** (Cambridge, MA: MIT Press, 2012); Masafumi Oizumi, Larissa Albantakis, and Giulio Tononi, "From the Phenomenology to the Mechanisms of Consciousness: Integrated Information Theory 3.0," **PLoS Computational Biology** 10, no. 5 (May 2014).

28. Scott Aaronson, "Why I Am Not an Integrated Information Theorist (or, The Unconscious Expander)," **Shtetl-Optimized.** https://www.scottaaronson.com/blog/?p=1799.

29. Michael Graziano, **Consciousness and the Social Brain** (New York: Oxford University Press, 2013); Taylor Webb and Michael

Graziano, "The attention schema theory: A mechanistic account of subjective awareness," **Frontiers in Psychology** 6 (2015): 500.

30. The human perception of color is more complex than suggested by my brief description. Our eyes have receptors whose sensitivities vary across a range of light's frequencies. Some are most sensitive to the highest visible frequencies, some to the lowest, and some to frequencies that lie between the two. The colors our brains perceive arise from a blending of the responses of the various receptors.

31. As in the previous endnote, this is a simplification, as "red" is the brain's interpretation of a blended union of responses to various frequencies received by its visual receptors. Nevertheless, the simplified description communicates the essential point: our sensation of color is a useful but coarse representation of the physical data carried to our eyes through electromagnetic waves.

32. David Premack and Guy Woodruff, "Does the chimpanzee have a theory of mind?" **Cognition and Consciousness in Nonhuman Species,** special issue of **Behavioral and Brain Sciences** 1, no. 4 (1978): 515–26.

33. Daniel Dennett, **The Intentional Stance** (Cambridge, MA: MIT Press, 1989).

34. See, for example, Dennett's multiple drafts model in Daniel Dennett, **Consciousness Explained** (Boston: Little, Brown & Co., 1991), Baar's global workspace theory in Bernard J.

Baars, **In the Theater of Consciousness** (New York: Oxford University Press, 1997), and Hameroff and Penrose's orchestrated reduction theory in Stuart Hameroff and Roger Penrose, "Consciousness in the universe: A review of the 'Orch OR' theory." **Physics of Life Reviews** 11 (2014): 39–78.

35. While all of quantum mechanics can be traced back to Schrödinger's equation, in the decades since the theory was introduced, many physicists have developed the mathematical formalism far further. The successful prediction I refer to emerges from calculations in a field of quantum mechanics known as quantum electrodynamics, which melds quantum mechanics with Maxwell's theory of electromagnetism.

36. An alternate way of expressing this is that according to quantum mechanics the electron, prior to being measured, does not have a position in the conventional sense of the term.

37. As pointed out in note 5 of chapter 3, there is a version of quantum mechanics in which particles do retain sharp and definite trajectories, thus offering a potential resolution to the quantum measurement problem. To date, this approach, called Bohmian mechanics or de Broglie–Bohm mechanics, is pursued by a small group of researchers worldwide. Although it is a darkhorse candidate, I would not write off Bohmian mechanics as an approach that could develop into a dominant perspective in the future. Another approach to the quantum measurement

problem is the Many Worlds interpretation, in which all of the potential outcomes allowed by the quantum mechanical evolution are realized upon measurement. And a third proposal is the Ghirardi–Rimini–Weber (GRW) theory, which introduces a new and fundamental physical process that rarely but randomly collapses the probability wave for an individual particle. For small collections of particles, the process happens too infrequently to impact the results of successful quantum experiments. But for large collections of particles, the process happens far more quickly, creating a domino-like effect that selects precisely one outcome to be realized in the macroworld. For further details, see, for example, **The Fabric of the Cosmos,** chapter 7.

38. Fritz London and Edmond Bauer, **La théorie de l'observation en mécanique quantique,** No. 775 of **Actualités scientifiques et industrielles; Exposés de physique générale, publiés sous la direction de Paul Langevin** (Paris: Hermann, 1939), as translated in John Archibald Wheeler and Wojciech Zurek, **Quantum Theory and Measurement** (Princeton: Princeton University Press, 1983), 220.

39. Eugene Wigner, **Symmetries and Reflections** (Cambridge, MA: MIT Press, 1970).

40. Aristotle described an action as "voluntary" if the action started within a given agent and emerged from that agent's own deliberations—a perspective that, with substantial refinements, has had significant influence. See Aristotle,

Nicomachean Ethics, trans. C. D. C. Reeve (Indianapolis, IN: Hackett Publishing, 2014), 35–41. Aristotle did not include deterministic laws of physics among the external forces with the capacity to render an action involuntary, but those (including me) who do consider such fundamental albeit impersonal influences, find that his notion of "voluntary" does not align with their intuition regarding free will.

41. As in note 17 of this chapter, when I refer to the particles constituting a macroscopic object that is a shorthand for the complete physical state of the object. Classically, this state is provided by the positions and velocities of the object's fundamental constituents. Quantum mechanically, the state is provided by the wave function describing the object's constituents. Now, my emphasis on particles might leave you wondering about fields. As the technically trained reader may know, in quantum field theory we learn that the influence of a field is transmitted by particles (e.g., the influence of the electromagnetic field is transmitted by photons); moreover, quantum field theory also shows that a macroscopic field can be described mathematically as a particular configuration of particles—a so-called **coherent state** of particles. So my reference to "particles" is meant to subsume fields as well. The informed reader will also note that certain quantum features, such as quantum entanglement, can render the state of an object a more subtle notion in the quantum as opposed

to the classical setting. For much of what we will discuss, we can ignore such subtleties; the lawful, unitary progression of the physical world is, fundamentally, all we will need.

42. More precisely, the likelihood of the rock's particles conspiring to push off the bench is so ridiculously small that on the timescales of interest the statistical possibility of the rock's saving me can be ignored.

43. The philosophical literature contains many compatibilist proposals. Of these, the approach I describe is closest to that proposed and developed by Daniel Dennett in **Freedom Evolves** (New York: Penguin Books, 2003) and also in **Elbow Room** (Cambridge, MA: MIT Press, 1984), to which I refer the reader for a more in-depth discussion. I have been ruminating on these ideas since first being prompted to think about them decades ago by Luise Vosgerchian, one of my most influential teachers. Vosgerchian, who was a professor of music at Harvard, had a deep interest in how scientific discoveries relate to aesthetic sensibilities, and she asked me to write about human freedom and creativity from the standpoint of modern physics.

44. Artificial intelligence and machine learning make the point more forcefully still. Researchers have developed algorithms for playing games like chess or Go that can update themselves based on analyzing the success or failure of previous moves. Inside the computer hosting such an algorithm, all we have are particles moving this

way and that under the full control of physical law. And yet the algorithm improves. The algorithm learns. The algorithm's moves become creative. So creative, in fact, that with just hours of such internal updating, the most refined systems can advance from playing at the level of a beginner to triumphing over world-class masters. See David Silver, Thomas Hubert, Julian Schrittwieser, et al., "A general reinforcement learning algorithm that masters chess, shogi, and Go through self-play," **Science** 362 (2018): 1140–44.

45. The issue here is that if "I" am my configuration of particles, when that configuration shifts, both in arrangement and composition, am I still me? It is a version of another of philosophy's towering questions—personal identity through time—and so has generated a broad range of viewpoints and responses. I am partial to Robert Nozick's approach in which, to use somewhat technical language, we identify my future self by minimizing a distance function over the space of candidates for that role, seeking the person who "most closely continues" the existence I have had until this moment. Specifying the distance function is of course essential, and Nozick notes that people who hold different emphases on the defining aspects of personhood may make different choices. In many cases the intuitive notion of who "most closely continues" me is adequate, but one can construct artificial but puzzling examples. For instance, imagine a transporter

malfunction that produces two identical copies of me at a target destination. Which collection of particles is "really" me? In this case, Nozick suggests that without a unique closest continuer I may no longer exist. However, as I am comfortable with non-unique minimizations of distance functions, my perspective is that both copies would be me. For the notion of "I" used in the chapter, the intuitive notion of personal identity aligns with Nozick's notion, since the various collections of particles which we would intuitively label, say, "Brian Greene" throughout my lifetime are indeed closest continuers. See Robert Nozick, **Philosophical Explanations** (Cambridge, MA: Belknap Press, 1983), 29–70.

46. A question this discussion raises is whether you should bear the consequences of behavior that fellow citizens or society deems unacceptable. Philosophers have long debated questions at the intersection of free will, moral responsibility, and the role of punishment. The issues are complex and thorny. In a nutshell, here is my view: For the same reasons given in the chapter, your actions—good or bad—are your responsibility, even in the absence of free will. You are your particles, and if your particles do the wrong thing, you have done the wrong thing. The real issue, then, is what should the consequences be? Putting aside the fact that consequences of actions are also not freely willed, the question is whether you should suffer punishment. The only answer I find coherent, or, really, the only

beginning to an answer I find coherent, is that punishment should be based on its capacity for protecting societal interests, including deterring future instances of unacceptable behavior. Again, free will is compatible with learning; the Roomba learns, as do people. Today's experiences are causally related to tomorrow's actions. So if punishment prevents or dissuades you and/or others from subsequently undertaking unacceptable actions, then through punishment we have guided society toward a more satisfactory outcome. Similar considerations are relevant to the "test cases" often raised in these discussions in which unacceptable behaviors are due to extenuating circumstances (brain tumors, coercion, schizophrenia, neural implants controlled by nefarious aliens, and so on) that would seem to release the perpetrator from responsibility. The view that follows from the above and the discussion in the chapter is that such individuals **are** responsible for their actions. Their particles **did** unacceptable things. And they are their particles. However, subject to the precise details in any given situation, because of the extenuating circumstances there may be no opportunity for punishment to have any benefit. If your unacceptable behavior was due to a brain tumor, punishing you will likely have no role in deterring similar behavior caused by similar circumstances in the future. And if we can remove the tumor, you no longer pose any threat, so punishment will offer society no additional protection. Briefly put, punishment must serve a pragmatic purpose.

Chapter 6: Language and Story

1. Alice Calaprice, ed., **The New Quotable Einstein** (Princeton: Princeton University Press, 2005), 149.
2. Max Wertheimer, **Productive Thinking,** enlarged ed. (New York: Harper and Brothers, 1959), 228.
3. Ludwig Wittgenstein, **Tractatus Logico-Philosophicus** (New York: Harcourt, Brace & Company, 1922), 149.
4. Toni Morrison, Nobel Prize lecture, 7 December 1993. https://www.nobelprize.org/prizes/literature/1993/morrison/lecture/.
5. As Darwin wrote, "Primeval man, or rather some early progenitor of man, probably first used his voice in producing true musical cadences, that is in singing" and added "This power would have been especially exerted during the courtship of the sexes,—would have expressed various emotions, such as love, jealousy, triumph,—and would have served as a challenge to rivals." Charles Darwin, **The Descent of Man** (New York: D. Appleton and Company, 1871), 56.
6. In the April 1869 edition of the **Quarterly Review,** Wallace, in reference to the forces driving evolution—"the laws of variation, multiplication and survival"—argued that, as noted in the chapter, "we must therefore admit

the possibility, that in the development of the human race, a Higher Intelligence has guided the same laws for nobler ends." Alfred Russel Wallace, "Sir Charles Lyell on geological climates and the origin of species," **Quarterly Review** 126 (1869): 359–94.

7. Joel S. Schwartz, "Darwin, Wallace, and the **Descent of Man," Journal of the History of Biology** 17, no. 2 (1984): 271–89.

8. Charles Darwin, letter to Alfred Russel Wallace, 27 March 1869. https://www.darwinproject.ac .uk/letter/?docId=letters/DCP-LETT-6684.xml ;query=child;brand=default.

9. Dorothy L. Cheney and Robert M. Seyfarth, **How Monkeys See the World: Inside the Mind of Another Species** (Chicago: University of Chicago Press, 1992). A recording of these alarm calls can be heard on the BBC webpage: https://www.bbc.co.uk/sounds/play/p016dgw1.

10. Bertrand Russell, **Human Knowledge** (New York: Routledge, 2009), 57–58.

11. R. Berwick and N. Chomsky, **Why Only Us?** (Cambridge, MA: MIT Press, 2015). Although some have questioned whether the proposal's need for comparatively rapid biological change creates tension with the understanding of evolution, Chomsky has argued that it fits squarely into the modern neo-Darwinian perspective that embraces biological episodes, such as the formation of the eye, which deviate from the traditional view that all things evolution are slow and gradual.

12. S. Pinker and P. Bloom, "Natural language and natural selection," **Behavioral and Brain Sciences** 13, no. 4 (1990): 707–84; Steven Pinker, **The Language Instinct** (New York: W. Morrow and Co., 1994); Steven Pinker, "Language as an adaptation to the cognitive niche," in **Language Evolution: States of the Art,** ed. S. Kirby and M. Christiansen (New York: Oxford University Press, 2003), 16–37.

13. For example, as linguist and developmental psychologist Michael Tomasello has noted, "For sure, all of the world's languages have things in common . . . But these commonalities come not from any universal grammar, but rather from universal aspects of human cognition, social interaction, and information processing—most of which were in existence in humans before anything like modern languages arose." Michael Tomasello, "Universal Grammar Is Dead," **Behavioral and Brain Sciences** 32, no. 5 (October 2009): 470–71.

14. Simon E. Fisher, Faraneh Vargha-Khadem, Kate E. Watkins, Anthony P. Monaco, and Marcus E. Pembrey, "Localisation of a gene implicated in a severe speech and language disorder," **Nature Genetics** 18 (1998): 168–70. C. S. L. Lai, et al., "A novel forkhead-domain gene is mutated in a severe speech and language disorder," **Nature** 413 (2001): 519–23.

15. Johannes Krause, Carles Lalueza-Fox, Ludovic Orlando, et al., "The Derived FOXP2 Variant of

Modern Humans Was Shared with Neandertals," **Current Biology** 17 (2007): 1908–12.

16. Fernando L. Mendez et al. "The Divergence of Neandertal and Modern Human Y Chromosomes," **American Journal of Human Genetics** 98, no. 4 (2016): 728–34.

17. Guy Deutscher, **The Unfolding of Language: An Evolutionary Tour of Mankind's Greatest Invention** (New York: Henry Holt and Company, 2005), 15.

18. Dean Falk, "Prelinguistic evolution in early hominins: Whence motherese?" **Behavioral and Brain Sciences** 27 (2004): 491–541; Dean Falk, **Finding Our Tongues: Mothers, Infants and the Origins of Language** (New York: Basic Books, 2009).

19. R. I. M. Dunbar, "Gossip in Evolutionary Perspective," **Review of General Psychology** 8, no. 2 (2004): 100–10; Robin Dunbar, **Grooming, Gossip, and the Evolution of Language** (Cambridge, MA: Harvard University Press, 1997).

20. N. Emler, "The Truth About Gossip," **Social Psychology Section Newsletter** 27 (1992): 23–37; R. I. M. Dunbar, N. D. C. Duncan, and A. Marriott, "Human Conversational Behavior," **Human Nature** 8, no. 3 (1997): 231–46.

21. Daniel Dor, **The Instruction of Imagination** (Oxford: Oxford University Press, 2015).

22. For the role of firemaking and cooking, see Richard Wrangha, **Catching Fire: How Cooking Made Us Human** (New York:

Basic Books; 2009); for group parenting of the young, see Sarah Hrdy, **Mothers and Others: The Evolutionary Origins of Mutual Understanding** (Cambridge, MA: Belknap Press, 2009); for learning and cooperation, see Kim Sterelny, **The Evolved Apprentice: How Evolution Made Humans Unique** (Cambridge, MA: MIT Press, 2012).

23. R. Berwick and N. Chomsky, **Why Only Us?** (Cambridge, MA: MIT Press, 2015), chapter 2.

24. David Damrosch, **The Buried Book: The Loss and Rediscovery of the Great Epic of Gilgamesh** (New York: Henry Holt and Company, 2007).

25. Andrew George, trans., **The Epic of Gilgamesh: The Babylonian Epic Poem and Other Texts in Akkadian and Sumerian** (London: Penguin Classics, 2003).

26. For an introduction to the perspective and principles of evolutionary psychology, see John Tooby and Leda Cosmides, "The Psychological Foundations of Culture," in **The Adapted Mind: Evolutionary Psychology and the Generation of Culture,** ed. Jerome H. Barkow, Leda Cosmides, and John Tooby (Oxford: Oxford University Press, 1992), 19–136; David Buss, **Evolutionary Psychology: The New Science of the Mind** (Boston: Allyn & Bacon, 2012).

27. S. J. Gould and R. C. Lewontin, "The Spandrels of San Marco and the Panglossian Paradigm: A Critique of the Adaptationist Programme,"

Proceedings of the Royal Society B 205, no. 1161 (21 September 1979): 581–98.

28. Steven Pinker, **How the Mind Works** (New York: W. W. Norton, 1997), 530; Brian Boyd, **On the Origin of Stories** (Cambridge, MA: Belknap Press, 2010); Brian Boyd, "The evolution of stories: from mimesis to language, from fact to fiction," **WIREs Cognitive Science** 9 (2018): e1444.

29. Patrick Colm Hogan, **The Mind and Its Stories** (Cambridge: Cambridge University Press, 2003); Lisa Zunshine, **Why We Read Fiction: Theory of Mind and the Novel** (Columbus: Ohio State University Press, 2006).

30. Jonathan Gottschall, **The Storytelling Animal** (Boston and New York: Mariner Books, Houghton Mifflin Harcourt, 2013), 63.

31. Keith Oatley, "Why fiction may be twice as true as fact," **Review of General Psychology** 3 (1999): 101–17.

32. For absorbing accounts of Jouvet's work, see Barbara E. Jones, "The mysteries of sleep and waking unveiled by Michel Jouvet," **Sleep Medicine** 49 (2018): 14–19; Isabelle Arnulf, Colette Buda, and Jean-Pierre Sastre, "Michel Jouvet: An explorer of dreams and a great storyteller," **Sleep Medicine** 49 (2018): 4–9.

33. Kenway Louie and Matthew A. Wilson, "Temporally Structured Replay of Awake Hippocampal Ensemble Activity During Rapid Eye Movement Sleep," **Neuron** 29 (2001): 145–56.

34. The outlandish narratives we often associate with dreams—violating physical law, logical progression, and internal coherence—might suggest that the act of dreaming has little relevance for real-world encounters. However, the prevalence of such bizarre dreams may be much less than our anecdotal assessments suggest. Instead, a significant fraction of our dreams may have realistic content. Antti Revonsuo, Jarno Tuominen, and Katja Valli, "The Avatars in the Machine—Dreaming as a Simulation of Social Reality," **Open MIND** (2015): 1–28; Serena Scarpelli, Chiara Bartolacci, Aurora D'Atri, et al., "The Functional Role of Dreaming in Emotional Processes," **Frontiers in Psychology** 10 (March 2019): 459.

35. Alfred North Whitehead, **Science and the Modern World** (New York: Free Press, 1953), 10.

36. Joyce Carol Oates, "Literature as Pleasure, Pleasure as Literature," **Narrative.** https://www .narrativemagazine.com/issues/stories-week -2015-2016/story-week/literature-pleasure -pleasure-literature-joyce-carol-oates.

37. Jerome Bruner, "The Narrative Construction of Reality," **Critical Inquiry** 18, no. 1 (Autumn 1991): 1–21.

38. Jerome Bruner, **Making Stories: Law, Literature, Life** (New York: Farrar, Straus and Giroux, 2002), 16.

39. Brian Boyd, "The evolution of stories: from mimesis to language, from fact to fiction," **WIREs Cognitive Science** 9 (2018): 7–8, e1444.

40. John Tooby and Leda Cosmides, "Does Beauty Build Adapted Minds? Toward an Evolutionary Theory of Aesthetics, Fiction and the Arts," **SubStance** 30, no. 1/2, issue 94/95 (2001): 6–27.

41. Ernest Becker, **The Denial of Death** (New York: Free Press, 1973), 97.

42. Joseph Campbell, **The Hero with a Thousand Faces** (Novato, CA: New World Library, 2008), 23.

43. Michael Witzel, **The Origins of the World's Mythologies** (New York: Oxford University Press, 2012).

44. Karen Armstrong, **A Short History of Myth** (Melbourne: The Text Publishing Company, 2005), 3.

45. Marguerite Yourcenar, **Oriental Tales** (New York: Farrar, Straus and Giroux, 1985).

46. Scott Leonard and Michael McClure, **Myth and Knowing** (New York: McGraw-Hill Higher Education, 2004), 283–301.

47. Michael Witzel, **The Origins of the World's Mythologies** (New York: Oxford University Press, 2012), 79.

48. Dan Sperber, **Rethinking Symbolism** (Cambridge: Cambridge University Press, 1975); Dan Sperber, **Explaining Culture: A Naturalistic Approach** (Oxford: Blackwell Publishers Ltd., 1996).

49. Pascal Boyer, "Functional Origins of Religious Concepts: Ontological and Strategic Selection in Evolved Minds," **Journal of the Royal**

Anthropological Institute 6, no. 2 (June 2000): 195–214. See also M. Zuckerman, "Sensation seeking: A comparative approach to a human trait," **Behavioral and Brain Sciences** 7 (1984): 413–71.

50. Bertrand Russell emphasized the role of language in facilitating thought, noting that "language serves not only to express thoughts, but to make possible thoughts which could not exist without it" (Bertrand Russell, **Human Knowledge** [New York: Routledge, 2009], 58). He described how certain "fairly elaborate thoughts" require words, and as an example notes the apparent impossibility of, without language, having any "thought at all closely corresponding to what is asserted in the sentence 'the ratio of the circumference of a circle to the diameter is approximately 3.14159.'" Less precise constructs but ones beyond the bounds of experience, such as talking trees or crying clouds or happy pebbles, are amenable to wordless incarnations in the human mind, but the combinatorial and hierarchical nature of language is particularly suited for creating them. Daniel Dennett has emphasized the role of language in the human capacity for inventing unions of qualities that individually exist in the real but in combination take us into the realm of the fantastic (Daniel Dennett, **Breaking the Spell: Religion as a Natural Phenomenon** [New York: Penguin Publishing Group, 2006], 121). As we will discuss in chapter 8, certain

types of art are particularly adept at facilitating the flow of ideas in the other direction: from thoughts articulated in words to language-free experiential feelings.

51.   Justin L. Barrett, **Why Would Anyone Believe in God?** (Lanham, MD: AltaMira, 2004); Stewart Guthrie, **Faces in the Clouds: A New Theory of Religion** (New York: Oxford University Press, 1993).

Chapter 7: Brains and Belief

1.   The Qafzeh excavation was begun in 1934, undertaken by French archaeologist René Neuville, and was carried on by a team led by anthropologist Bernard Vandermeersch. In the words of Vandermeersch and his team, the burial arrangement of Qafzeh 11 "attested a funerary offering and not an accidental incorporation. All these observations strongly support the interpretation of a deliberate, ceremonial burial." See Hélène Coqueugniot et al., "Earliest cranio-encephalic trauma from the Levantine Middle Palaeolithic: 3D reappraisal of the Qafzeh 11 skull, consequences of pediatric brain damage on individual life condition and social care," **PloS One** 9 (23 July 2014): 7 e102822.

2.   Erik Trinkaus, Alexandra Buzhilova, Maria Mednikova, and Maria Dobrovolskaya, **The People of Sunghir: Burials, Bodies and Behavior in the Earlier Upper Paleolithic** (New York: Oxford University Press, 2014).

3. Edward Burnett Tylor, **Primitive Culture,** vol. 2 (London: John Murray 1873; Dover Reprint Edition, 2016), 24.

4. Mathias Georg Guenther, **Tricksters and Trancers: Bushman Religion and Society** (Bloomington, IN: Indiana University Press, 1999), 180–98.

5. Peter J. Ucko and Andrée Rosenfeld, **Paleolithic Cave Art** (New York: McGraw-Hill, 1967), 117–23, 165–74.

6. David Lewis-Williams, **The Mind in the Cave: Consciousness and the Origins of Art** (New York: Thames & Hudson, 2002), 11. Although many works were created on more accessible surfaces, too, the existence of a substantial collection that presented significant difficulties to execute gives this perspective relevance.

7. Salomon Reinach, **Cults, Myths and Religions,** trans. Elizabeth Frost (London: David Nutt, 1912), 124–38.

8. The proposal gained wide currency, but the subsequent discovery of a mismatch between the animals whose bones have been dug up in the vicinity of various caves and those depicted on the walls of those caves casts doubt. If you're looking for a little extra luck hunting bison, you're going to paint a bison. Or so you would think. But the data failed to bear out this expectation. See Jean Clottes, **What Is Paleolithic Art? Cave Paintings and the Dawn of Human Creativity** (Chicago: University of Chicago Press, 2016).

9.	Benjamin Smith, personal communication, 13 March 2019.

10.	Pascal Boyer, **Religion Explained: The Evolutionary Origins of Religious Thought** (New York: Basic Books, 2007), 2.

11.	For a detailed discussion see, for example, **The Adapted Mind: Evolutionary Psychology and the Generation of Culture,** Jerome H. Barkow, Leda Cosmides, and John Tooby, eds. (Oxford: Oxford University Press, 1992); David Buss, **Evolutionary Psychology: The New Science of Mind** (Boston: Allyn & Bacon, 2012).

12.	For other accessible contributions to the cognitive science of religion, see, for example, Justin L. Barrett, **Why Would Anyone Believe in God?** (Lanham, MD: AltaMira Press, 2004); Scott Atran, **In Gods We Trust: The Evolutionary Landscape of Religion** (Oxford: Oxford University Press, 2002); Todd Tremlin, **Minds and Gods: The Cognitive Foundations of Religion** (Oxford: Oxford University Press, 2006).

13.	Pascal Boyer, **Religion Explained: The Evolutionary Origins of Religious Thought** (New York: Basic Books, 2007), 46–47; Daniel Dennett, **Breaking the Spell: Religion as a Natural Phenomenon** (New York: Penguin Books, 2006), 122–23; Richard Dawkins, **The God Delusion** (New York: Houghton Mifflin Harcourt, 2006), 230–33.

14.	First described by Darwin, kin selection (or inclusive fitness) was developed in R. A. Fisher,

The Genetical Theory of Natural Selection (Oxford: Clarendon Press, 1930); J. B. S. Haldane, **The Causes of Evolution** (London: Longmans, Green & Co., 1932); and W. D. Hamilton, "The Genetical Evolution of Social Behaviour," **Journal of Theoretical Biology** 7, no. 1 (1964): 1–16. More recently, the utility of inclusive fitness in understanding evolutionary development has been challenged in M. A. Nowak, C. E. Tarnita, and E. O. Wilson, "The evolution of eusociality," **Nature** 466 (2010): 1057–62, with a critical response signed by 136 researchers: P. Abbot, J. Abe, J. Alcock, et al., "Inclusive fitness theory and eusociality," **Nature** 471 (2010): E1–E4.

15. David Sloan Wilson, **Does Altruism Exist? Culture, Genes and the Welfare of Others** (New Haven: Yale University Press, 2015); David Sloan Wilson, **Darwin's Cathedral: Evolution, Religion and the Nature of Society** (Chicago: University of Chicago Press, 2002).

16. For one example, Steven Pinker in "The Believing Brain," World Science Festival public program, New York City, Gerald Lynch Theatre, 2 June 2018, https://www.worldsciencefestival.com/videos/believing-brain-evolution-neuroscience-spiritual-instinct/46:50-49:16.

17. Charles Darwin, **The Descent of Man and Selection in Relation to Sex** (New York: D. Appleton and Company, 1871), 84. Kindle. Darwin's comment nods in the direction of a long-simmering debate in evolutionary theory

on the process of **group selection.** Standard evolutionary theory is based on natural selection operating on individual organisms: those organisms better able to survive and reproduce will be more successful in passing their genetic material on to subsequent individuals. Group selection is a similar type of selection but one that acts on entire groups: those groups better able to survive (as entire groups of individuals) and reproduce (in the sense of gaining greater numbers and splintering off new groups) will be more successful in passing dominant traits on to subsequent groups. (Darwin's remark focuses upon cooperating individuals contributing to a group's success, manifested by the group's population growing larger as opposed to the group yielding greater numbers of similar groups, but still relies on the fundamental interplay between behaviors beneficial to the individual and those beneficial to the group.) There is no controversy over whether group selection can happen in principle. The controversy is whether it happens in practice. The issue is one of timescales: The general expectation is that the typical timescale during which an individual will either reproduce or die is far shorter than the corresponding timescales during which a group will either divide or dissolve. And if this is the case, as critics of group selection argue, group selection is too slow to matter. In response, David Sloan Wilson, a longtime proponent of group selection (in a yet more generalized form known as **multilevel**

selection), has argued that much of the debate comes down to different but ultimately equivalent accounting methods (different ways of partitioning the entire population) and is thus less contentious than the ongoing disagreements have made it appear (see David Sloan Wilson, **Does Altruism Exist? Culture, Genes and the Welfare of Others** [New Haven: Yale University Press, 2015], 31–46).

18. The importance of the emotional basis for religious commitment is examined in R. Sosis, "Religion and intra-group cooperation: Preliminary results of a comparative analysis of utopian communities," **Cross-Cultural Research** 34 (2000): 70–87; R. Sosis and C. Alcorta, "Signaling, solidarity, and the sacred: The evolution of religious behavior," **Evolutionary Anthropology** 12 (2003): 264–74.

19. Robert Axelrod and William D. Hamilton, "The Evolution of Cooperation," **Science** 211 (March 1981): 1390–96; Robert Axelrod, **The Evolution of Cooperation,** rev. ed. (New York: Perseus Books Group, 2006).

20. Jesse Bering, **The Belief Instinct** (New York: W. W. Norton, 2011).

21. Sheldon Solomon, Jeff Greenberg, and Tom Pyszczynski, **The Worm at the Core: On the Role of Death in Life** (New York: Random House Publishing Group, 2015), 122.

22. Abram Rosenblatt, Jeff Greenberg, Sheldon Solomon, et al., "Evidence for Terror

Management Theory I: The Effects of Mortality Salience on Reactions to Those Who Violate or Uphold Cultural Values," **Journal of Personality and Social Psychology** 57 (1989): 681–90. For a review, see Sheldon Solomon, Jeff Greenberg, and Tom Pyszczynski, "Tales from the Crypt: On the Role of Death in Life," **Zygon** 33, no. 1 (1998): 9–43.

23. Tom Pyszczynski, Sheldon Solomon, and Jeff Greenberg, "Thirty Years of Terror Management Theory," **Advances in Experimental Social Psychology** 52 (2015): 1–70.

24. Pascal Boyer, **Religion Explained: The Evolutionary Origins of Religious Thought** (New York: Basic Books, 2007), 20.

25. William James, **The Varieties of Religious Experience: A Study in Human Nature** (New York: Longmans, Green, and Co., 1905), 485.

26. Stephen Jay Gould, **The Richness of Life: The Essential Stephen Jay Gould** (New York: W. W. Norton, 2006), 232–33.

27. Stephen J. Gould, in **Conversations About the End of Time** (New York: Fromm International, 1999). For a study of the impact of mortality awareness on belief in supernatural entities, see, for example, A. Norenzayan and I. G. Hansen, "Belief in supernatural agents in the face of death," **Personality and Social Psychology Bulletin** 32 (2006): 174–87.

28. Karl Jaspers, **The Origin and Goal of History** (Abingdon, UK: Routledge, 2010), 2.

29. Wendy Doniger, trans., **The Rig Veda** (New York: Penguin Classics, 2005), 25–26.

30. His Holiness the Dalai Lama, Houston, Texas, 21 September 2005. While I have been unable to locate a transcript of the conversation, at the very least this is a close paraphrase of his response.

31. As with the historical roots of all the major religions, there is scholarly debate on precisely when various texts were written, when they reached canonical form, and so on. The dates I have quoted are compatible with some scholarly opinion, but as there is not universal agreement, they should be viewed as providing only a rough sketch.

32. David Buss, **Evolutionary Psychology: The New Science of Mind** (Boston: Allyn & Bacon, 2012), 90–95, 205–206, 405–409.

33. For an in-depth, accessible, and lively discussion of human belief and the various factors that influence it, see Michael Shermer, **The Believing Brain: From Ghosts and Gods to Politics and Conspiracies** (New York: St. Martin's Griffin, 2011). Although the influence that emotion can have on belief may seem manifest, until recently scholarly focus has tended to emphasize the influence of belief on emotion, a point stressed in N. Frijda, A. S. R. Manstead, and S. Bem, "The influence of emotions on belief," in **Emotions and Beliefs: How Feelings Influence Thoughts** (Studies in Emotion and

Social Interaction), ed. N. Frijda, A. Manstead, and S. Bem (Cambridge: Cambridge University Press, 2000), 1–9. A study of the impact of emotion on the establishment of beliefs in contexts where none were previously held, as well as the influence of emotion on the willingness to change beliefs is described in N. Frijda and B. Mesquita, "Beliefs through emotions," in **Emotions and Beliefs: How Feelings Influence Thoughts** (Studies in Emotion and Social Interaction), ed. N. Frijda, A. Manstead, and S. Bem (Cambridge: Cambridge University Press, 2000), 45–77.

34. Pascal Boyer, **Religion Explained: The Evolutionary Origins of Religious Thought** (New York: Basic Books, 2007), 303.

35. Karen Armstrong, **A Short History of Myth** (Melbourne: The Text Publishing Company, 2005), 57.

36. Ibid.

37. Guy Deutscher, **The Unfolding of Language: An Evolutionary Tour of Mankind's Greatest Invention** (New York: Henry Holt and Company, 2005).

38. William James, **The Varieties of Religious Experience: A Study in Human Nature** (New York: Longmans, Green and Co., 1905), 498.

39. Ibid., 506–507.

Chapter 8: Instinct and Creativity

1. Howard Chandler Robbins Landon, **Beethoven: A Documentary Study** (New York: Macmillan Publishing Co., Inc., 1970), 181.

2. Friedrich Nietzsche, **Twilight of the Idols,** trans. Duncan Large (Oxford: Oxford University Press, 1998, reissue 2008), 9.

3. George Bernard Shaw, **Back to Methuselah** (Scotts Valley, CA: CreateSpace Independent Publishing Platform, 2012), 277.

4. David Sheff, "Keith Haring, An Intimate Conversation," **Rolling Stone** 589 (August 1989): 47.

5. Josephine C. A. Joordens et al., "**Homo erectus** at Trinil on Java used shells for tool production and engraving," **Nature** 518 (12 February 2015): 228–31.

6. More precisely, what matters is that one's genes are propagated to the next generation, a goal that can be achieved by having progeny or by ensuring that other individuals who share a substantial portion of one's genes have progeny.

7. White-bearded manakin courtship rituals are richly described in Richard Prum, **The Evolution of Beauty: How Darwin's Forgotten Theory on Mate Choice Shapes the Animal World and Us** (New York: Doubleday, 2017), 1544–45, Kindle. Firefly flashing and mate choice is reviewed in S. M. Lewis and C. K. Cratsley, "Flash signal evolution, mate choice, and predation in fireflies," **Annual Review of**

Entomology 53 (2008): 293–321. Bowerbird constructions are described and illustrated in Peter Rowland, **Bowerbirds** (Collingwood, Australia: CSIRO Publishing, 2008), especially pages 40–47.

8.   The resistance to sexual selection was also due, in part, to the selective power ceded to choosy females, a proposal that was off-putting to Victorian biologists, almost all male. See, for example, H. Cronin, **The Ant and the Peacock: Altruism and Sexual Selection from Darwin to Today** (Cambridge: Cambridge University Press, 1991). Note too that there are examples of species in which males play the role of chooser as well as species in which both males and females are engaged in this role.

9.   Charles Darwin, **The Descent of Man and Selection in Relation to Sex,** ill. ed. (New York: D. Appleton and Company, 1871), 59.

10.   Wallace offered alternate explanations for male bodily ornaments, such as males possessing excess "vigor," which, with no other available outlet, would power the emergence of vibrant colors, long tails, prolonged calls, and so on. He also argued that attractive bodily adornments necessarily correlated with health and strength and thus offered outward fitness indicators, rendering sexual selection nothing more than a particular instance of natural selection. See Alfred Russel Wallace, **Natural Selection and Tropical Nature** (London: Macmillan and Co., 1891). Ornithologist Richard Prum argues that

researchers have unjustifiably discounted intrinsic aesthetic sensibilities in favor of adaptive
explanations, a controversial position he lays out
in **The Evolution of Beauty: How Darwin's
Forgotten Theory on Mate Choice Shapes the
Animal World and Us** (New York: Doubleday,
2017).

11. The male-female asymmetry in the arena of
    reproductive strategy was studied and illumi-
    nated by Robert Trivers in "Parental Investment
    and Sexual Selection," in **Sexual Selection and
    the Descent of Man: The Darwinian Pivot,**
    ed. Bernard G. Campbell (Chicago: Aldine
    Publishing Company, 1972), 136–79.

12. Geoffrey Miller, **The Mating Mind: How
    Sexual Choice Shaped the Evolution of
    Human Nature** (New York: Anchor, 2000);
    Denis Dutton, **The Art Instinct** (New York:
    Bloomsbury Press, 2010). The perspective is
    closely related to an earlier proposal of Amotz
    Zahavi, the **handicap principle,** which envi-
    sions that some animals advertise their fitness
    through conspicuous-consumption-like displays
    that can take the form of extravagant body parts
    or behaviors. A peacock that can afford to carry
    around a beautiful but ungainly tail assures
    potential partners of his strength and fitness,
    as weaker brethren would be unable to survive
    with such an excessive, survival-challenging trait.
    The idea, then, is that early human artists may
    have leveraged the adaptive irrelevance of their
    art into an analogous public display of strength

and fitness, advancing reproductive opportunities and hence passing on the tendency for art to be used as a means for attracting mates. See Amotz Zahavi, "Mate selection—A selection for a handicap," **Journal of Theoretical Biology** 53, no. 1 (1975): 205–14.

13. Brian Boyd, "Evolutionary Theories of Art," in **The Literary Animal: Evolution and the Nature of Narrative,** ed. Jonathan Gottschall and David Sloan Wilson (Evanston, IL: Northwestern University Press, 2005), 147.

The criticisms of sexual selection as the explanation of human artistic activity mentioned in this paragraph have been spelled out in a number of works. Here is a sampling: If sexual selection is the explanation for the arts, wouldn't we expect art to be a male-driven undertaking that is fine-tuned for sexual access, i.e., an activity pursued most vigorously by males at the apogee of their reproductive drive and directed exclusively to potential female partners? (Brian Boyd, **On the Origin of Stories** [Cambridge: Belknap Press, 2010], 76; Ellen Dissanayake, **Art and Intimacy** [Seattle: University of Washington Press, 2000], 136.) Intelligence and creativity are not necessarily trustworthy indicators of physical fitness—the combination of physical weakness and creative prowess is not uncommon. (James R. Roney, "Likeable but Unlikely, a Review of the Mating Mind by Geoffrey Miller," **Psycoloquy** 13, no. 10 (2002), article 5.) Is there evidence that a male's artistic forays provide a better means of advertising fitness than other activities like flaunting social connections, displaying wealth, winning sporting events, and so on? (Stephen Davies, **The Artful Species: Aesthetics, Art, and Evolution** [Oxford: Oxford University Press, 2012], 125.)

14. Steven Pinker, **How the Mind Works** (New York: W. W. Norton, 1997), 525.

15. Ellen Dissanayake, **Art and Intimacy: How the Arts Began** (Seattle: University of Washington Press, 2000), 94.

16. Noël Carroll, "The Arts, Emotion, and Evolution," in **Aesthetics and the Sciences of Mind,** ed. Greg Currie, Matthew Kieran, Aaron Meskin, and Jon Robson (Oxford: Oxford University Press, 2014).

17. Glenn Gould in **The Glenn Gould Reader,** ed. Tim Page (New York: Vintage Books, 1984), 240.

18. Brian Boyd, **On the Origin of Stories** (Cambridge, MA: Belknap Press, 2010), 125.

19. Jane Hirshfield, **Nine Gates: Entering the Mind of Poetry** (New York: Harper Perennial, 1998), 18.

20. Saul Bellow, Nobel lecture, 12 December 1976, from **Nobel Lectures, Literature 1968–1980,** ed. Sture Allén (Singapore: World Scientific Publishing Co., 1993).

21. Joseph Conrad, **The Nigger of the "Narcissus"** (Mineola, NY: Dover Publications, Inc., 1999), vi.

22. Yip Harburg, "Yip at the 92nd Street

YM-YWHA, December 13, 1970," transcript 1-10-3, p. 3, tapes 7-2-10 and 7-2-20.

23. Yip Harburg, "E. Y. Harburg, Lecture at UCLA on Lyric Writing, February 3, 1977," transcript, pp. 5–7, tape 7-3-10.

24. Marcel Proust, **Remembrance of Things Past,** vol. 3: **The Captive, The Fugitive, Time Regained** (New York: Vintage, 1982), 260, 931.

25. Ibid., 260.

26. George Bernard Shaw, **Back to Methuselah** (Scotts Valley, CA: Create Space Independent Publishing Platform, 2012), 278.

27. Ellen Greene, "Sappho 58: Philosophical Reflections on Death and Aging," in **The New Sappho on Old Age: Textual and Philosophical Issues,** ed. Ellen Greene and Marilyn B. Skinner, Hellenic Studies Series 38 (Washington, DC: Center for Hellenic Studies, 2009); Ellen Greene, ed., **Reading Sappho: Contemporary Approaches** (Berkeley: University of California Press, 1996).

28. Joseph Wood Krutch, "Art, Magic, and Eternity," **Virginia Quarterly Review** 8, no. 4, (Autumn 1932); https://www.vqronline.org/essay/art -magic-and-eternity.

29. For an alternate perspective (as in note 5 to chapter 1) some authors have suggested that mortality anxiety and its attendant impact through death denial, as described by Ernest Becker, is a modern influence, largely spurred by the rise of longevity and the decline of religion. See, for example, Philippe Ariès, **The Hour of**

Our Death, trans. Helen Weaver (New York: Alfred A. Knopf, 1981).

30. W. B. Yeats, **Collected Poems** (New York: Macmillan Collector's Library Books, 2016), 267.

31. Herman Melville, **Moby-Dick** (Hertfordshire, UK: Wordsworth Classics, 1993) 235.

32. Edgar Allan Poe as quoted in J. Gerald Kennedy, **Poe, Death, and the Life of Writing** (New Haven: Yale University Press, 1987), 48.

33. Tennessee Williams, **Cat on a Hot Tin Roof** (New York: New American Library, 1955), 67–68.

34. Fyodor Dostoevsky, **Crime and Punishment,** trans. Michael R. Katz (New York: Liveright, 2017), 318.

35. Sylvia Plath, **The Collected Poems,** ed. Ted Hughes (New York: Harper Perennial, 1992), 255.

36. Douglas Adams, **Life, the Universe and Everything** (New York: Del Rey, 2005), 4–5.

37. Pablo Casals, from Bach Festival: Prades 1950, as quoted in Paul Elie, **Reinventing Bach** (New York: Farrar, Straus and Giroux, 2012), 447.

38. Joseph Conrad, **The Nigger of the "Narcissus"** (Mineola, NY: Dover Publications, Inc., 1999), vi.

39. Helen Keller, Letter to New York Symphony Orchestra, 2 February 1924, digital archives of American Foundation for the Blind, filename HK01-07_B114_F08_015_002.tif.

Chapter 9: Duration and Impermanence

1. Some prominent thinkers have suggested that human evolution has drawn to a close. For instance, Stephen Jay Gould noted that, from the standpoint of biology, humans today are essentially the same as humans living as far back as fifty thousand years ago (Stephen Jay Gould, "The spice of life," **Leader to Leader** 15 [2000]: 14–19). Other researchers, studying the human genome, have argued to the contrary that the rate of human evolution is accelerating (see, for example, John Hawks, Eric T. Wang, Gregory M. Cochran, et al., "Recent acceleration of human adaptive evolution," **Proceedings of the National Academy of Sciences** 104, no. 52 [December 2007]: 20753–58; Wenqing Fu, Timothy D. O'Connor, Goo Jun, et al., "Analysis of 6,515 exomes reveals the recent origin of most human protein-coding variants," **Nature** 493 [10 January 2013]: 216–20). Studies of various populations have provided evidence of relatively recent genetic evolution. Examples include the height of Dutch men, whose exceptional average increase may reflect the effects of sexual and natural selection (Gert Stulp, Louise Barrett, Felix C. Tropf, and Melinda Mill, "Does natural selection favour taller stature among the tallest people on earth?" **Proceedings of the Royal Society B** 282, no. 1806 [7 May 2015]: 20150211) and adaptations to high-altitude environments (Abigail Bigham et al., "Identifying

signatures of natural selection in Tibetan and Andean populations using dense genome scan data," **PLoS Genetics** 6, no. 9 [9 September 2010]: e1001116).

2. Choongwon Jeong and Anna Di Rienzo, "Adaptations to local environments in modern human populations," **Current Opinion in Genetics & Development** 29 (2014), 1–8; Gert Stulp, Louise Barrett, Felix C. Tropf, and Melinda Mill, "Does natural selection favour taller stature among the tallest people on earth?" **Proceedings of the Royal Society B** 282, no. 1806 (7 May 2015): 20150211 (and see note 1, above).

3. A cautionary consideration of this assumption is provided by Steven Carlip, "Transient Observers and Variable Constants, or Repelling the Invasion of the Boltzmann's Brains," **Journal of Cosmology and Astroparticle Physics** 06 (2007): 001. Note that one possible variation we will consider is that the value of the dark energy might change. As we discuss in this chapter, it was only in the late 1990s that astronomical observations convinced the community of physicists that Einstein's elimination of the cosmological constant in 1931 ("Away with the cosmological term!") was premature. Premature too was labeling the cosmological constant "constant." It is quite possible that the value of Einstein's cosmological term varies over time—a possibility, as we will see, that comes with profound implications for the future.

4. For a different perspective on the future of intelligence, see David Deutsch, **The Beginning of Infinity** (New York: Viking, 2011).

5. Physical eschatology, the physics of the far future, has received less attention than the physics of the far past. Nevertheless, there have been numerous studies. A comprehensive list of technical references is contained in Milan M. Ćirković, "Resource Letter: PEs-1, Physical Eschatology," **American Journal of Physics** 71 (2003): 122. In the discussion that follows, the seminal paper by Freeman Dyson, "Time without end: Physics and biology in an open universe," **Reviews of Modern Physics** 51 (1979): 447–60, has been particularly influential, as has the paper by Fred C. Adams and Gregory Laughlin, "A dying universe: The long-term fate and evolution of astrophysical objects," **Reviews of Modern Physics** 69 (1997): 337–72, which further develops the subject, including new results on planetary, stellar, and galactic dynamics, discussed as well in their excellent general-level book **The Five Ages of the Universe: Inside the Physics of Eternity** (New York: Free Press, 1999). The subject owes its modern origins to the paper by M. J. Rees, "The collapse of the universe: An eschatological study," **Observatory** 89 (1969): 193–98, as well as the paper by Jamal N. Islam, "Possible Ultimate Fate of the Universe," **Quarterly Journal of the Royal Astronomical Society** 18 (March 1977): 3–8.

6. I.-J. Sackmann, A. I. Boothroyd, and K. E.

Kraemer, "Our Sun. III. Present and Future," **Astrophysical Journal** 418 (1993): 457; Klaus-Peter Schroder and Robert C. Smith, "Distant future of the Sun and Earth revisited," **Monthly Notices of the Royal Astronomical Society** 386, no. 1 (2008): 155–63.

7. The expert reader will note that the Pauli exclusion principle would have already played a role in the sun's evolution. Prior to the ignition of helium fusion in the sun's core, the density would have been high enough for the exclusion principle's electron degeneracy pressure to become relevant. Indeed, the "spectacular and momentary eruption" I mentioned as marking the transition to helium fusion arises because of special properties of the gas of degenerate electrons populating the core (the gas does not expand or cool in response to the heat generated by the onset of helium fusion, leading to a colossal nuclear reaction, not that different from a helium bomb).

8. Alan Lindsay Mackay, **The Harvest of a Quiet Eye: A Selection of Scientific Quotations** (Bristol, UK: Institute of Physics, 1977): 117.

9. The initial recognition of the essential role of Pauli's exclusion principle in the structure of white dwarfs was made in R. H. Fowler, "On Dense Matter," **Monthly Notices of the Royal Astronomical Society** 87, no. 2 (1926): 114–22. The recognition of the important inclusion of relativistic effects was made in Subrahmanyan Chandrasekhar, "The Maximum Mass of

Ideal White Dwarfs," **Astrophysical Journal** 74 (1931): 81–82. The result, known as the Chandrasekhar limit, shows that the contraction of any star with mass less than about 1.4 times that of the sun will be similarly halted by the resistance due to Pauli's exclusion principle. Subsequent work revealed that in the case of more massive stars, the force of stellar contraction can drive electrons to meld with protons, forming neutrons. The process allows stars to contract further, but at some point the neutrons will be so tightly packed that Pauli's exclusion principle becomes relevant anew, once again halting further contraction. The result is a neutron star.

10. While on average galactic separations are growing, there are galaxies that are sufficiently close that their mutual gravitational attraction drives them to approach each other. As we will discuss, such is the case, for example, with the Milky Way and Andromeda galaxies.

11. S. Perlmutter et al., "Measurements of Ω and Λ from 42 High-Redshift Supernovae," **Astrophysical Journal** 517, no. 2 (1999): 565; B. P. Schmidt et al., "The High-Z Supernova Search: Measuring Cosmic Deceleration and Global Curvature of the Universe Using Type IA Supernovae," **Astrophysical Journal** 507 (1998): 46.

12. For completeness, note that the explanations of accelerated spatial expansion that are taken seriously all point the finger at gravity. But broadly

speaking they do so in two different ways. Either the behavior of the force of gravity over cosmological distances differs from our expectation based on Einstein's and Newton's descriptions, or the sources giving rise to gravity differ from our expectation based on the conventional understanding of matter and energy. While both approaches are viable, the second has been more fully developed and widely applied (to explain not just the quickening expansion of space but also detailed observations of the cosmic microwave background radiation), and so is the approach we follow.

13. The dark energy density is about 5×10^{-10} joules per cubic meter or about 5×10^{-10} watt-seconds per cubic meter. To run a 100-watt bulb for one second requires 2×10^{11} times the dark energy contained in a single cubic centimeter. Such energy can thus run a 100-watt bulb for about 5×10^{-12} seconds, or five trillionths of a second.

14. If the value of dark energy does not change over time, then it is identical to Einstein's cosmological constant—a Hail Mary Einstein threw into his calculations in 1917 when he realized that the equations of general relativity were unable to account for the consensus view that on large scales the universe was static. The challenge Einstein had encountered is that stasis requires equilibrium, but gravity seemingly pulls in only one direction. With no counterbalancing force, a static universe seemed impossible. Happily,

Einstein then realized that by inserting one new term into his equations—the cosmological constant—general relativity also allowed for repulsive gravity that could counter ordinary attractive gravity and make a static universe possible. (Einstein did not realize that the balancing act was unstable—a small change in the size of the static universe, larger or smaller, would upset the balance, leading to expansion or contraction.) Within a little over a decade, however, Einstein learned that the universe is expanding. With that realization, Einstein famously erased the cosmological constant from his equations. But Einstein had let the repulsive-gravity genie out of general relativity's bottle. In time, repulsive gravity would serve cosmology grandly, granting the outward push of the big bang and, after that, offering an explanation for the accelerated expansion of space. As many have said, it all goes to show that even Einstein's bad ideas are good.

15. Robert R. Caldwell, Marc Kamionkowski, and Nevin N. Weinberg, "Phantom Energy and Cosmic Doomsday," **Physical Review Letters** 91 (2003): 071301.

16. Abraham Loeb, "Cosmology with hypervelocity stars," **Journal of Cosmology and Astroparticle Physics** 04 (2011): 023.

17. Energy within the earth is also a remnant of the heat produced when the pull of gravity crushed a cloud of dust and gas into the nascent planet. Additionally, heat is also generated as the earth

spins, because the motion stresses layers of deep rock that need a constant force to keep up with the rotational velocity.

18. Fred C. Adams and Gregory Laughlin, "A dying universe: The long-term fate and evolution of astrophysical objects," **Reviews of Modern Physics** 69 (1997): 337–72; Fred C. Adams and Greg Laughlin, **The Five Ages of the Universe: Inside the Physics of Eternity** (New York: Free Press, 1999), 50–52. Similar considerations are relevant to planets or moons that have always been too far from their host star for surface conditions hospitable for life to arise. Internal processes within such bodies, their astrogeology, can yield energy capable of sustaining life well below their surfaces. Saturn's moon Enceladus is a prime candidate. At such a great distance from the sun, its icy surface is an inauspicious home for life. But the various gravitational pulls exerted by Saturn and its other moons, slightly stretching Enceladus this way and compressing it that way, create stresses and strains that heat its interior, melting ice and possibly sustaining reservoirs of liquid water. It's not entirely far-fetched to imagine that we might one day drill a small hole in Enceladus's frozen crust, lower down a probe, and come eye to eye with a native if ocean-dwelling Enceladan.

19. For a demonstration of this, see my segment on **The Late Show with Stephen Colbert** in which a stack of five balls were dropped, propelling the lightest more than thirty feet in the air

(surely the only Guinness World Record I will ever hold). https://www.youtube.com/watch?v =75szwX09pg8.

20. Dyson gives a simple back-of-the-envelope estimate of the rate at which planets are ejected from solar systems as well as the rate at which stars are ejected from galaxies: Freeman Dyson, "Time without end: Physics and biology in an open universe," **Reviews of Modern Physics** 51 (1979): 450. Adams and Laughlin provide fuller explanations and calculations, as well as original research contributions to some of these processes (for example, the implications of small stars wandering through our solar system). F. C. Adams and G. Laughlin, "A dying universe: The long-term fate and evolution of astrophysical objects," **Reviews of Modern Physics** 69 (1997): 343–47; Fred C. Adams and Greg Laughlin, **The Five Ages of the Universe: Inside the Physics of Eternity** (New York: Free Press, 1999), 50–51.

21. For a video demonstration of the rubber sheet metaphor, using spandex, and a brief discussion of the point made in the next paragraph regarding gravitational waves and the decay of planetary orbits, see https://www.youtube.com/watch?v=uRijc-AN-F0.

22. R. A. Hulse and J. H. Taylor, "Discovery of a pulsar in a binary system," **Astrophysical Journal** 195 (1975): L51.

23. The possibility that a slowly decaying orbit

might indicate energy being lost through gravitational radiation was raised in R. V. Wagoner, "Test for the existence of gravitational radiation," **Astrophysical Journal** 196 (1975): L63.

24. J. H. Taylor, L. A. Fowler, and P. M. McCulloch, "Measurements of general relativistic effects in the binary pulsar PSR 1913+16," **Nature** 277 (1979): 437.

25. Freeman Dyson, "Time without end: Physics and biology in an open universe," **Reviews of Modern Physics** 51 (1979): 451; Fred C. Adams and Gregory Laughlin, "A dying universe: The long-term fate and evolution of astrophysical objects," **Reviews of Modern Physics** 69 (1997): 344–47.

26. Fred C. Adams and Gregory Laughlin, "A dying universe: The long-term fate and evolution of astrophysical objects," **Reviews of Modern Physics** 69 (1997): 347–49.

27. Neutrons, when isolated, have a short lifetime of about fifteen minutes. However, because neutrons are heavier than protons, their decay process involves the production of a proton (and an electron and an antineutrino). For a neutron to decay within an atom, the nucleus would need to accommodate the proton produced, but often that requirement can't be met. Protons already in the nucleus fill out the available quantum slots, which according to Pauli and his exclusion principle cannot be shared, shoring up the neutron's stability in this context. If protons

decay, being lighter than neutrons they would not produce neutrons, and a similar stabilizing process would not come into play.

28. Howard Georgi and Sheldon Glashow, "Unity of All Elementary-Particle Forces," **Physical Review Letters** 32, no. 8 (1974): 438.

29. A 50 percent decay rate over 10^{30} years implies that in a sample of 10^{30} protons there's a 50 percent chance that within a single year one of them will fall apart.

30. Howard Georgi, personal communication, Harvard University, 28 December 1997.

31. If protons do not disintegrate in the manner envisioned by theories like grand unification or string theory that step beyond the established laws of particle physics—the standard model of particle physics—the rollout toward the future that I have described would need various modifications. For instance, we usually think of solids, like iron, as objects that hold their shape, unlike liquids, whose shape is fluid. But over sufficiently long timescales even iron would act like a fluid, its constituent atoms tunneling through all barriers normally erected by physical and chemical processes. Over the course of about 10^{65} years, a lump of iron floating in space would rearrange its atoms, "melting" into a spherical blob—as would all other matter that still exists. Beyond shape reconfigurations, over longer durations the identity of matter would change: atoms lighter than iron would gradually fuse together, while atoms heavier than iron

would fission apart. Iron is the most stable of all atomic configurations, and would thus be the end product of all such nuclear processes. The timescale for such processes to conclude is about $10^{1,500}$ years. Over yet longer timescales, matter would quantum tunnel into black holes, which at this temporal scale would immediately evaporate by Hawking radiation. Note, though, that even in the standard model of particle physics—no exotic or hypothetical extensions— it is believed that protons will decay, just on a much longer timescale than the 10^{38} years we assumed in the chapter. For instance, there is an exotic quantum process fully within the standard model that physicists have studied theoretically (known as an instanton, making use of the so-called sphaleron solution to the electroweak field equations) that would result in protons disintegrating. The process relies on a quantum tunneling event, so the timescale for this to occur is long, with estimates putting it at roughly 10^{150} years in the future, but much less than the $10^{1,500}$ years noted above. Physicists have studied other exotic processes that would also cause the proton to decay with various timescales that are mostly within, roughly, 10^{200} years. So by that future era, it is likely that any remaining complex matter will have fallen apart. See Freeman Dyson, "Time without end: Physics and biology in an open universe," **Reviews of Modern Physics** 51 (1979): 451–52, for the estimates on the fluidity of solid matter and the transformation of matter into iron. For technical references on quantum tunneling leading to proton decay, see G. 't Hooft, "Computation of the quantum effects due to a four-dimensional pseudoparticle," **Physical Review D** 14 (1976): 3432, and F. R. Klinkhamer and N. S. Manton, "A saddle-point solution in the Weinberg-Salam theory," **Physical Review D** 30 (1984): 2212.

32. Freeman Dyson, "Time without end: Physics and biology in an open universe," **Reviews of Modern Physics** 51 (1979): 447–60.

33. Dyson computes the necessary rate of energy dissipation **D** for a Thinker whose "complexity" is **Q** (which is the rate of entropy production per unit of the Thinker's subjective time, roughly the entropy production per Thinker's thought), operating at a temperature T, and finds $D \propto QT^2$.

34. More precisely, in the language I'm using, Dyson assumes that if we have an ensemble of Thinkers, all tuned to function at different temperatures, then the rate of each Thinker's metabolic processes, whatever they might be, scale linearly with the temperature. In technical terms, Dyson is positing what he calls the **biological scaling hypothesis,** which says the following: if you have a replica of a given environment, quantum mechanically identical to the original save for the temperature of the new environment being T_{new} while that of the original environment was $T_{original}$, and if you make a replica of a living system so that its quantum

mechanical Hamiltonian, up to a unitary transformation, is given by $H_{new} = (T_{new} / T_{original}) H_{original}$, then the copy is in fact alive and has subjective experiences identical to the original, except that all its internal functions are reduced by a factor of $T_{new} / T_{original}$.

35. For the mathematically inclined reader, note that if temperature, T, is a function of time, t, according to $T(t) \sim t^{-p}$, the integral of the expression in note 33, QT^2, will converge for $p > \frac{1}{2}$, while the total number of thoughts (the integral of $T(t)$) will diverge for $p < 1$. So, with $\frac{1}{2} < p < 1$, the Thinker can carry out an infinite number of thoughts while requiring a finite supply of energy.

36. For the mathematically inclined reader, the key issue here is that the maximum rate of waste disposal (assuming the Thinker expels waste through electron-based dipole radiation) is proportional to T^3, while the energy dissipated is proportional to T^2. This implies there is a lower bound on T to avoid waste heat building up faster than it can be expelled.

37. The computer scientists responsible for these influential results include Charles Bennett, Edward Fredkin, Rolf Landauer, and Tommaso Toffoli, among many others. For an insightful and accessible exposition, see Charles H. Bennett and Rolf Landauer, "The Fundamental Physical Limits of Computation," **Scientific American** 253, no. 1 (July 1985): 48–56.

38. More precisely, it is **virtually** impossible to undo

the computation. As the act of erasure is itself a physical process, in principle we could undo it by the same process we would use to undo the shattering of a glass: reverse the motion of every particle everywhere. But again, in any practical sense, that is not feasible.

39.　A number of authors have considered the impact of a cosmological constant on the future of life and mind. Long before the observational discovery of dark energy, John Barrow and Frank Tipler analyzed the physics of computation in a universe with a cosmological constant and argued that information processing necessarily draws to a close, bringing an end to life and mind (John D. Barrow and Frank J. Tipler, **The Anthropic Cosmological Principle** [Oxford: Oxford University Press, 1988], 668–69). Lawrence Krauss and Glenn Starkman revisited Dyson's analysis in a universe with a cosmological constant and arrived at a similar conclusion (Lawrence M. Krauss and Glenn D. Starkman, "Life, the Universe, and Nothing: Life and Death in an Ever-Expanding Universe," **Astrophysical Journal** 531 [2000]: 22–30). Krauss and Starkman also argued on general grounds that the discrete nature of states for a finite-sized quantum system would similarly imperil infinite thought in **any** expanding spacetime, even in the absence of a cosmological constant. However, Barrow and Hervik argued that by using temperature gradients generated by gravitational waves, information processing can in fact continue indefinitely in a universe that does not have a cosmological constant (John D. Barrow and Sigbjørn Hervik, "Indefinite information processing in ever-expanding universes," **Physics Letters B** 566, nos. 1–2 [24 July 2003]: 1–7). Freese and Kinney came to a similar conclusion, arguing that in a spacetime whose horizon size increases over time (unlike that of a universe with a cosmological constant in which the horizon size is fixed), the phase space continually acquires new modes (those whose wavelengths drop below that of the increasing horizon size), which give the system an ongoing supply of new degrees of freedom that can transport waste to the environment, thus allowing computation to proceed indefinitely far into the future (K. Freese and W. Kinney, "The ultimate fate of life in an accelerating universe," **Physics Letters B** 558, nos. 1–2 [10 April 2003}: 1–8).

40. K. Freese and W. Kinney, "The ultimate fate of life in an accelerating universe," **Physics Letters B** 558, nos. 1–2 [10 April 2003]: 1–8.

Chapter 10: The Twilight of Time

1. The fact that processes with minuscule probabilities can leverage long durations to pry their way into reality is something we have encountered in earlier chapters. In one explanation of what may have ignited the big bang, I noted that the cosmic unfolding might have long waited for the highly unlikely configuration of

a uniform inflaton field to fill a small region, where it would source repulsive gravity and set the expansion of space in motion. For another important and general example, I also emphasized that the second law of thermodynamics is not a law in the conventional sense but is instead a statistical tendency. Entropic decreases are extraordinarily rare, but if you wait long enough, even the most unlikely of things will happen.

2. Freeman Dyson in Jon Else, dir., **The Day After Trinity** (Houston: KETH, 1981).

3. Personal communication with John Wheeler, Princeton University, 27 January 1998.

4. W. Israel, "Event Horizons in Static Vacuum Space-Times," **Physical Review** 164 (1967): 1776; W. Israel, "Event Horizons in Static Electrovac Space-Times," **Communications in Mathematical Physics** 8 (1968): 245; B. Carter, "Axisymmetric Black Hole Has Only Two Degrees of Freedom," **Physical Review Letters** 26 (1971): 331.

5. Jacob D. Bekenstein, "Black Holes and Entropy," **Physical Review D** 7 (15 April 1973): 2333. For a beautiful and accessible mathematical summary of Bekenstein's calculation, see Leonard Susskind, **The Black Hole War: My Battle with Stephen Hawking to Make the World Safe for Quantum Mechanics** (New York: Little, Brown and Co., 2008), 151–54.

6. More precisely, the area increases by one square unit if the unit is chosen to be one-quarter of a squared Planck length.

7. The electron's magnetic properties, which are highly sensitive to quantum fluctuations in empty space, provide the most impressive agreement between observations and mathematical predictions. The mathematical calculations are nothing short of heroic. In the late 1940s, Richard Feynman introduced a graphical scheme for organizing such quantum calculations, using what are now known as **Feynman diagrams.** Each diagram stands for a mathematical contribution that requires careful evaluation, and at the conclusion of the calculation, all such terms need to be summed. In determining quantum contributions to the magnetic properties of electrons (the electron dipole moment), the researchers needed to evaluate more than twelve thousand Feynman diagrams. The spectacular agreement between such calculations and experimental measurements ranks among the greatest of all triumphs emerging from our understanding of quantum physics (see Tatsumi Aoyama, Masashi Hayakawa, Toichiro Kinoshita, and Makiko Nio, "Tenth-order electron anomalous magnetic moment: Contribution of diagrams without closed lepton loops," **Physical Review D** 91 [2015]: 033006).

8. Although I am using charcoal as an analogy, it is worth noting one essential difference between radiation emitted from familiar burning and radiation emitted from a black hole. When charcoal glows, the radiation is emitted directly from burning the material constituting the

charcoal; the radiation, therefore, carries an imprint of the charcoal's specific material makeup. By contrast, the material constituting a black hole has all been crushed into the black hole's singularity—and the more massive the black hole, the larger the separation between the singularity and the black hole's event horizon—so the radiation emitted from the event horizon would not appear to carry an imprint of the black hole's material makeup. This difference is one way of understanding the origin of what is known as the **black hole information paradox.** If the radiation emitted by a black hole is insensitive to the specific ingredients from which the black hole formed, then by the time the black hole has fully transformed into radiation, the information contained in those ingredients will have been lost. Such a loss of information would disrupt the quantum mechanical progression of the universe, and so physicists have spent decades trying to establish that the information is not lost. Most physicists now agree that we have strong arguments supporting the contention that the information is indeed preserved, but there are numerous important details still at the forefront of research.

9.　Hawking's formula shows that the black-body radiation emitted by a Schwarzschild black hole (an uncharged, nonrotating black hole) of mass M is given by $T_{Hawking} = hc^3/16\pi^2 GMk_b$ (h is Planck's constant, c is the speed of light,

G is Newton's constant, and k_b is Boltzmann's constant). S. W. Hawking, "Particle Creation by Black Holes," **Communications in Mathematical Physics** 43 (1975): 199–220.

10. Don N. Page, "Particle emission rates from a black hole: Massless particles from an uncharged, nonrotating hole," **Physical Review D** 13 no. 2 (1976), 198–206. The numbers quoted update Page's calculation based on more recent assessments of particle properties, especially nonzero masses for neutrinos.

11. More precisely, a ball whose radius is no greater than the so-called Schwarzschild radius, whose mathematical form in terms of the mass, **M,** is $R_{Schwarzschild} = 2GM/c^2$.

12. Note that I am referring to what might be called a black hole's **effective average density:** its total mass divided by the total volume contained within a sphere equal in radius to that of its event horizon. The notion is intuitively useful but is, as the expert reader will recognize, at best heuristic. When a black hole forms, the radial direction within its event horizon becomes timelike, and so the notion of the black hole's interior spatial volume becomes a more subtle notion (and, in fact, it becomes divergent). Moreover, the black hole's mass does not uniformly fill any such volume, so the average density we have computed is not physically realized by the black hole itself. Nevertheless, the average density of a black hole, as we have

defined it, gives an intuitive sense of why larger black holes yield less extreme external environments and give rise to Hawking radiation with lower temperatures.

13. In the previous chapter, we noted that the accelerated expansion of space gives rise to a tiny and constant background temperature of about 10^{-30} K. The temperature of a black hole with mass greater than about 10^{23} times the mass of the sun would be smaller than the ambient temperature of space in the far future. However, such a black hole would be larger than the cosmological horizon itself.

14. According to the mathematics, as photons pass through the Higgs field they experience no drag resistance at all, rendering them massless and the Higgs field invisible.

15. Peter Higgs in "What Is Space?"—the first episode of the four-part **NOVA** documentary **The Fabric of the Cosmos,** based on the book of the same name. Other physicists who developed similar ideas to Higgs around the same time include Robert Brout and François Englert, and Gerald Guralnik, C. Richard Hagen, and Tom Kibble. Higgs and Englert shared the Nobel Prize for their work.

16. There's less significance in this particular number than there might appear. The value 246 (or, more precisely, 246.22 GeV, where GeV stands for the conventional unit of gigaelectron volts) depends on the mathematical conventions that physicists generally invoke. But less standard

conventions would yield equivalent physics with different numerical values.

17. Sidney Coleman, "Fate of the False Vacuum," **Physical Review D** 15 (1977): 2929; Erratum, **Physical Review D** 16 (1977): 1248.

18. More precisely, the sphere would spread slowly at first and then increase its speed rapidly toward that of light.

19. A. Andreassen, W. Frost, and M. D. Schwartz, "Scale Invariant Instantons and the Complete Lifetime of the Standard Model," **Physical Review D** 97 (2018): 056006.

20. The possibility that our universe might have emerged from a high-entropy uniform bath of particles bumping and jostling in the void, in which a rare spontaneous drop to lower entropy resulted in the orderly structures we witness, was raised by Ludwig Boltzmann in two papers (Ludwig Boltzmann, "On Certain Questions of the Theory of Gases," **Nature** 51 [1895]: 1322, 413–15; Ludwig Boltzmann, **"Entgegnung auf die wärmetheoretischen Betrachtungen des Hrn. E. Zermelo," Annalen der Physik** 57 [1896]: 773–84). Later, Arthur Eddington pointed out that because less significant drops in entropy are more likely to happen, it is far more probable that such a fluctuation would not yield an entire universe filled with stars, planets, and people—a dramatic drop in entropy—but would instead yield only "mathematical physicists" (observers engaging in the very thought experiments he was exploring)

within an otherwise disorganized environment (A. Eddington, "The End of the World: From the Standpoint of Mathematical Physics," **Nature** 127, no. 1931 [3203]: 447–53). Much later, the notion of "mathematical physicists" was further stripped down to an even more modest entropic drop—only giving rise to the observers' cogitating components, referred to as "Boltzmann brains" (as far as I know, the first explicit use of the term was in A. Albrecht and L. Sorbo, "Can the Universe Afford Inflation?" **Physical Review D** 70 [2004]: 063528).

21.   For reasons emphasized in the chapter, my focus will be on the spontaneous creation of structures that can think—Boltzmann brains—but the spontaneous creation of entire new universes or the spontaneous recreation of conditions that set off inflationary cosmological expansion are also worthy of attention. To avoid overburdening the chapter, I consider such possibilities in notes 22 and 34.

22.   The expert reader will recognize that I am gliding over both subtlety and controversy. There isn't universal consensus on how to calculate the probabilities of the various spontaneous cosmological fluctuations to which I refer. Leonard Susskind and collaborators advocated one approach in L. Dyson, M. Kleban, and L. Susskind, "Disturbing Implications of a Cosmological Constant," **Journal of High Energy Physics** 0210 (2002): 011, based on an earlier idea of Susskind's known as "horizon

complementarity." Recall that because the expansion of space is accelerating, we are surrounded by a distant cosmological horizon. Locations farther than the cosmological horizon recede from us faster than the speed of light, so there is no possibility for us to be influenced by anything located at or beyond that distance. Susskind, motivated by such isolation (and by his earlier work on black holes, which have their own variety of horizon), advocates considering only physical processes that take place within our "causal patch"—you can think of this as the region of space lying within our cosmological horizon—effectively discarding all physics in the potentially infinite expanse of space that lies beyond. More precisely, Susskind argues that physics outside our causal patch is redundant with the physics within our causal patch (much as the wave and particle descriptions in quantum mechanics are two complementary ways of discussing the same physics, interior-patch and exterior-patch physics would also be complementary ways of discussing the same physics). With this assumption, reality is considered to be a finite patch of space, with a fixed cosmological constant, Λ, yielding a temperature $T \sim \sqrt{\Lambda}$—somewhat like the canonical case of hot gas in a box studied in elementary statistical mechanics. Calculating the relative probabilities of two different macrostates then amounts to taking ratios of the number of microstates associated to each. That is, the likelihood of a given configuration is proportional to (the exponential of) its entropy. With this approach, Susskind and collaborators note that the coming together of particles within our patch to yield conditions necessary for an inflationary big bang is extraordinarily less likely (because it has low entropy) than particles coming together to directly yield the world as we know it, from stars to people (because such a configuration has higher entropy). An alternative approach to calculating likelihoods is suggested in A. Albrecht and L. Sorbo, "Can the Universe Afford Inflation?" **Physical Review D** 70 (2004): 063528, which is based on inflation arising from a local quantum tunneling event. This approach yields radically different probabilities. Albrecht and Sorbo consider fluctuations to lower entropy—a region that will subsequently inflate—within a background environment that itself has high entropy; this ensures that the full configuration still has high entropy, thereby enhancing likelihoods. Susskind and collaborators consider the entropy only within the fluctuation itself, reasoning that because the region will subsequently inflate, everything outside the region lies beyond its cosmological horizon and so can be ignored. The lower total entropy Susskind and collaborators assign to the fluctuation drastically decreases its likelihood of happening.

23. In note 9 of chapter 2, I explained that the entropy of a system is more properly defined as the natural logarithm of the number of

accessible quantum states. So, if a system has entropy S, the number of such states is e^S. If we assume that a system spends nearly equal time in any of the microstates compatible with its macrostate, then the probability P of a fluctuation from an initial state of entropy S_1 to a state of final entropy S_2 is given by the ratio of the number of microstates associated to each, hence $P = e^{S_2}/e^{S_1} = e^{(S_2 - S_1)}$. For clarity, write $S_2 = S_1 - D$, where D stands for the "drop" in entropy from the initial value of S_1. Then $P = e^{(S_1 - D - S_1)} = e^{-D}$, where we see the exponential decrease in likelihood as a function of the drop in entropy. What then is the probability of forming a Boltzmann brain? Well, at temperature T, the particles in our thermal bath have energies very equal to T (using units with $k_B = 1$), and so to build a brain of mass M we need to siphon off about M/T such particles (using units with c = 1). Since the entropy of the bath tracks the number of particles, the drop D is essentially equal to M/T and so the probability is about $e^{-M/T}$. For a particularly relevant example, we can set our sights on the very distant future and take T to be equal to the temperature of the thermal bath arising from the cosmological horizon, about 10^{-30} K, which is about 10^{-41} GeV (where a GeV, gigaelectron volt, is about equal to the energy equivalent of the mass of a proton). Since a brain has about 10^{27} protons, M/T is about $10^{27}/10^{-41} = 10^{68}$. The probability of spontaneously forming a brain is thus about equal to $e^{-10^{68}}$. The time necessary to have a reasonable chance of such a rare event taking place is proportional to $1/(e^{-10^{68}})$, namely $e^{10^{68}}$, which in this chapter, for ease, we approximate as $10^{10^{68}}$.

24.   Although time may well be unlimited, there is a natural but finite timescale of relevance known as the "recurrence time." I discuss this in endnote 34, so here suffice it to say that the recurrence time is so long that the number of Boltzmann brains that will arise before we reach that limit is—even with the tiny rate of formation—vast.

25.   The particularly diligent reader will recognize that we are implicitly invoking the principle of indifference described in note 8 of chapter 3. That is, when I consider the origin of my brain, I am assigning equal likelihood to each incarnation that has the same physical configuration. Since almost all of these would have formed in the Boltzmannian manner, it is highly unlikely that the usual story I tell of how my brain came to be is true. However, as in note 8 of chapter 3, one can mount a challenge against the use of the principle of indifference in situations that bear no resemblance to those in which the principle has been empirically verified (coin tosses, dice rolls, and the vast assortment of chancy situations we encounter in everyday life). Nevertheless, many leading cosmologists are not satisfied with this approach and thus consider the Boltzmann

Brain puzzles I describe in the chapter to be a serious concern.

26. See David Albert, **Time and Chance** (Cambridge, MA: Harvard University Press, 2000), 116; Brian Greene, **The Fabric of the Cosmos** (New York: Vintage, 2005), 168.

27. Let me mention two other related approaches for resolving the problem. One is to imagine that over time the "constants" of nature drift in such a way that the physical processes necessary to form Boltzmann brains are suppressed. See, for example, Steven Carlip, "Transient Observers and Variable Constants, or Repelling the Invasion of the Boltzmann's Brains," **Journal of Cosmology and Astroparticle Physics** 06 (2007): 001. Another, argued by Sean Carroll and collaborators, is that the fluctuations necessary to form Boltzmann brains do not arise under a careful quantum mechanical treatment (K. K. Boddy, S. M. Carroll, and J. Pollack, "De Sitter Space Without Dynamical Quantum Fluctuations," **Foundations of Physics** 46, no. 6 [2016]: 702).

28. See, for instance, A. Ceresole, G. Dall'Agata, A. Giryavets, et al., "Domain walls, near-BPS bubbles, and probabilities in the landscape," **Physical Review D** 74 (2006): 086010. Physicist Don Page has taken a different approach to formulating the Boltzmann brain problem, noting that in any finite volume of space undergoing accelerated expansion, such

as ours, there will be—over unlimited time—an unlimited number of spontaneously created brains. To avoid our brains being atypical members in this expanding volume, Page suggests that our region does not have unlimited time but is instead heading toward some variety of destruction. His calculations (Don N. Page, "Is our universe decaying at an astronomical rate?" **Physics Letters B** 669 [2008]: 197–200) indicate that the maximum lifetime of our universe might be as low as twenty billion years. A number of other physicists (see, for example, R. Bousso and B. Freivogel, "A Paradox in the Global Description of the Multiverse," **Journal of High Energy Physics** 6 [2007]: 018; A. Linde, "Sinks in the Landscape, Boltzmann Brains, and the Cosmological Constant Problem," **Journal of Cosmology and Astroparticle Physics** 0701 [2007]: 022; A. Vilenkin, "Predictions from Quantum Cosmology," **Physical Review Letters** 74 [1995]: 846) have suggested other ways to avoid the Boltzmann brain problem using different mathematical formalisms for calculating the probability that they will form. In short, there remains much disagreement on how to calculate the probability of these kinds of processes, no doubt a fruitful source of controversy to drive further research.

29. Kimberly K. Boddy and Sean M. Carroll, "Can the Higgs Boson Save Us from the Menace of the Boltzmann Brains?" 2013, arXiv:1308.468.

30. At least, that's the story as told by Einstein's

equations. Determining whether that powerful crunch would truly be the end or whether some variety of exotic process would rear up at the last moment will require a full quantum treatment of gravity. The current general consensus is that tunneling to a negative value yields a terminal state—in that realm, a true end to time.

31. Paul J. Steinhardt and Neil Turok, "The cyclic model simplified," **New Astronomy Reviews** 49 (2005): 43–57; Anna Ijjas and Paul Steinhardt, "A New Kind of Cyclic Universe" (2019): arXiv:1904.0822[gr-qc].

32. Alexander Friedmann, trans. Brian Doyle, "On the Curvature of Space," **Zeitschrift für Physik** 10 (1922): 377–386; Richard C. Tolman, "On the problem of the entropy of the universe as a whole," **Physical Review** 37 (1931): 1639–60; Richard C. Tolman, "On the theoretical requirements for a periodic behavior of the universe," **Physical Review** 38 (1931): 1758–71.

33. More than likely, however, the case would not be clear-cut. The reason is that the inflationary paradigm can also accommodate the lack of primordial gravitational waves: models that reduce the energy scale of inflation would produce waves too weak to be observed. Some researchers would argue vociferously that such models are unnatural and are thus less convincing than the cyclic model. But that's a qualitative judgment on which different

researchers will have different opinions. The potential data I reference (or, really, the lack thereof) would surely open a heated debate in the physics community between proponents of these two cosmological theories, but it is not likely that the inflationary scenario would be abandoned.

34. While it would have taken us too far afield in the chapter, I will note here that there is a version of cyclic cosmology that may emerge from more standard cosmological scenarios as well. While differing substantially from the cyclic approach just described, this cosmology involves sequential episodes, but with enormously longer timescales and arising via a completely different mechanism. The essential physics was derived toward the end of the nineteenth century by mathematician Henri Poincaré, and is now called the **Poincaré Recurrence Theorem.** To get the gist of the theorem, think about shuffling a deck of cards. Because there are only finite many different orders of the cards (a huge number, yes, but definitely finite), if you continue to shuffle them, sooner or later the order of the cards must repeat. Poincaré realized that if you have, say, molecules of steam randomly bouncing around a container, a similar sort of repetition is also guaranteed to happen. For example, imagine I place a tight cluster of steam molecules in one corner of a container and then let them disperse. They will quickly fill the

container and for a spectacularly long time they will maintain a uniform appearance as they continue to move randomly around the available space. But if we wait long enough, the molecules will, by chance, happen to migrate into more ordered, lower-entropy configurations. Poincaré went further. He argued that the molecules will, through their random motions, come arbitrarily close to the very configuration from which they began: a tightly clustered group in a corner of the container. The reasoning, while technical, is similar to the way in which we concluded that the order of an endlessly shuffled deck of cards must repeat. An endless list of random particle positions and speeds necessarily repeats as well. Now, you might be skeptical of this claim—after all, unlike the case with shuffled cards, there are infinitely many different configurations for the steam molecules in the container. But Poincaré took care of this complication by not arguing for an exact recreation of an earlier configuration, but rather for an arbitrarily close approximate recreation. The more precise the desired recreation, the longer you will have to wait for it to happen, but choose any tolerance you like and the particles will recreate the earlier configuration within that specification.

Although Poincaré's reasoning was classical, in the 1950s his theorem was extended to quantum mechanics. If you start a closed system off with particular probabilities for its particles to be found at particular locations, and allow it to evolve for a sufficiently long time, the probabilities will come arbitrarily close to their initial values, a cycle that will also repeat indefinitely. Essential to Poincaré's argument, whether classical or quantum, is the fact that the steam is confined to a container. Otherwise, the molecules would continue to disperse outward, never to return. Since the universe is not a closed container, you might think his theorem does not have cosmological relevance. However, as we discussed in note 22 of this chapter, Leonard Susskind has argued that a cosmological horizon does indeed act like the walls of a container: it confines the part of the universe with which we can interact to a finite size, making Poincaré's theorem applicable. And so, much as the steam in the container will, over extraordinarily long periods, return arbitrarily close to any given configuration, so too for the conditions within our cosmological horizon: any given configuration of particles and fields will, to any given precision, be realized over and over. It is a literal version of an eternal return. Based on the size of our cosmological horizon, we can calculate the timescale necessary for recurrences, and the result is the longest timescale we've yet encountered—roughly $10^{10^{120}}$ years.

One can't help but think about such recurrences in earthly terms. Each of the hundred billion people who've lived and died were configurations of particles. If those configurations will be realized once again, well—as you can see, this line of thinking heads toward places that science generally avoids with a vengeance. But before getting too carried away, note that, as we have seen, spontaneous drops of entropy can threaten the very basis of rational understanding. If a random reconfiguration of particles and fields sparks a new cosmological unfolding—a new big bang—that ultimately yields stars, planets, and people, that's one thing. However, if it turns out that there is a greater probability to spontaneously recreate conditions like those of today's universe—with no big bang and no cosmological unfolding— we will find ourselves in the same morass we encountered with Boltzmann brains. Even if our universe did emerge in the cosmological manner we've described in previous chapters, looking to the far future we would conclude that the vast majority of observers like us (some who would have the same memories as us and thus claim to be us) would not have arisen through that cosmological sequence. Yet each will think they did. As with Boltzmann brains, we will have run into an epistemological quagmire. You might suggest that this would not undercut **our** grasp on reality—you and I and everything familiar could have emerged from a bona fide cosmological unfolding. The disturbing insight, though, is that everyone in the future can latch on to the very same consoling story, and yet most of them would be wrong. Given that the vast majority of observers across the timeline would not have emerged from the standard cosmological evolution, we would need a convincing argument that we are not among the deluded. And that's an argument that physicists have attempted to formulate, but as yet no such argument has achieved broad acceptance. Part of the issue is that we don't yet fully understand the melding of quantum mechanics and gravity and so our calculational schemes are tentative. Faced with this situation, some physicists, Susskind most prominently, have suggested that the cosmological constant may not be truly constant. After all, if in the far future the cosmological constant dissipates away, the era of accelerated expansion would end and the cosmological horizon would disappear. With that, Poincaré and his recurrences would be neutered. The jury awaits observations that, optimistically, will provide insight into this potential future.

35. Since inflationary expansion begins with a tiny region of space that rapidly swells under the force of repulsive gravity, you might think the resulting realm would necessarily have a finite size. After all, however much you stretch something finite, it will remain finite. But the reality is more intricate. In the standard formulation of inflation, the comingling of space and time results in observers **within** an inflating region of space

residing in an expanse that is **infinite.** I explain this in some detail in chapter 2 of **The Hidden Reality,** to which I refer the interested reader for a fuller explanation. Also note that inflationary cosmology can yield a distinct but related multiverse: A common feature of many inflationary scenarios is that inflationary expansion is not a one-time event. Instead, distinct bursts of inflationary expansion can yield many—generally infinitely many—expanding universes, with our universe being but one among the vast collection. The collection of such universes is known as the inflationary multiverse and arises from what is known as eternal inflation. Aspects of the multiverse description I give in this chapter apply as well to the inflationary multiverse. For details see chapter 3 of **The Hidden Reality.**

36. To avoid interactions at their boundaries, you can surround each such region with a large enough buffer, ensuring that no region has had any contact with any other.

37. Jaume Garriga and Alexander Vilenkin, "Many Worlds in One," **Physical Review D** 64, no. 4 (2001): 043511. See also J. Garriga, V. F. Mukhanov, K. D. Olum, and A. Vilenkin, "Eternal Inflation, Black Holes, and the Future of Civilizations," **International Journal of Theoretical Physics** 39, no. 7 (2000): 1887–1900, as well as the general-level book, Alex Vilenkin, **Many Worlds in One** (New York: Hill and Wang, 2006).

Chapter 11: The Nobility of Being

1. The role of evolution in the shaping of ethics was discussed in E. O. Wilson, **Sociobiology: The New Synthesis** (Cambridge, MA: Harvard University Press, 1975), initiating a new paradigm for analyzing human behavior in general and human morality in particular. For one detailed proposal laying out potential stages in the evolution of human morality, see P. Kitcher, "Biology and Ethics," in **The Oxford Handbook of Ethical Theory** (Oxford: Oxford University Press, 2006), 163–85, and P. Kitcher, "Between Fragile Altruism and Morality: Evolution and the Emergence of Normative Guidance," **Evolutionary Ethics and Contemporary Biology** (2006): 159–77.

2. T. Nagel, **Mortal Questions** (Cambridge: Cambridge University Press, 1979), 142–46.

3. See, for example, J. Haidt, "The Emotional Dog and Its Rational Tail: A Social Intuitionist Approach to Moral Judgment," **Psychological Review** 108, no. 4 (2001): 814–34, and Jonathan Haidt, **The Righteous Mind: Why Good People Are Divided by Politics and Religion** (New York: Pantheon Books, 2012).

4. Jorge Luis Borges, "The Immortal," in **Labyrinths: Selected Stories and Other Writings** (New York: New Directions Paperbook, 2017), 115. Other books referenced in this paragraph are Jonathan Swift, **Gulliver's Travels** (New York: W. W. Norton, 1997); Karel

Čapek, **The Makropulos Case,** in **Four Plays: R. U. R.; The Insect Play; The Makropulos Case; The White Plague** (London: Bloomsbury, 2014).

5. Bernard Williams, **Problems of the Self** (Cambridge: Cambridge University Press, 1973).

6. Aaron Smuts, "Immortality and Significance," **Philosophy and Literature** 35, no. 1 (2011): 134–49.

7. Samuel Scheffler, **Death and the Afterlife** (New York: Oxford University Press, 2016), 59–60.

8. As Wolf writes, "Our confidence in the continuation of the human race plays an enormous, if mostly tacit, role in the way we conceive of our activities and understand their value." Samuel Scheffler, "The Significance of Doomsday," **Death and the Afterlife** (New York: Oxford University Press, 2016), 113.

9. Harry Frankfurt, "How the Afterlife Matters," in Samuel Scheffler, **Death and the Afterlife** (New York: Oxford University Press, 2016), 136.

10. Adherents to the Many Worlds view of quantum mechanics may cast this description in a different light. If all possible outcomes happen in one world or another, this world was foreordained. But the fact that self-aware collections are among the possible outcomes is rendered no less extraordinary.

Bibliography

Aaronson, Scott. "Why I Am Not an Integrated Information Theorist (or, The Unconscious Expander)." **Shtetl-Optimized.** https://www.scottaaronson.com/blog/?p=1799.

Abbot, P., J. Abe, J. Alcock, et al. "Inclusive fitness theory and eusociality." **Nature** 471 (2010): E1–E4.

Adams, Douglas. **Life, the Universe and Everything.** New York: Del Rey, 2005.

Adams, Fred C., and Gregory Laughlin. "A dying universe: The long-term fate and evolution of astrophysical objects." **Reviews of Modern Physics** 69 (1997): 337–72.

———. **The Five Ages of the Universe: Inside the Physics of Eternity.** New York: Free Press, 1999.

Albert, David. **Time and Chance.** Cambridge, MA: Harvard University Press, 2000.

Alberts, Bruce, et al. **Molecular Biology of the Cell,** 5th ed. New York: Garland Science, 2007.

Albrecht, A., and L. Sorbo. "Can the Universe Afford Inflation?" **Physical Review D** 70 (2004): 063528.

Albrecht, A., and P. Steinhardt. "Cosmology for Grand Unified Theories with Radiatively Induced Symmetry Breaking." **Physical Review Letters** 48 (1982): 1220.

Andreassen, A., W. Frost, and M. D. Schwartz. "Scale Invariant Instantons and the Complete Lifetime of the Standard Model." **Physical Review D** 97 (2018): 056006.

Aoyama, Tatsumi, Masashi Hayakawa, Toichiro Kinoshita, and Makiko Nio. "Tenth-order electron anomalous magnetic moment: Contribution of diagrams without closed lepton loops." **Physical Review D** 91 (2015): 033006.

Aquinas, T. **Truth,** volume II. Translated by James V. McGlynn, S.J. Chicago: Henry Regnery Company, 1953.

Ariès, Philippe. **The Hour of Our Death.** Translated by Helen Weaver. New York: Alfred A. Knopf, 1981.

Aristotle, **Nicomachean Ethics.** Translated by C. D. C. Reeve. Indianapolis, IN: Hackett Publishing, 2014.

Armstrong, Karen. **A Short History of Myth.** Melbourne: The Text Publishing Company, 2005.

Arnulf, Isabelle, Colette Buda, and Jean-Pierre Sastre. "Michel Jouvet: An explorer of dreams and a great storyteller." **Sleep Medicine** 49 (2018): 4–9.

Atran, Scott. **In Gods We Trust: The Evolutionary Landscape of Religion.** Oxford: Oxford University Press, 2002.

Augustine. **Confessions.** Translated by F. J. Sheed. Indianapolis, IN: Hackett Publishing, 2006.

Auton, A., L. Brooks, R. Durbin, et al. "A global reference for human genetic variation." **Nature** 526, no. 7571 (October 2015): 68–74.

Axelrod, Robert. **The Evolution of Cooperation,** rev. ed. New York: Perseus Books Group, 2006.

Axelrod, Robert, and William D. Hamilton. "The Evolution of Cooperation." **Science** 211 (March 1981): 1390–96.

Baars, Bernard J. **In the Theater of Consciousness.** New York: Oxford University Press, 1997.

Barrett, Justin L. **Why Would Anyone Believe in God?** Lanham, MD: AltaMira, 2004.

Barrow, John D., and Sigbjørn Hervik. "Indefinite information processing in ever-expanding universes." **Physics Letters B** 566, nos. 1–2 (24 July 2003): 1–7.

Barrow, John D., and Frank J. Tipler. **The Anthropic Cosmological Principle.** Oxford: Oxford University Press, 1988.

Becker, Ernest. **The Denial of Death.** New York: Free Press, 1973.

Bekenstein, Jacob D. "Black Holes and Entropy." **Physical Review D** 7 (15 April 1973): 2333.

Bellow, Saul. Nobel lecture, December 12, 1976. In **Nobel Lectures, Literature 1968–1980,** ed. Sture Allén. Singapore: World Scientific Publishing Co., 1993.

Bennett, Charles H., and Rolf Landauer. "The Fundamental Physical Limits of Computation." **Scientific American** 253, no. 1 (July 1985).

Bering, Jesse. **The Belief Instinct.** New York: W. W. Norton, 2011.

Berwick, R., and N. Chomsky. **Why Only Us?** Cambridge, MA: MIT Press, 2015.

Bierce, Ambrose. **The Devil's Dictionary.** Mount Vernon, NY: The Peter Pauper Press, 1958.

Bigham, Abigail, et al. "Identifying signatures of natural selection in Tibetan and Andean populations using dense genome scan data." **PLoS Genetics** 6, no. 9 (9 September 2010): e1001116.

Blackmore, Susan. **The Meme Machine.** Oxford: Oxford University Press, 1999.

Boddy, Kimberly K., and Sean M. Carroll. "Can the Higgs Boson Save Us from the Menace of the Boltzmann Brains?" 2013. arXiv:1308.468.

Boddy, K. K., S. M. Carroll, and J. Pollack. "De Sitter Space Without Dynamical Quantum Fluctuations." **Foundations of Physics** 46, no. 6 (2016): 702.

Boltzmann, Ludwig. "On Certain Questions of the Theory of Gases." **Nature** 51, no. 1322 (1895): 413–15.

———. **"Entgegnung auf die wärmetheoretischen Betrachtungen des Hrn. E. Zermelo." Annalen der Physik** 57 (1896): 773–84.

Borges, Jorge Luis. "The Immortal." In **Labyrinths: Selected Stories and Other Writings.** New York: New Directions Paperbook, 2017.

Born, Max. **"Zur Quantenmechanik der Stoßvorgänge." Zeitschrift für Physik** 37, no. 12 (1926): 863–67.

Bousso, R., and B. Freivogel. "A Paradox in the Global Description of the Multiverse." **Journal of High Energy Physics** 6 (2007): 018.

Boyd, Brian. "The evolution of stories: from mimesis

to language, from fact to fiction." **WIREs Cognitive Science** 9, no. 1 (2018), e1444–46.

———. "Evolutionary Theories of Art," in **The Literary Animal: Evolution and the Nature of Narrative.** Edited by Jonathan Gottschall and David Sloan Wilson. Evanston, IL: Northwestern University Press, 2005, 147.

———. **On the Origin of Stories.** Cambridge, MA: Belknap Press, 2010.

Boyer, Pascal. "Functional Origins of Religious Concepts: Ontological and Strategic Selection in Evolved Minds." **Journal of the Royal Anthropological Institute** 6, no. 2 (June 2000): 195–214.

———. **Religion Explained: The Evolutionary Origins of Religious Thought.** New York: Basic Books, 2007.

Bruner, Jerome. **Making Stories: Law, Literature, Life.** New York: Farrar, Straus and Giroux, 2002.

———. "The Narrative Construction of Reality." **Critical Inquiry** 18, no. 1 (Autumn 1991): 1–21.

Buss, David. **Evolutionary Psychology: The New Science of the Mind.** Boston: Allyn & Bacon, 2012.

Cairns-Smith, A. G. **Seven Clues to the Origin of Life.** Cambridge: Cambridge University Press, 1990.

Calaprice, Alice, ed. **The New Quotable Einstein.** Princeton, NJ: Princeton University Press, 2005.

Caldwell, Robert R., Marc Kamionkowski, and Nevin N. Weinberg. "Phantom Energy and Cosmic Doomsday." **Physical Review Letters** 91 (2003): 071301.

Campbell, Joseph. **The Hero with a Thousand Faces.** Novato, CA: New World Library, 2008.

Camus, Albert. **Lyrical and Critical Essays.** Translated by Ellen Conroy Kennedy. New York: Vintage Books, 1970.

————. **The Myth of Sisyphus.** Translated by Justin O'Brien. London: Hamish Hamilton, 1955.

Čapek, Karel. **The Makropulos Case.** In **Four Plays: R. U. R.; The Insect Play; The Makropulos Case; The White Plague.** London: Bloomsbury, 2014.

Carlip, Steven. "Transient Observers and Variable Constants, or Repelling the Invasion of the Boltzmann's Brains." **Journal of Cosmology and Astroparticle Physics** 06 (2007): 001.

Carnot, Sadi. **Reflections on the Motive Power of Fire.** Mineola, NY: Dover Publications, Inc., 1960.

Carroll, Noël. "The Arts, Emotion, and Evolution." In **Aesthetics and the Sciences of Mind,** ed. Greg Currie, Matthew Kieran, Aaron Meskin, and Jon Robson. Oxford: Oxford University Press, 2014.

Carroll, Sean. **The Big Picture: On the Origins of Life, Meaning, and the Universe Itself.** New York: Dutton, 2016.

Carter, B. "Axisymmetric Black Hole Has Only Two Degrees of Freedom." **Physical Review Letters** 26 (1971): 331.

Casals, Pablo. Bach Festival: Prades 1950. As referenced by Paul Elie. **Reinventing Bach.** New York: Farrar, Straus and Giroux, 2012.

Cavosie, A. J., J. W. Valley, and S. A. Wilde. "The Oldest Terrestrial Mineral Record: Thirty Years of Research on Hadean Zircon from Jack Hills, Western Australia," in **Earth's Oldest Rocks,** ed. M. J. Van Kranendonk. New York: Elsevier, 2018, 255–78.

Ceresole, A., G. Dall'Agata, A. Giryavets, et al. "Domain walls, near-BPS bubbles, and probabilities in the landscape." **Physical Review D** 74 (2006): 086010.

Chalmers, David J. "Facing Up to the Problem of Consciousness." **Journal of Consciousness Studies** 2, no. 3 (1995): 200–19.

———. **The Conscious Mind: In Search of a Fundamental Theory.** Oxford: Oxford University Press, 1997.

Chandrasekhar, Subrahmanyan. "The Maximum Mass of Ideal White Dwarfs." **Astrophysical Journal** 74 (1931): 81–82.

Cheney, Dorothy L., and Robert M. Seyfarth. **How Monkeys See the World: Inside the Mind of Another Species.** Chicago: University of Chicago Press, 1992.

Ćirković, Milan M. "Resource Letter: PEs-1: Physical Eschatology." **American Journal of Physics** 71 (2003): 122.

Cloak, F. T., Jr. "Cultural Microevolution." **Research Previews** 13 (November 1966): 7–10.

Clottes, Jean. **What Is Paleolithic Art? Cave Paintings and the Dawn of Human Creativity.** Chicago: University of Chicago Press, 2016.

Coleman, Sidney. "Fate of the False Vacuum." **Physical Review D** 15 (1977): 2929; erratum, **Physical Review D** 16 (1977): 1248.

Conrad, Joseph. **The Nigger of the "Narcissus."** Mineola, NY: Dover Publications, Inc., 1999.

Coqueugniot, Hélène, et al. "Earliest cranio-encephalic trauma from the Levantine Middle Palaeolithic: 3D reappraisal of the Qafzeh 11 skull, consequences of pediatric brain damage on individual life condition and social care." **PloS One** 9 (23 July 2014): 7 e102822.

Crick, F. H. C., Leslie Barnett, S. Brenner, and R. J. Watts-Tobin. "General nature of the genetic code for proteins," **Nature** 192 (Dec. 1961): 1227–32.

Cronin, H. **The Ant and the Peacock: Altruism and Sexual Selection from Darwin to Today.** Cambridge: Cambridge University Press, 1991.

Crooks, G. E. "Entropy production fluctuation theorem and the nonequilibrium work relation for free energy differences." **Physical Review E** 60 (1999): 2721.

Damrosch, David. **The Buried Book: The Loss and Rediscovery of the Great Epic of Gilgamesh.** New York: Henry Holt and Company, 2007.

Darwin, Charles. **The Descent of Man and Selection in Relation to Sex.** New York: D. Appleton and Company, 1871.

———. **The Expression of the Emotions in Man and Animals.** Oxford: Oxford University Press, 1998.

———. Letter to Alfred Russel Wallace, 27 March 1869. https://www.darwinproject.ac.uk/letter/?docId=letters/DCP-LETT-6684.xml;query=child ;brand=default.

―――. **The Origin of Species.** New York: Pocket Books, 2008.

Davies, Stephen. **The Artful Species: Aesthetics, Art, and Evolution.** Oxford: Oxford University Press, 2012.

Dawkins, Richard. **The God Delusion.** New York: Houghton Mifflin Harcourt, 2006.

―――. **The Selfish Gene.** Oxford: Oxford University Press, 1976.

De Caro, M., and D. Macarthur. **Naturalism in Question.** Cambridge, MA: Harvard University Press, 2004.

Deamer, David. **Assembling Life: How Can Life Begin on Earth and Other Habitable Planets?** Oxford: Oxford University Press, 2018.

Dehaene, Stanislas. **Consciousness and the Brain.** New York: Penguin Books, 2014.

Dehaene, Stanislas, and Jean-Pierre Changeux. "Experimental and Theoretical Approaches to Conscious Processing." **Neuron** 70, no. 2 (2011): 200–227.

Dennett, Daniel. **Breaking the Spell: Religion as a Natural Phenomenon.** New York: Penguin Books, 2006.

―――. **Consciousness Explained.** Boston: Little, Brown and Co., 1991.

―――. **Elbow Room.** Cambridge, MA: MIT Press, 1984.

―――. **Freedom Evolves.** New York: Penguin Books, 2003.

―――. **The Intentional Stance.** Cambridge, MA: MIT Press, 1989.

Deutsch, David. **The Beginning of Infinity: Explanations That Transform the World.** New York: Viking, 2011.

Deutscher, Guy. **The Unfolding of Language: An Evolutionary Tour of Mankind's Greatest Invention.** New York: Henry Holt and Company, 2005.

Dickinson, Emily. **The Poems of Emily Dickinson,** reading ed., ed. R. W. Franklin. Cambridge, MA: The Belknap Press of Harvard University Press, 1999.

Dissanayake, Ellen. **Art and Intimacy: How the Arts Began.** Seattle: University of Washington Press, 2000.

Distin, Kate. **The Selfish Meme: A Critical Reassessment.** Cambridge: Cambridge University Press, 2005.

Doniger, Wendy, trans. **The Rig Veda.** New York: Penguin Classics, 2005.

Dor, Daniel. **The Instruction of Imagination.** Oxford: Oxford University Press, 2015.

Dostoevsky, Fyodor. **Crime and Punishment.** Translated by Michael R. Katz. New York: Liveright, 2017.

Dunbar, R. I. M. "Gossip in Evolutionary Perspective." **Review of General Psychology** 8, no. 2 (2004): 100–110.

———. **Grooming, Gossip, and the Evolution of Language.** Cambridge, MA: Harvard University Press, 1997.

Dunbar, R. I. M., N. D. C. Duncan, and A. Marriott. "Human Conversational Behavior." **Human Nature** 8, no. 3 (1997): 231–46.

Dupré, John. "The Miracle of Monism," in **Naturalism in Question,** ed. Mario de Caro and David Macarthur. Cambridge, MA: Harvard University Press, 2004.

Durant, Will. **The Life of Greece.** Vol. 2 of **The Story of Civilization.** New York: Simon & Schuster, 2011. Kindle, 8181–82.

Dutton, Denis. **The Art Instinct.** New York: Bloomsbury Press, 2010.

Dyson, Freeman. "Time without end: Physics and biology in an open universe." **Reviews of Modern Physics** 51 (1979): 447–60.

Dyson, L., M. Kleban, and L. Susskind. "Disturbing Implications of a Cosmological Constant." **Journal of High Energy Physics** 0210 (2002): 011.

Eddington, A. "The End of the World: From the Standpoint of Mathematical Physics." **Nature** 127, no. 3203 (1931): 447–53.

Einstein, Albert. **Autobiographical Notes.** La Salle, IL: Open Court Publishing, 1979.

Elgendi, Mohamed, et al. "Subliminal Priming-State of the Art and Future Perspectives." **Behavioral Sciences** (Basel, Switzerland) 8, no. 6 (30 May 2018): 54.

Ellenberger, Henri. **The Discovery of the Unconscious.** New York: Basic Books, 1970.

Else, Jon, dir. **The Day After Trinity.** Houston: KETH, 1981.

Emerson, Ralph Waldo. **The Conduct of Life.** Boston and New York: Houghton Mifflin Company, 1922.

Emler, N. "The Truth About Gossip." **Social Psychology Section Newsletter** 27 (1992): 23–37.

England, J. L. "Statistical physics of self-replication." **Journal of Chemical Physics** 139 (2013): 121923.

Epicurus. **The Essential Epicurus.** Translated by Eugene O'Connor. Amherst, NY: Prometheus Books, 1993.

Falk, Dean. **Finding Our Tongues: Mothers, Infants and the Origins of Language.** New York: Basic Books, 2009.

———. "Prelinguistic evolution in early hominins: Whence motherese?" **Behavioral and Brain Sciences** 27 (2004): 491–541.

Fisher, R. A. **The Genetical Theory of Natural Selection.** Oxford: Clarendon Press, 1930.

Fisher, Simon E., Faraneh Vargha-Khadem, Kate E. Watkins, Anthony P. Monaco, and Marcus E. Pembrey. "Localisation of a gene implicated in a severe speech and language disorder." **Nature Genetics** 18 (1998): 168–70.

Fowler, R. H. "On Dense Matter." **Monthly Notices of the Royal Astronomical Society** 87, no. 2 (1926): 114–22.

Freese, K., and W. Kinney. "The ultimate fate of life in an accelerating universe." **Physics Letters B** 558, nos. 1–2 (10 April 2003): 1–8.

Friedmann, Alexander. Translated by Brian Doyle. "On the Curvature of Space." **Zeitschrift für Physik** 10 (1922): 377–86.

Frijda, N., A. S. R. Manstead, and S. Bem. "The influence of emotions on belief," in **Emotions and Beliefs: How Feelings Influence Thoughts** (Studies in Emotion and Social Interaction), ed. N. Frijda,

A. Manstead, and S. Bem. Cambridge: Cambridge University Press, 2000, 1–9.

Frijda, N., and B. Mesquita. "Beliefs through emotions," in **Emotions and Beliefs: How Feelings Influence Thoughts** (Studies in Emotion and Social Interaction), ed. N. Frijda, A. Manstead, and S. Bem. Cambridge: Cambridge University Press, 2000, 45–77.

Fu, Wenqing, Timothy D. O'Connor, Goo Jun, et al. "Analysis of 6,515 exomes reveals the recent origin of most human protein-coding variants." **Nature** 493 (10 January 2013): 216–20.

Garriga, Jaume, and Alexander Vilenkin. "Many Worlds in One." **Physical Review D** 64, no. 4 (2001): 043511.

Garriga, J., V. F. Mukhanov, K. D. Olum, and A. Vilenkin. "Eternal Inflation, Black Holes, and the Future of Civilizations." **International Journal of Theoretical Physics** 39, no. 7 (2000): 1887–1900.

George, Andrew, trans. **The Epic of Gilgamesh: The Babylonian Epic Poem and Other Texts in Akkadian and Sumerian.** London: Penguin Classics, 2003.

Georgi, Howard, and Sheldon Glashow. "Unity of All Elementary-Particle Forces." **Physical Review Letters** 32, no. 8 (1974): 438.

Gottschall, Jonathan. **The Storytelling Animal.** Boston and New York: Mariner Books, Houghton Mifflin Harcourt, 2013.

Gould, Stephen J. **Conversations About the End of Time.** New York: Fromm International, 1999.

———. "The spice of life." **Leader to Leader** 15 (2000): 14–19.

———. **The Richness of Life: The Essential Stephen Jay Gould.** New York: W. W. Norton, 2006.

Gould, S. J., and R. C. Lewontin. "The Spandrels of San Marco and the Panglossian Paradigm: A Critique of the Adaptationist Programme." **Proceedings of the Royal Society B,** 205, no. 1161 (21 September 1979): 581–98.

Graziano, M. **Consciousness and the Social Brain.** New York: Oxford University Press, 2013.

Greene, Brian. **The Elegant Universe.** New York: Vintage, 2000.

———. **The Fabric of the Cosmos.** New York: Alfred A. Knopf, 2005.

———. **The Hidden Reality.** New York: Alfred A. Knopf, 2011.

Greene, Ellen. "Sappho 58: Philosophical Reflections on Death and Aging." In **The New Sappho on Old Age: Textual and Philosophical Issues,** ed. Ellen Greene and Marilyn B. Skinner. Hellenic Studies Series 38. Washington, DC: Center for Hellenic Studies, 2009. https://chs.harvard.edu/CHS/article/display/6036.11-ellen-greene-sappho-58-philosophical-reflections-on-death-and-aging#n.1.

Greene, Ellen, ed. **Reading Sappho: Contemporary Approaches.** Berkeley: University of California Press, 1996.

Guenther, Mathias Georg. **Tricksters and Trancers: Bushman Religion and Society**. Bloomington, IN: Indiana University Press, 1999.

Guth, Alan H. "Inflationary universe: A possible

solution to the horizon and flatness problems." **Physical Review D** 23 (1981): 347.

———. **The Inflationary Universe.** New York: Basic Books, 1998.

Guthrie, Stewart. **Faces in the Clouds: A New Theory of Religion.** New York: Oxford University Press, 1993.

Haidt, Jonathan. "The Emotional Dog and Its Rational Tail: A Social Intuitionist Approach to Moral Judgment." **Psychological Review** 108, no. 4 (2001): 814–34.

———. **The Righteous Mind: Why Good People Are Divided by Politics and Religion.** New York: Pantheon Books, 2012.

Haldane, J. B. S. **The Causes of Evolution.** London: Longmans, Green & Co., 1932.

Halligan, Peter, and John Marshall. "Blindsight and insight in visuo-spatial neglect." **Nature** 336, no. 6201 (December 22–29, 1988): 766–67.

Hameroff, S., and R. Penrose. "Consciousness in the universe: A review of the 'Orch OR' theory." **Physics of Life Reviews** 11 (2014): 39–78.

Hamilton, W. D. "The Genetical Evolution of Social Behaviour." **Journal of Theoretical Biology** 7, no. 1 (1964): 1–16.

Harburg, Yip. "E. Y. Harburg, Lecture at UCLA on Lyric Writing, February 3, 1977." Transcript, pp. 5–7, tape 7-3-10.

———. "Yip at the 92nd Street YM-YWHA, December 13, 1970." Transcript #1-10-3, p. 3, tapes 7-2-10 and 7-2-20.

Hawking, S. W. "Particle Creation by Black Holes."

Communications in Mathematical Physics 43 (1975): 199–220.

Hawking, Stephen, and Leonard Mlodinow. **The Grand Design.** New York: Bantam Books, 2010.

Hawks, John, Eric T. Wang, Gregory M. Cochran, et al. "Recent acceleration of human adaptive evolution." **Proceedings of the National Academy of Sciences** 104, no. 52 (December 2007): 20753–58.

Heisenberg, Werner. **Physics and Philosophy: The Revolution in Modern Science.** London: Penguin Books, 1958.

Hirshfield, Jane. **Nine Gates: Entering the Mind of Poetry.** New York: Harper Perennial, 1998.

Hogan, Patrick Colm. **The Mind and Its Stories.** Cambridge: Cambridge University Press, 2003.

Hrdy, Sarah. **Mothers and Others: The Evolutionary Origins of Mutual Understanding.** Cambridge, MA: Belknap Press, 2009.

Hulse, R. A., and J. H. Taylor. "Discovery of a pulsar in a binary system." **Astrophysical Journal** 195 (1975): L51.

Ijjas, Anna, and Paul Steinhardt. "A New Kind of Cyclic Universe" (2019). arXiv:1904.0822[gr-qc].

Islam, Jamal N. "Possible Ultimate Fate of the Universe." **Quarterly Journal of the Royal Astronomical Society** 18 (March 1977): 3–8.

Israel, W. "Event Horizons in Static Electrovac Space-Times." **Communications in Mathematical Physics** 8 (1968): 245.

———. "Event Horizons in Static Vacuum Space-Times." **Physical Review** 164 (1967): 1776.

Jackson, Frank. "Epiphenomenal Qualia." **Philosophical Quarterly** 32 (1982): 127–36.

———. "Postscript on Qualia." In **Mind, Method, and Conditionals: Selected Essays.** London: Routledge, 1998, 76–79.

James, William. **The Varieties of Religious Experience: A Study in Human Nature.** New York: Longmans, Green, and Co., 1905.

Jarzynski, C. "Nonequilibrium equality for free energy differences." **Physical Review Letters** 78 (1997): 2690–93.

Jaspers, Karl. **The Origin and Goal of History.** Abingdon, UK: Routledge, 2010.

Jeong, Choongwon, and Anna Di Rienzo. "Adaptations to local environments in modern human populations." **Current Opinion in Genetics & Development** 29 (2014): 1–8.

Jones, Barbara E. "The mysteries of sleep and waking unveiled by Michel Jouvet." **Sleep Medicine** 49 (2018): 14–19.

Joordens, Josephine C. A., et al. "**Homo erectus** at Trinil on Java used shells for tool production and engraving." **Nature** 518 (12 February 2015): 228–31.

Jørgensen, Timmi G., and Ross P. Church. "Stellar escapers from M67 can reach solar-like Galactic orbits." arxiv.org: arXiv:1905.09586.

Joyce, G. F., and J. W. Szostak. "Protocells and RNA Self-Replication." **Cold Spring Harbor Perspectives in Biology** 10, no. 9 (2018).

Jung, Carl. "The Soul and Death." In **Complete Works of C. G. Jung,** ed. Gerald Adler and R. F. C. Hull. Princeton: Princeton University Press, 1983.

Kachman, Tal, Jeremy A. Owen, and Jeremy L. England. "Self-Organized Resonance During Search of a Diverse Chemical Space." **Physical Review Letters** 119, no. 3 (2017): 038001–1.

Kafka, Franz. **The Blue Octavo Notebooks.** Translated by Ernst Kaiser and Eithne Wilkens, edited by Max Brod. Cambridge, MA: Exact Change, 1991.

Keller, Helen. Letter to New York Symphony Orchestra, 2 February 1924. Digital archives of American Foundation for the Blind, filename HK01-07_B114_F08_015_002.tif.

Kennedy, J. Gerald. **Poe, Death, and the Life of Writing.** New Haven: Yale University Press, 1987.

Kierkegaard, Søren. **The Concept of Dread.** Translated and with introduction and notes by Walter Lowrie. Princeton: Princeton University Press, 1957.

Kitcher, P. "Between Fragile Altruism and Morality: Evolution and the Emergence of Normative Guidance." **Evolutionary Ethics and Contemporary Biology** (2006): 159–77.

———. "Biology and Ethics." In **The Oxford Handbook of Ethical Theory.** Oxford: Oxford University Press, 2006.

Klinkhamer, F. R., and N. S. Manton. "A saddle-point solution in the Weinberg-Salam theory." **Physical Review D** 30 (1984): 2212.

Koch, Christof. **Consciousness: Confessions of a Romantic Reductionist.** Cambridge, MA: MIT Press, 2012.

Kragh, Helge. "Naming the Big Bang." **Historical Studies in the Natural Sciences** 44, no. 1 (February 2014): 3–36.

Krause, Johannes, Carles Lalueza-Fox, Ludovic Orlando, et al. "The Derived FOXP2 Variant of Modern Humans Was Shared with Neandertals." **Current Biology** 17 (2007): 1908–12.

Krauss, Lawrence M., and Glenn D. Starkman. "Life, the Universe, and Nothing: Life and Death in an Ever-Expanding Universe." **Astrophysical Journal** 531 (2000): 22–30.

Krutch, Joseph Wood. "Art, Magic, and Eternity." **Virginia Quarterly Review** 8, no. 4 (Autumn 1932).

Lai, C. S. L., et al. "A novel forkhead-domain gene is mutated in a severe speech and language disorder." **Nature** 413 (2001): 519–23.

Landon, H. C. Robbins. **Beethoven: A Documentary Study.** New York: Macmillan Publishing Co., Inc., 1970.

Laurent, John. "A Note on the Origin of 'Memes'/ 'Mnemes.'" **Journal of Memetics** 3 (1999): 14–19.

Lemaître, Georges. **"Rencontres avec Einstein." Revue des questions scientifiques** 129 (1958): 129–32.

Leonard, Scott, and Michael McClure. **Myth and Knowing.** New York: McGraw-Hill Higher Education, 2004.

Lewis, David. **Papers in Metaphysics and Epistemology,** vol. 2. Cambridge: Cambridge University Press, 1999.

———. "What Experience Teaches." **Proceedings of the Russellian Society** 13 (1988): 29–57.

Lewis, S. M., and C. K. Cratsley. "Flash signal

evolution, mate choice, and predation in fireflies." **Annual Review of Entomology** 53 (2008): 293–321.

Lewis-Williams, David. **The Mind in the Cave: Consciousness and the Origins of Art.** New York: Thames & Hudson, 2002.

Linde, A. "A new inflationary universe scenario: A possible solution of the horizon, flatness, homogeneity, isotropy and primordial monopole problems." **Physics Letters B** 108 (1982): 389.

———. "Sinks in the Landscape, Boltzmann Brains, and the Cosmological Constant Problem." **Journal of Cosmology and Astroparticle Physics** 0701 (2007): 022.

Loeb, Abraham. "Cosmology with hypervelocity stars." **Journal of Cosmology and Astroparticle Physics** 04 (2011): 023.

Loewi, Otto. "An Autobiographical Sketch." **Perspectives in Biology and Medicine** 4, no. 1 (Autumn 1960): 3–25.

Louie, Kenway, and Matthew A. Wilson. "Temporally Structured Replay of Awake Hippocampal Ensemble Activity during Rapid Eye Movement Sleep." **Neuron** 29 (2001): 145–56.

Mackay, Alan Lindsay. **The Harvest of a Quiet Eye: A Selection of Scientific Quotations.** Bristol: Institute of Physics, 1977.

Maddox, Brenda. **Rosalind Franklin: The Dark Lady of DNA.** New York: Harper Perennial, 2003.

Marcel, Anthony J. "Conscious and Unconscious Perception: Experiments on Visual Masking and

Word Recognition." **Cognitive Psychology** 15 (1983): 197–237.

Martin, W., and M. J. Russell. "On the origin of biochemistry at an alkaline hydrothermal vent." **Philosophical Transactions of the Royal Society B** 367 (2007): 1887–925.

Matthaei, J. Heinrich, Oliver W. Jones, Robert G. Martin, and Marshall W. Nirenberg. "Characteristics and Composition of RNA Coding Units." **Proceedings of the National Academy of Sciences** 48, no. 4 (1962): 666–77.

Melville, Herman. **Moby-Dick.** Hertfordshire, U.K.: Wordsworth Classics, 1993.

Mendez, Fernando L., et al. "The Divergence of Neandertal and Modern Human Y Chromosomes." **American Journal of Human Genetics** 98, no. 4 (2016): 728–34.

Miller, Geoffrey. **The Mating Mind: How Sexual Choice Shaped the Evolution of Human Nature.** New York: Anchor, 2000.

Mitchell, P. "Coupling of phosphorylation to electron and hydrogen transfer by a chemi-osmotic type of mechanism." **Nature** 191 (1961): 144–48.

Morrison, Toni. Nobel Prize lecture, 7 December 1993. https://www.nobelprize.org/prizes/literature/1993/morrison/lecture/.

Müller, Max, trans. **The Upanishads.** Oxford: The Clarendon Press, 1879.

Nabokov, Vladimir. **Speak, Memory: An Autobiography Revisited.** New York: Alfred A. Knopf, 1999.

Naccache, L., and S. Dehaene. "The Priming Method: Imaging Unconscious Repetition Priming Reveals an Abstract Representation of Number in the Parietal Lobes." **Cerebral Cortex** 11, no. 10 (2001): 966–74.

———. "Unconscious Semantic Priming Extends to Novel Unseen Stimuli." **Cognition** 80, no. 3 (2001): 215–29.

Nagel, Thomas. **Mortal Questions.** Cambridge: Cambridge University Press, 1979.

———. "What Is It Like to Be a Bat?" **Philosophical Review** 83, no. 4 (1974): 435–50.

Nelson, Philip. **Biological Physics: Energy, Information, Life.** New York: W. H. Freeman and Co., 2014.

Nemirow, Laurence. "Physicalism and the cognitive role of acquaintance." In **Mind and Cognition,** ed. W. Lycan. Oxford: Blackwell, 1990, 490–99.

———. "Review of Nagel's Mortal Questions." **Philosophical Review** 89 (1980): 473–77.

Newton, Isaac. Letter to Henry Oldenburg, 6 February 1671. http://www.newtonproject.ox.ac.uk/view/texts/normalized/NATP00003.

Nietzsche, Friedrich. **Twilight of the Idols.** Translated by Duncan Large. Oxford: Oxford University Press, 1998.

Norenzayan, A., and I. G. Hansen. "Belief in supernatural agents in the face of death." **Personality and Social Psychology Bulletin** 32 (2006): 174–87.

Nowak, M. A., C. E. Tarnita, and E. O. Wilson. "The evolution of eusociality." **Nature** 466, no. 7310 (2010): 1057–62.

Nozick, Robert. **Philosophical Explanations.** Cambridge, MA: Belknap Press, 1983.

———. "Philosophy and the Meaning of Life." In **Life, Death, and Meaning: Key Philosophical Readings on the Big Questions,** ed. David Benatar. Lanham, MD: The Rowman & Littlefield Publishing Group, 2010, 65–92.

Nussbaumer, Harry. "Einstein's conversion from his static to an expanding universe." **European Physics Journal—History** 39 (2014): 37–62.

Oates, Joyce Carol. "Literature as Pleasure, Pleasure as Literature." **Narrative.** https://www.narrativemagazine.com/issues/stories-week-2015-2016/story-week/literature-pleasure-pleasure-literature-joyce-carol-oates.

Oatley, K. "Why fiction may be twice as true as fact." **Review of General Psychology** 3 (1999): 101–17.

Oizumi, Masafumi, Larissa Albantakis, and Giulio Tononi. "From the Phenomenology to the Mechanisms of Consciousness: Integrated Information Theory 3.0." **PLoS Computational Biology** 10, no. 5 (May 2014).

Page, Don N. "Is our universe decaying at an astronomical rate?" **Physics Letters B** 669 (2008): 197–200.

———. "The Lifetime of the Universe." **Journal of the Korean Physical Society** 49 (2006): 711–14.

———. "Particle emission rates from a black hole: Massless particles from an uncharged, nonrotating hole." **Physical Review D** 13, no. 2 (1976): 198–206.

Page, Tim, ed. **The Glenn Gould Reader**. New York: Vintage, 1984.

Parker, Eric, Henderson J. Cleaves, Jason P. Dworkin, et al. "Primordial synthesis of amines and amino acids in a 1958 Miller H_2S-rich spark discharge experiment." **Proceedings of the National Academy of Sciences** 108, no. 14 (April 2011): 5526–31.

Perlmutter, Saul, et al. "Measurements of Ω and Λ from 42 High-Redshift Supernovae." **Astrophysical Journal** 517, no. 2 (1999): 565.

Perunov, Nikolay, Robert A. Marsland, and Jeremy L. England. "Statistical Physics of Adaptation." **Physical Review X** (June 2016): 021036-1.

Pichardo, Bárbara, Edmundo Moreno, Christine Allen, et al. "The Sun was not born in M67." **The Astronomical Journal** 143, no. 3 (2012): 73–84.

Pinker, Steven. **How the Mind Works.** New York: W. W. Norton, 1997.

———. "Language as an adaptation to the cognitive niche." In **Language Evolution: States of the Art,** ed. S. Kirby and M. Christiansen. New York: Oxford University Press, 2003.

———. **The Language Instinct.** New York: W. Morrow and Co., 1994.

Pinker, S., and P. Bloom. "Natural language and natural selection." **Behavioral and Brain Sciences** 13, no. 4 (1990): 707–84.

Plath, Sylvia. **The Collected Poems.** Edited by Ted Hughes. New York: Harper Perennial, 1992.

Prebble, John, and Bruce Weber. **Wandering in the Gardens of the Mind: Peter Mitchell and** the Making of Glynn. Oxford: Oxford University Press, 2003.

Premack, David, and Guy Woodruff. "Does the chimpanzee have a theory of mind?" **Cognition and Consciousness in Nonhuman Species,** special issue of **Behavioral and Brain Sciences** 1, no. 4 (1978): 515–26.

Proust, Marcel. **Remembrance of Things Past.** Vol. 3: **The Captive, The Fugitive, Time Regained.** New York: Vintage, 1982.

Prum, Richard. **The Evolution of Beauty: How Darwin's Forgotten Theory on Mate Choice Shapes the Animal World and Us.** New York: Doubleday, 2017.

Pyszczynski, Tom, Sheldon Solomon, and Jeff Greenberg. "Thirty Years of Terror Management Theory." **Advances in Experimental Social Psychology** 52 (2015): 1–70.

Rank, Otto. **Art and Artist: Creative Urge and Personality Development.** Translated by Charles Francis Atkinson. New York: Alfred A. Knopf, 1932.

———. **Psychology and the Soul.** Translated by William D. Turner. Philadelphia: University of Pennsylvania Press, 1950.

Rees, M. J. "The collapse of the universe: An eschatological study." **Observatory** 89 (1969): 193–98.

Reinach, Salomon. **Cults, Myths and Religions.** Translated by Elizabeth Frost. London: David Nutt, 1912.

Revonsuo, Antti, Jarno Tuominen, and Katja Valli. "The Avatars in the Machine—Dreaming as a

Simulation of Social Reality." **Open MIND** (2015): 1–28.

Rodd, F. Helen, Kimberly A. Hughes, Gregory F. Grether, and Colette T. Baril. "A possible non-sexual origin of mate preference: Are male guppies mimicking fruit?" **Proceedings of the Royal Society B** 269 (2002): 475–81.

Roney, James R. "Likeable but Unlikely, a Review of the Mating Mind by Geoffrey Miller." **Psycoloquy** 13, no. 10 (2002): article 5.

Rosenblatt, Abram, Jeff Greenberg, Sheldon Solomon, et al. "Evidence for Terror Management Theory I: The Effects of Mortality Salience on Reactions to Those Who Violate or Uphold Cultural Values." **Journal of Personality and Social Psychology** 57 (1989): 681–90.

Rowland, Peter. **Bowerbirds.** Collingwood, Australia: CSIRO Publishing, 2008.

Russell, Bertrand. **Why I Am Not a Christian.** New York: Simon and Schuster, 1957.

———. **Human Knowledge**. New York: Routledge, 2009.

Ryan, Michael. **A Taste for the Beautiful.** Princeton: Princeton University Press, 2018.

Sackmann I.-J., A. I. Boothroyd, and K. E. Kraemer. "Our Sun. III. Present and Future." **Astrophysical Journal** 418 (1993): 457.

Sartre, Jean-Paul. **The Wall and Other Stories.** Translated by Lloyd Alexander. New York: New Directions Publishing, 1975.

Scarpelli, Serena, Chiara Bartolacci, Aurora D'Atri, et al. "The Functional Role of Dreaming in

Emotional Processes." **Frontiers in Psychology** 10 (Mar. 2019): 459.

Scheffler, Samuel. **Death and the Afterlife.** New York: Oxford University Press, 2016.

Schmidt, B. P., et al. "The High-Z Supernova Search: Measuring Cosmic Deceleration and Global Curvature of the Universe Using Type IA Supernovae." **Astrophysical Journal** 507 (1998): 46.

Schrödinger, Erwin. **What Is Life?** Cambridge: Cambridge University Press, 2012.

Schroder, Klaus-Peter, and Robert C. Smith, "Distant future of the Sun and Earth revisited." **Monthly Notices of the Royal Astronomical Society** 386, no. 1 (2008): 155–63.

Schvaneveldt, R. W., D. E. Meyer, and C. A. Becker. "Lexical ambiguity, semantic context, and visual word recognition." **Journal of Experimental Psychology: Human Perception and Performance** 2, no. 2 (1976): 243–56.

Schwartz, Joel S. "Darwin, Wallace, and the **Descent of Man.**" **Journal of the History of Biology** 17, no. 2 (1984): 271–89.

Shakespeare, William. **Measure for Measure.** Edited by J. M. Nosworthy. London: Penguin Books, 1995.

Shaw, George Bernard. **Back to Methuselah.** Scotts Valley, CA: CreateSpace Independent Publishing Platform, 2012.

Sheff, David. "Keith Haring, An Intimate Conversation." **Rolling Stone** 589 (August 1989): 47.

Shermer, Michael. **The Believing Brain: From Ghosts**

and Gods to Politics and Conspiracies. New York: St. Martin's Griffin, 2011.

Silver, David, Thomas Hubert, Julian Schrittwieser, et al. "A general reinforcement learning algorithm that masters chess, shogi, and Go through self-play." **Science** 362 (2018): 1140–44.

Smuts, Aaron. "Immortality and Significance." **Philosophy and Literature** 35, no. 1 (2011): 134–49.

Solomon, Sheldon, Jeff Greenberg, and Tom Pyszczynski. "Tales from the Crypt: On the Role of Death in Life." **Zygon** 33, no. 1 (1998): 9–43.

———. **The Worm at the Core: On the Role of Death in Life.** New York: Random House Publishing Group, 2015.

Sosis, R. "Religion and intra-group cooperation: Preliminary results of a comparative analysis of utopian communities." **Cross-Cultural Research** 34 (2000): 70–87.

Sosis, R., and C. Alcorta. "Signaling, solidarity, and the sacred: The evolution of religious behavior." **Evolutionary Anthropology** 12 (2003): 264–74.

Spengler, Oswald. **Decline of the West.** New York: Alfred A. Knopf, 1986.

Sperber, Dan. **Explaining Culture: A Naturalistic Approach.** Oxford: Blackwell Publishers Ltd., 1996.

———. **Rethinking Symbolism.** Cambridge: Cambridge University Press, 1975.

Stapledon, Olaf. **Star Maker.** Mineola, NY: Dover Publications, 2008.

Steinhardt, Paul J., and Neil Turok. "The cyclic model

simplified." **New Astronomy Reviews** 49 (2005): 43–57.

Sterelny, Kim. **The Evolved Apprentice: How Evolution Made Humans Unique.** Cambridge, MA: MIT Press, 2012.

Stroud, Barry. "The Charm of Naturalism," **Proceedings and Addresses of the American Philosophical Association** 70, no. 2 (November 1996).

Stulp, G., L. Barrett, F. C. Tropf, and M. Mills. "Does natural selection favour taller stature among the tallest people on earth?" **Proceedings of the Royal Society B** 282: 20150211.

Susskind, Leonard. **The Black Hole War: My Battle with Stephen Hawking to Make the World Safe for Quantum Mechanics.** New York: Little, Brown and Co., 2008.

Swift, Jonathan. **Gulliver's Travels.** New York: W. W. Norton, 1997.

Szent-Györgyi, Albert. "Biology and Pathology of Water." **Perspectives in Biology and Medicine** 14, no. 2 (1971): 239–49.

't Hooft, G. "Computation of the quantum effects due to a four-dimensional pseudoparticle." **Physical Review D** 14 (1976): 3432.

Thoreau, Henry David. **The Journal 1837–1861.** New York: New York Review Books Classics, 2009.

Time 41, no. 14 (April 5, 1943): 42.

Tolman, Richard C. "On the problem of the entropy of the universe as a whole." **Physical Review** 37 (1931): 1639–60.

———. "On the theoretical requirements for a

periodic behavior of the universe." **Physical Review** 38 (1931): 1758–71.

Tomasello, Michael. "Universal Grammar Is Dead." **Behavioral and Brain Sciences** 32, no. 5 (October 2009): 470–71.

Tononi, Giulio. **Phi: A Voyage from the Brain to the Soul.** New York: Pantheon, 2012.

Tooby, John, and Leda Cosmides. "Does Beauty Build Adapted Minds? Toward an Evolutionary Theory of Aesthetics, Fiction and the Arts." **SubStance** 30, no. 1/2, issue 94/95 (2001): 6–27.

———. "The Psychological Foundations of Culture." In **The Adapted Mind: Evolutionary Psychology and the Generation of Culture,** ed. Jerome H. Barkow, Leda Cosmides, and John Tooby. Oxford: Oxford University Press, 1992, 19–136.

Tremlin, Todd. **Minds and Gods: The Cognitive Foundations of Religion.** Oxford: Oxford University Press, 2006.

Trinkaus, Erik, Alexandra Buzhilova, Maria Mednikova, and Maria Dobrovolskaya. **The People of Sunghir: Burials, Bodies and Behavior in the Earlier Upper Paleolithic.** New York: Oxford University Press, 2014.

Trivers, Robert. "Parental Investment and Sexual Selection." In **Sexual Selection and the Descent of Man: The Darwinian Pivot,** ed. Bernard G. Campbell. Chicago: Aldine Publishing Company, 1972.

Tylor, Edward Burnett. **Primitive Culture,** vol. 2. London: John Murray, 1873; Dover Reprint Edition, 2016, 24.

Ucko, Peter J., and Andrée Rosenfeld. **Paleolithic Cave Art.** New York: McGraw-Hill, 1967, 117–23, 165–74.

Valley, John W., William H. Peck, Elizabeth M. King, and Simon A. Wilde. "A Cool Early Earth." **Geology** 30 (2002): 351–54.

Vilenkin, A. "Predictions from Quantum Cosmology." **Physical Review Letters** 74 (1995): 846.

Vilenkin, Alex. **Many Worlds in One.** New York: Hill and Wang, 2006.

Wagoner, R. V. "Test for the existence of gravitational radiation." **Astrophysical Journal** 196 (1975): L63.

Wallace, Alfred Russel. **Natural Selection and Tropical Nature.** London: Macmillan and Co., 1891.

———. "Sir Charles Lyell on geological climates and the origin of species." **Quarterly Review** 126 (1869): 359–94.

Watson, J. D., and F. H. C. Crick. "Molecular Structure of Nucleic Acids: A Structure for Deoxyribose Nucleic Acid." **Nature** 171 (1953): 737–38.

Webb, Taylor, and M. Graziano. "The attention schema theory: A mechanistic account of subjective awareness." **Frontiers in Psychology** 6 (2015): 500.

Wertheimer, Max. **Productive Thinking,** enlarged ed. New York: Harper and Brothers, 1959.

Wheeler, John Archibald, and Wojciech Zurek. **Quantum Theory and Measurement.** Princeton: Princeton University Press, 1983.

Whitehead, Alfred North. **Science and the Modern World.** New York: The Free Press, 1953.

Wigner, Eugene. **Symmetries and Reflections.** Cambridge, MA: MIT Press, 1970.

Wilkins, Maurice. **The Third Man of the Double Helix.** Oxford: Oxford University Press, 2003.

Williams, Bernard. **Problems of the Self.** Cambridge: Cambridge University Press, 1973.

Williams, Tennessee. **Cat on a Hot Tin Roof.** New York: New American Library, 1955.

Wilson, David Sloan. **Darwin's Cathedral: Evolution, Religion and the Nature of Society.** Chicago: University of Chicago Press, 2002.

———. **Does Altruism Exist? Culture, Genes and the Welfare of Others.** New Haven: Yale University Press, 2015.

Wilson, E. O. **Sociobiology: The New Synthesis.** Cambridge, MA: Harvard University Press, 1975.

Wilson, K. G. "Critical phenomena in 3.99 dimensions." **Physica** 73 (1974): 119.

Wittgenstein, Ludwig. **Tractatus Logico-Philosophicus.** New York: Harcourt, Brace & Company, 1922.

Witzel, Michael. **The Origins of the World's Mythologies.** New York: Oxford University Press, 2012.

Woosley, S. E., A. Heger, and T. A. Weaver. "The evolution and explosion of massive stars." **Reviews of Modern Physics** 74 (2002): 1015–71.

Wrangha, Richard. **Catching Fire: How Cooking Made Us Human.** New York: Basic Books, 2009.

Yeats, W. B. **Collected Poems.** New York: Macmillan Collector's Library Books, 2016.

Yourcenar, Marguerite. **Oriental Tales.** New York: Farrar, Straus and Giroux, 1985.

Zahavi, Amotz. "Mate selection—a selection for a

handicap." **Journal of Theoretical Biology** 53, no. 1 (1975): 205–14.

Zuckerman, M. "Sensation seeking: A comparative approach to a human trait." **Behavioral and Brain Sciences** 7 (1984): 413–71.

Zunshine, Lisa. **Why We Read Fiction: Theory of Mind and the Novel.** Columbus: Ohio State University Press, 2006.

Zunshine, Lisa. Why We Read Fiction: Theory of Mind and the Novel. Columbus: Ohio State University Press, 2006.

Index

A Note About the Author

Brian Greene is a professor of physics and mathematics at Columbia University and is renowned for a number of groundbreaking discoveries in string theory. He is the author of the **New York Times** best-selling books **The Elegant Universe, The Fabric of the Cosmos,** and **The Hidden Reality.** Greene hosted two award-winning **NOVA** miniseries based on his books and is also a cofounder of the World Science Festival. With his wife and children, he lives in Andes, New York, and in New York City.